כתבי האקדמיה הלאומית הישראלית למדעים

PUBLICATIONS OF THE ISRAEL ACADEMY
OF SCIENCES AND HUMANITIES

SECTION OF SCIENCES

———

"AROMATICITY, PSEUDO-AROMATICITY, ANTI-AROMATICITY"

THE JERUSALEM SYMPOSIA ON QUANTUM CHEMISTRY AND BIOCHEMISTRY

III

JERUSALEM 1971

THE ISRAEL ACADEMY OF SCIENCES AND HUMANITIES

"AROMATICITY, PSEUDO-AROMATICITY, ANTI-AROMATICITY"

*Proceedings of an International Symposium
held in Jerusalem, 31 March–3 April 1970*

Edited by

ERNST D. BERGMANN

*Department of Organic Chemistry, the Hebrew University
Jerusalem, Israel*

and

BERNARD PULLMAN

*Université de Paris, Institut de Biologie Physico-Chimique
(Fondation Edmond de Rothschild)
Paris, France*

JERUSALEM 1971

THE ISRAEL ACADEMY OF SCIENCES AND HUMANITIES

Library of Congress Catalog Card Number 79–134838

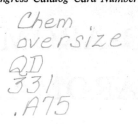
Sole Distributors

ACADEMIC PRESS, INC.

111 Fifth Avenue, New York, N.Y. 10003

ISBN 0–12–091040–3

Printed in Israel
at the Jerusalem Academic Press

Preface

THIS THIRD Jerusalem Symposium has tried to tackle a problem that, since Kekulé's paper on the structure of benzene, almost hundred years ago, has stimulated the thought and the experimental skill of scientists: the definition of aromaticity. The survey of the existing experimental data and of new observations, the recapitulation of the quantum-mechanical approaches used so far and the presentation of new approaches and their results might have been expected to give a simple answer to so simple a question; in fact, they only made clearer the limitations of the notion of aromaticity. Some of the participants have advocated the elimination of this notion entirely from the lexicon of the chemist; however, we believe the general conclusion to have been that 'aromaticity' is useful and will continue to be employed as long as we know the inherent limitations of this notion. If this Symposium has made this point clear, it has served its purpose.

Apart from the papers read by the participants and apart from their contribution to the discussions, there were three more factors that combined to make the Symposium a success: the generosity of the Edmond de Rothschild Foundation; the hospitality of the Israel Academy of Sciences and Humanities, under whose auspices and in whose premises the Symposium took place; and, last but not least, the *genius loci* of Jerusalem, the serenity of its sky and the lucidity of its air.

We are thus looking forward with confidence to the future Jerusalem Symposia on Quantum Chemistry and Biochemistry.

BERNARD PULLMAN
ERNST D. BERGMANN

Contents

Review of the Theoretical and Experimental Means for the Determination of Aromaticity

by E.D. BERGMANN *and* I. AGRANAT

Department of Organic Chemistry, the Hebrew University of Jerusalem

THE PROBLEM of aromaticity has always been one of the most difficult and yet one of the most fascinating problems in chemistry. Ever since Kekulé's intuitive idea on the structure of the benzene molecule (I) in 1865 [1], aromatic chemistry [2] has been a challenge both to the theoretician and to the synthesist.

I

In recent years, the attention of those interested in aromaticity was directed mainly towards cyclic conjugated non-benzenoid aromatic compounds [3–6]. Some of them possessed features similar to those of benzene, a phenomenon which deprived this hydrocarbon of its unique status and tended to create the necessity for a new and broader definition of aromaticity.

Even today, the concept of aromaticity is not defined unequivocally and is used with different meanings [7]. This vagueness is, in our opinion, due first of all to the fact that aromaticity has two meanings, which are basically different one from another: Classically, a compound is considered aromatic if it has 'a chemistry like that of benzene', while, according to the modern definition, a compound is aromatic if it has 'a low ground-state enthalpy' [8].

Originally, the concept of aromaticity developed as a means of characterizing a certain type of organic molecules that was inclined to substitution and disinclined to addition reactions and was thermally stable [9]. It is true that it has been known for some time that the aromatic compounds possess typical physical properties, such as the anisotropy of the diamagnetic susceptibility [10], but the emphasis has been on the chemical activity rather than on the physical properties in the ground state. Only relatively recently has it been discovered that aromatic molecules possess very low enthalpy in the ground state [11]. With the advent of quantum chemistry, this empirical finding was given a theoretical grounding. Quantitative approximation methods, the valence bond method and the MO method were developed, and these permitted the calculation of the resonance energy [12] of a conjugated system — which is both a ground state property, and can also be measured experimentally. There has thus been a continuous process of transforming the meaning of aromaticity from the chemical definition, which emphasizes the energy content of the molecule in the excited state, to the physical viewpoint, which underlines the properties of the molecules in the ground state [13].

This ambiguity in the concept of aromaticity is especially pronounced in the series of the non-benzenoid aromatics, e.g., the fulvenes. In the benzenoid aromatics, there is a quite good correlation between the two definitions. However, the cyclopentadienyl anion $C_5H_5^-$, II, prepared by Thiele in 1900 [14] but not recognized as such, which is the prototype of the non-benzenoid aromatic compounds, possesses a low enthalpy in the ground state, but is very reactive chemically and very far from behaving like benzene.

II

The dichotomy is quite common and is most probably the source for the mushrooming of the various prefixes attached to the word aromaticity; to mention the most common:

1. Pseudo-aromaticity [12, 15–16], which is a basic concept in Craig's rules (to which we shall return later), but in the broader sense refers to all non-benzenoid aromatic compounds.
2. Quasi-aromaticity [17].
3. Anti-aromaticity [18–20].
4. Non-aromaticity [18].
5. Homo-aromaticity [21–22].

And perhaps even

6. Pseudo-anti-aromaticity [23].

These were attempts to modify the concept with the purpose of trying to bridge the gap between the two fundamental definitions mentioned. The observations made, e.g., in the fulvene series, especially regarding the variation of properties with structure, have made desirable a clear-cut, 'black-and-white' definition of aromaticity [24–27].

The leading role in the theory of non-benzenoid aromatic compounds has been played by Hückel's $4n + 2$ aromaticity rule [28–33]. Contrary to the 'aromatic sextet' of Armit and Robinson [34–35], which was developed essentially as an empirical generalization, the Hückel rule was derived from theoretical considerations. Considering the fact that it has been so successful both in explaining a vast variety of observed facts and in predicting new ones [13, 33], it is rather curious that its theoretical foundation, the HMO method, is so weak [36]. Even for the monocyclic conjugated hydrocarbons, where the rule can formally be applied, there is an obvious discrepancy between theory and experiment: while HMO calculations indicated that systems with $4n\,\pi$ electrons should have lower resonance energies per atom than the corresponding $4n + 2$ π-electron systems, they did not imply that the difference between the two systems would be as large as the experiment appears to suggest.

A first approximation to a better understanding has been put forward in terms of the PMO method introduced by Coulson and Longuet-Higgins [37–41] and then extended into a general theory by Dewar [42], who applied it to the problem of aromaticity [43–45]. The theoretical foundation of the Hückel rule was further weakened with the introduction of the bond alternation concept [46–47], for example, in $4n+2$ electron rings, with n sufficiently large.

Some other attempts were also made to broaden the Hückel rule. Examples are the periphery modification of Platt [48–49] and the polycyclic modification of Volpin [50]. The Platt model has recently been extended to include anti-aromatic systems [51–52], and applied to dicyclopenta [*ef, kl*] heptalene (III), a non-alternant hydrocarbon containing no benzenoid moiety at all [53]. Nevertheless, the validity of the $4n + 2$ rule, even in LCAO, has been attacked by Musher, who defined aromaticity differently, viz., as the group of physical and chemical properties exhibited by roughly planar cyclic molecules in which the bond angles at the carbon atoms are roughly 120° (240°) [54].

III

In contrast to the annulenes, the application of the Hückel rule to the non-alternant hydrocarbons [55] in general, and to the simple non-alternant aromatic compounds — the monocyclic systems pentafulvene (IV), heptafulvene (V), pentafulvalene (VI), sesquifulvalene (VII) and heptafulvalene (VIII) and the bicyclic systems pentalene (IX), azulene (X) and heptalene (XI) — in particular, was only intuitive.

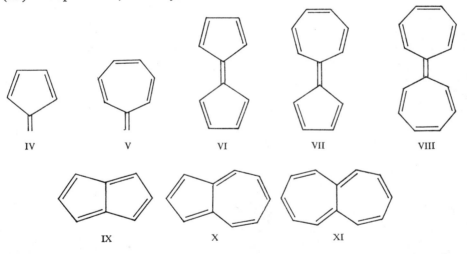

IV   V   VI   VII   VIII

IX   X   XI

To improve the theoretical foundation of the rule, it was deemed necessary to apply a MO theory that would take into account explicitly the effects of the interelectronic and internuclear repulsions. One attempt, based on Pople's method [56–57], was made by Fukui and his co-workers [58], who suggested guidelines for the synthesis of new aromatic compounds: they proposed to choose suitable substituents to be attached to the ring in such a way that the ring will acquire $4n+2$ electrons. The 'push-pull' approach towards the synthesis of stable cyclobutadienes [59–60], e. g., diethyl 2, 4-*bis*(diethylamino)-1, 3-cyclobutadiene-1, 3-dicarboxylate (XII) [61], and of 10-dimethylaminononafulvene (XIII) [62] is based on this suggestion. One may also mention the preparation of hetero-analogues of

11

the non-benzenoid aromatics, such as the hetero-fulvalenes [63–64]. Typical examples are the following: 2, 6-dimethyl-4-tetraphenylcyclopentadienylidenepyran (XIV) [65–66], 1-benzyl-4-cyclopentadienylidene-1, 4-dihydropyridine (XV) [67–68], 1, 4-diaza-5, 8-dithia-2, 3, 6-triphenylpentafulvalene (XVI) [69], 2, 6-dimethyl-4-(1-cyanocyclopentadienylidene) thiin (XVII) [64] and 10, 9-borazaronaphthalene (XVIII) [70–71].

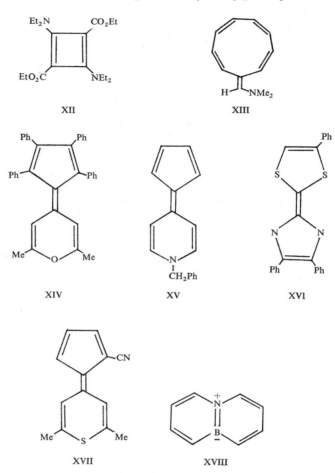

Various other attempts were based on SCF–MO calculations by Nakajima [72–74], Julg [75] and Daudel [76]. These attempts were carried out with various degrees of sophistication and were all related specifically to certain non-alternant systems, and not to the Hückel rule in a general sense.

A most important attempt to define more clearly the meaning of the Hückel rule as a criterion for aromaticity has been made by Dewar and co-workers [36, 77–79]. He applied Mulliken and Parr's definition of resonance energy and showed that both the magnitude and the sign of the resonance energy might serve as a criterion for aromaticity. The results confirmed the Hückel rule in the annulenes for $n < 5$, taking into account realistic geometries of the molecules, and pointed to the existence of a critical value of n for which a $4n + 2$ annulene would not be aromatic.

Difficulties encountered with the resonance energy criterion prompted the idea of defining aromaticity by the aromatic ring current concept [25–33, 80]. This suggestion was derived from the explanation given to the chemical shifts of the ring protons of benzene relative to ethylene [81]. An aromatic compound would be defined as a compound which will sustain an induced ring current. The magnitude of the ring current will be a function of the delocalization of π electrons around the ring, and therefore a quantitative measure of aromaticity. The concept has been successfully applied by Sondheimer in the annulene series [82] and by Boekelheide to aromatic molecules having substituents within the cavity of the π-electron cloud [83–84], e.g., to pyrene. However, this concept is not free of limitations [54, 85–86]. It was discovered that the shift to a lower field of the ring protons of an aromatic molecule, as compared to the protons of the respective ethylene, is due not only to the ring current, but also to local, induced atomic currents [87–88]. Moreover, it must be realized that the aromatic ring current is not linked directly either to the resonance energy or to the reactivity of the molecule [85]. Finally, the theoretical basis of the ring current criterion has been criticized by Musher, who argued that the 'chemical shift of aromatic hydrocarbons, generally attributed to π-electron ring currents, can be correctly represented as a sum of contributions from localized electrons of both π and σ characters' [86]. This controversy is still unsettled [54, 89–90].

Very recently, Dauben proposed the diamagnetic susceptibility exaltation as a criterion for aromaticity [91–92], as it appears to reflect unequivocally the presence of appreciable cyclic π-electron delocalization. For example, benzene (I) and pentafulvene (IV) have values of 13.7 and 1.1, respectively, while heptalene (XI) has a value of -6. The importance of this criterion lies in the fact that it is related to a theoretically well-defined quantity, the London diamagnetism, though, of course, it is empirical in character.

Closely related is the Faraday effect ring current concept as a criterion for aromaticity, proposed by Labarre and co-workers [93–94].

Two other approaches, both originating from NMR techniques, should be mentioned here. The first method tries to determine the extent of bond alternation from NMR coupling constants data. It is based on the correlation between vicinal olefinic coupling constants and π-bond order [80]. The following example is taken from the triapentafulvalene series [95]:

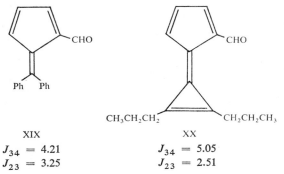

XIX	XX
$J_{34} = 4.21$	$J_{34} = 5.05$
$J_{23} = 3.25$	$J_{23} = 2.51$

We see a greater bond alternating in 1-formyl-6, 6-diphenylpentafulvene (XIX) than in 3-formyl-1, 2-dipropyltriapentafulvalene (XX). The latter is 'more aromatic'. Another example is found in the work of Bertelli and co-workers [96] on the NMR spectra of tropone (XXI), 8, 8-dicyanoheptafulvene (XXII) and related compounds. The spectra point to a bond alternation, which is characteristic of polyenes and polyenones, and thus exclude any

13

aromatic character of these substances. The method has recently been reviewed for the pentafulvene series [97].

The second method determines the energy barriers for restricted rotations. Using the same example of 3-formyl-1, 2-dipropyltriapentafulvene (XX) [98], the activation free energy for free rotation was calculated from the temperature dependence of the α-methylene signal. The observed facility of internal rotation indicated a strikingly low π-bond order of the pinch, and thus a significant 'cyclopropenium-cyclopentadienide aromaticity'. Another example is 6-dimethylaminopentafulvene (XXIII), in which, from the study of the energy barriers for restricted rotation about the C_6–N and C_1–C_6 bonds, the delocalization energy was estimated as 25–32 Kcal [99].

| XXI | XXII | XXIII |

Formally, the best method for determining bond alternations is, of course, X-ray diffraction. It is true that it is time consuming and that one may not always be justified in translating its results into conclusions related to the solute state (in which the chemist is usually interested), but it is regrettable that more use has not yet been made of this method.

An example from our own studies is that of 1, 2, 3, 4-tetrachloro-5, 6-diphenyltriapentafulvalene (XXIV) [100–101]. The X-ray investigation [102] indicated a marked bond alternation, but it is noteworthy that in the three-membered ring the bond lengths are considerably shorter than the theoretical values. Similar results were obtained by Kitahara for the 1, 2, 3, 4-tetrachloro-5, 6-dipropyltriapentafulvalene (XXV) [103–104].

Three further examples may be mentioned. For 1-methyl-2-fluorenylidene-1, 2-dihydropyridine (XXVI) the X-ray diffraction pattern shows that the bond length of the 'pinch' is large for a formal double bond, and that the distances in the dihydropyridine ring clearly reflect the effects of electron delocalization [105]. An X-ray analysis of 1, 6-methano [10]

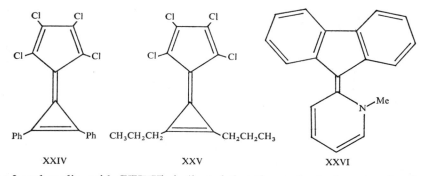

| XXIV | XXV | XXVI |

annulene-2-carboxylic acid (XXVII) indicated that the perimeter is not quite planar, but that the bonds are typical benzenoid aromatic bonds, their lengths varying only between 1.38 and 1.42 Å [106]. The structure reported for [18] annulene (XXVIII) in the solid state indicated only a small deviation from planarity, but two types of bonds of different lengths [107–108].

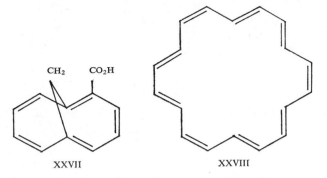

CH₂   CO₂H

XXVII   XXVIII

The aromaticity of non-alternant systems may also be determined by the dipole moment, which can be calculated to various approximations by the existing quantum-mechanical theories. For hexaphenyl-triapentafulvalene (XXIX) [109], which is a hydrocarbon and thus should have a negligible dipole moment, the experimental value is $\mu = 6.3$ D, while the best calculated value is that of Nakajima [109], 6.04 D, assuming the molecule is planar.

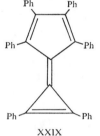

Ph   Ph

Ph   Ph

Ph   Ph

XXIX

Recently, this approach has been criticized by Dewar [44, 111], who claims that 'the existence of large dipole moments cannot in itself be taken as evidence that a molecule has a highly delocalized aromatic character' [44]. However, in the example cited, there undoubtedly exists a certain contribution of a dipolar, aromatic structure in the ground state. It is true that the evaluation of observed dipole moments and their comparison with theoretical values can only lead to qualitative results, as the measured moment is a global property of the molecule and its breakdown to contributions of the different parts of the molecule is not always unequivocal. It is essential to find proper reference compounds; only thus has it been possible to show, e.g., that tropone (XXI) has no aromatic character [112]. In particular, one has to take into consideration that the dipole moment of a polar-substituted parent system is not necessarily the sum of the moments of the parent system and of the polar substituents [113]. The attempt [114] to derive the moment (3 D) of heptafulvene (V) from the moment (7.5 D) of 8, 8-dicyanoheptafulvene (XXII) is not theoretically valid.

This question of the moments of the fundamental members of the fulvene series (which are generally unstable) has been thrown into relief by the report of Brown [115] that the dipole moment of pentafulvene (IV), determined by microwave spectroscopy, is only 0.49 D, whilst the value obtained from solution measurements is about 1.1 D [97]. The new value shows the absence of any significant contribution of a dipolar, aromatic character in the ground state of the hydrocarbon. This is all the more surprising as the isomer of pentafulvene (IV), 3, 4-dimethylenecyclobutene (XXX), has a moment of 0.62 D [116], so that it became useful to introduce a new notion, 'pseudo-alternant' compound [117–118].

15

XXX

Among the various other indices for aromaticity, the one proposed by Julg [119], based on the bond lengths in the periphery, should be mentioned:

$$A = 1 - \frac{225}{n} \sum_{rs} \left(1 - \frac{d_{rs}}{\bar{d}}\right),$$

where

$$\bar{d} = \frac{1}{n} \sum_{rs} d_{rs},$$

and d_{rs} is the bond length in the periphery. For benzene, $A = 1$, while for fulvene, $A = 0.62$. This formula would permit us to express the 'gradation' in aromatic character in quantitative terms.

Analogously, a description of delocalization, and thus implicitly of aromaticity, in terms of bond index has been put forward [120]. The delocalization index [121],

$$D = 1 - \frac{2}{m} \sum_{r=1}^{m} p - p_r,$$

where p is the average bond order, p_r the bond order of the rth bond and m the number of peripheral bonds, gives a value of $D = 1$ for benzene (I), $D = 0$ for the Kekulé structure of benzene and $D = 0.57$ for pentafulvene (IV), thus approximately the same figures as Julg's method.

The importance of bond alternation in the study of the problem of aromaticity has prompted Heilbronner and Binsch to introduce the concepts of first-order and second-order double bond fixation [122–124]. Binsch has suggested that a conjugated π-electron system could be called aromatic if it shows neither strong first-order nor second-order double bond fixation [125–126]. The strongest tendency for second-order double bond fixation was calculated for pentalene (IX), indicating that it consists of an equilibrating mixture of two isoenergetic forms of lower symmetry.

Binsch's structural criterion [126] has the advantage that it is closely related to a physical phenomenon: the tendency of π electrons to cluster in certain bonds. In contrast to the resonance energy, it is capable of pinpointing the lack of π-electron delocalization in a particular structural segment and of detecting the dynamic distortions in certain π systems which hamper electron delocalization.

In conclusion, we would like to return to the concept of pseudo-aromaticity and to analyse Craig's rules for aromaticity [12, 15, 16, 127], which are associated with this concept. Craig's rules, which are empirical guidelines for the classification of non-benzenoid aromatic compounds, relate aromatic character to high resonance energy: they try to predict whether the resonance energy between the canonical structures, representing the ground state of a conjugated hydrocarbon and possessing certain symmetry features, is high or low. Craig's rules distinguish between molecules with a valence bond (VB) wave function totally symmetric in the ground state and molecules with a valence bond wave function *not* totally symmetric in this state.

16

Craig's rules have been criticized [68, 72, 74, 128] for two reasons: they are often limited by the fact that in many cases the symmetry conditions necessary for their application are not fulfilled, and they have led to inaccurate predictions, e.g., in the case of pentafulvene itself. In our opinion it is important to realize that for Craig's rules to have any practical value they should be applied only to those molecules for which there exists a problem of resonance between Kekulé structures in the ground state, that is, only to non-classical conjugated hydrocarbons. Accordingly, the rules could be re-formulated in the following manner: An *aromatic* molecule is a non-classical conjugated molecule that is represented in the ground state by a totally symmetric VB wave function. A *pseudo-aromatic* molecule is a non-classical conjugated molecule that is represented in the ground state by a non-totally symmetric VB wave function.

In a classical conjugated molecule, the problem of resonance between distinct Kekulé structures does not exist, so that one cannot predict, on the basis of Craig's rules, whether such a molecule is aromatic, whatever may be the degree of symmetry of its VB wave function. The fulvenes, e.g., are classical conjugated molecules. It can be shown (using the detailed analytical procedure of Craig's rules) that they have totally symmetric VB wave functions. However, we cannot predict on the basis of Craig's rules that they should be aromatic. The interesting point is that the two properties, the classical character of a conjugated molecule and its representation by a totally symmetric VB wave function, are mutually dependent. A non-classical conjugated molecule may be represented by a VB wave function that is neither necessarily totally symmetric nor non-totally symmetric. However, it can be shown that each Kekulé structure that is transformed into itself by a symmetry operation does so with a positive sign.

In Craig's terminology,

$$A \rightarrow (-1)^{f+g} B,$$

$$A \rightarrow (-1)^{f+g} A, \qquad f + g \text{ being even.}$$

Consequently, for a conjugated molecule, the classical character is a sufficient, but not a necessary, condition for being represented in the ground state by a totally symmetric VB wave function.

We have tried to survey briefly the numerous attempts to define aromaticity and the discrepancies, overlaps and incompatibilities among the various definitions. We are not of the opinion that this state of affairs is a sufficient cause for condemning the word 'aromaticity' and eliminating it from the vocabulary of the chemist. On the contrary, 'a clash of doctrines', in Whitehead's words, 'is an opportunity and not a disaster'. But one should always remember the following remark, made by Ernst Nagel [129], on any attempt to define explicitly a theoretical notion by a few experimental concepts:

> In these cases in which a given theoretical notion is made to correspond to two or more experimental ideas, it would be absurd to maintain that the theoretical concept is explicitly defined by each of the two experimental ones in turn.

REFERENCES

1 A. Kekulé (1865) *Bull. Soc. Chim. France*, 3 : 98.

2 G. M. Badger (1969) *Aromatic Character and Aromaticity*, Cambridge University Press, New York.

3 D. Ginsburg (1959) *Non-Benzenoid Aromatic Compounds*, Interscience, New York.

4 K. Hafner (1964) *Angew. Chem.* (*Int. Edn*), 3 : 165.

5 D. Lloyd (1966) *Carbocyclic Non-Benzenoid Aromatic Compounds*, Elsevier, London.

6 P. J. Garratt & M. V. Sargent (1969) in: *Advances in Organic Chemistry — Methods and Results*, VI (eds. E. C. Taylor & H. Wynberg), Interscience, New York, p. 1.

7 R. Robinson (1959) in: *Non-Benzenoid Aromatic Compounds* (ed. D. Ginsburg), Interscience, New York, p. v.

8 D. Peters (1960) *J. Chem. Soc.*, p. 1274.

9 R. Robinson (1958) *Tetrahedron*, 3 : 323.

10 L. Pauling (1936) *J. Chem. Phys.*, 4 : 673.

11 G. W. Wheland (1955) *Resonance in Organic Chemistry*, Wiley, New York.

12 D. P. Craig (1959) in: *Non-Benzenoid Aromatic Compounds* (ed. D. Ginsburg), Interscience, New York, p. 1.

13 R. Breslow (1965) *Chem. Eng. News*, 43 (26) : 90.

14 J. Thiele (1900) *Ber. Dt. Chem. Ges.*, 33 : 666.

15 D. P. Craig (1951) *J. Chem. Soc.*, p. 3175.

16 Idem (1959) in: *Theoretical Organic Chemistry* (*Papers presented to the Kekulé Symposium*), Butterworths, London, p. 20.

17 D. M. G. Lloyd & D. R. Marshall (1964) *Chemy Ind.*, p. 1760.

18 M. J. S. Dewar (1965) in: *Advances in Chemical Physics*, VIII (ed. R. Daudel), Interscience, New York, p. 65.

19 R. Breslow (1968) *Chem. Britain*, 4 : 100.

20 Idem (1968) *Angew. Chem.* (*Int. Edn*), 7 : 565.

21 S. Winstein (1967) in: *Aromaticity* (*Special Publication, No. 21*), The Chemical Society, London, p. 5.

22 Idem (1969) *Q. Rev.*, 23 : 141.

23 D. T. Clark & D. R. Armstrong (1969) *Chem. Comm.*, p. 850.

24 A. P. Marchand (1965) *Chemy Ind.*, p. 161.

25 J. A. Elvidge & L. M. Jackman (1961) *J. Chem. Soc.*, p. 859.

26 J. A. Elvidge (1965) *Chem. Comm.*, p. 160.

27 G. G. Hall, A. Hardisson & L. M. Jackman (1963) *Tetrahedron*, Vol. 19, Suppl. 2, p. 101.

28 E. Hückel (1931) *Z. Physik.*, 70 : 204.

29 *Ibid.* (1932) 76 : 628.

30 Idem (1935) *International Conference on Physics, London 1934*, II, The Physical Society, London, p. 9.

31 Idem (1937) *Z. Elektrochem.*, 43 : 752.

32 Idem (1938) *Grundzüge der Theorie ungesättigter und aromatischer Verbindungen*, Verlag Chemie, Berlin.

33 A. Streitwieser Jr (1961) *Molecular Orbital Theory for Organic Chemists*, Wiley, New York.

34 J. W. Armit & R. Robinson (1925) *J. Chem. Soc.*, p. 1604.

35 R. Robinson (1967) in: *Aromaticity* (*Special Publication, No. 21*), The Chemical Society, London, p. 47.

36 A. L. H. Chung & M. J. S. Dewar (1965) *J. Chem. Phys.*, 42 : 756.

37 C. A. Coulson & H. C. Longuet-Higgins (1947) *Proc. Roy. Soc.*, A 191 : 39.

38 *Ibid.* (1947/8) A 192 : 16.

39 *Ibid.* (1948) A 193 : 447, 456.

40 *Ibid.* (1948/9) A 195 : 188.

41 H. C. Longuet-Higgins (1950) *J. Chem. Phys.*, 18 : 265, 275, 283.

42 M. J. S. Dewar (1952) *J. Am. Chem. Soc.*, 74 : 3341, 3345, 3350, 3353, 3355, 3357.

43 M. J. S. Dewar (1966) *Tetrahedron*, Suppl. 8, Part 1, p. 75.

44 M. J. S. Dewar (1967) in: *Aromaticity* (*Special Publication, No. 21*), The Chemical Society, London, p. 177.

45 Idem (1969) *The Molecular Orbital Theory of Organic Chemistry*, McGraw-Hill, New York, p. 191.

46 Y. Ooshika (1957) *J. Phys. Soc. Japan*, 12 : 1238.

47 H. C. Longuet-Higgins & L. Salem (1959) *Proc. Roy. Soc.*, A 251 : 172.

48 J. R. Platt (1954) *J. Chem. Phys.*, 22 : 1448.

49 W. Baker & J. F. W. McOmie (1959) in: *Non-Benzenoid Aromatic Compounds* (ed. D. Ginsburg), Interscience, New York, p. 477.

50 M. E. Vol'pin (1960) *Russ. Chem. Rev.*, 29 : 129.

51 B. M. Trost, S. F. Nelsen & D. R. Brittelli (1967) *Tetrahedron Lett.*, p. 3959.

52 B. M. Trost (1969) *J. Am. Chem. Soc.*, 91 : 918.

53 A. G. Anderson, A. A. MacDonald & A. F. Montana (1968) *J. Am. Chem. Soc.*, 90 : 2993.

54 J. I. Musher (1966) in: *Advances in Magnetic Resonance*, II (ed. J. S. Waugh), Academic Press, New York, p. 177.

55 R. Zahradnik (1965) *Angew. Chem.* (*Int. Edn*), 4 : 1039.

56 J. A. Pople (1953) *Trans. Faraday Soc.*, 49 : 1375.

57 A. Brickstock & J. A. Pople (1954) *ibid.*, 50 : 901.

58 K. Fukui, A. Imamura, T. Yonezawa & C. Nagata (1960) *Bull. Chem. Soc. Japan*, 33 : 1591.

59 J. D. Roberts (1958) *Developments in Aromatic Chemistry (Special Publication, No. 12)*, The Chemical Society, London, p. 111.

60 R. Breslow, D. Kivelevich, M. J. Mitchell, W. Fabian & K. Wendel (1965) *J. Am. Chem. Soc.*, 87 : 5132.

61 R. Gompper & G. Seybold (1968) *Angew. Chem. (Int. Edn)*, 7 : 824.

62 K. Hafner & H. Tappe (1969) *ibid.*, 8 : 593.

63 E. D. Bergmann (1968) *Chem. Rev.*, 68 : 41.

64 G. Seitz (1969) *Angew. Chem. (Int. Edn)*, 8 : 478.

65 D. Lloyd & F. I. Wasson (1963) *Chemy Ind.*, p. 1559.

66 Idem (1966) *J. Chem. Soc.* (C), p. 1086.

67 D. N. Kursanov, N. K. Baranetskaya & W. N. Setkina (1957) *Dokl. Akad. Nauk S.S.S.R.*, 113 : 116.

68 J. A. Berson, E. M. Evleth & Z. Hamlet (1965) *J. Am. Chem. Soc.*, 87 : 2887.

69 R. Gompper & R. Weiss (1968) *Angew. Chem. (Int. Edn)*, 7 : 296.

70 M. J. S. Dewar & R. Jones (1968) *J. Am. Chem. Soc.*, 90 : 2137.

71 F. A. Davis, M. J. S. Dewar, R. Jones & S. D. Worley (1969) *J. Am. Chem. Soc.*, 91 : 2094.

72 T. Nakajima (1964) in : *Molecular Orbitals in Chemistry, Physics and Biology* (eds. P. O. Löwdin & B. Pullman), Academic Press, New York, p. 451.

73 T. Nakajima & S. Katagiri (1962) *Bull. Chem. Soc. Japan*, 35 : 910.

74 Idem (1964) *Molec. Phys.*, 7 : 149.

75 A. Julg & P. François (1964) *C. r. Acad. Sci. (Paris)*, 258 : 2067.

76 O. Chalvet, R. Daudel & J. J. Kaufman (1964) *J. Phys. Chem.*, 68 : 490.

77 M. J. S. Dewar & G. J. Gleicher (1965) *J. Am. Chem. Soc.*, 87 : 685.

78 *Ibid.*, p. 692.

79 M. J. S. Dewar & C. de Llano (1969) *ibid.*, 91 : 789.

80 L. Salem (1966) *The Molecular Orbital Theory of Conjugated Systems*, Benjamin, New York, Chap. IV.

81 J. A. Pople (1956) *J. Chem. Phys.*, 24 : 1111.

82 F. Sondheimer, I. C. Calder, J. A. Elix, Y. Gaoni, P. J. Garratt, K. Grohmann, G. Di Maio, J. Mayer, M. V. Sargent & R. Wolovsky (1967) in : *Aromaticity (Special Publication, No. 21)*, The Chemical Society, London, p. 75.

83 V. Boekelheide & J. B. Phillips (1967) *J. Am. Chem. Soc.*, 89 : 1695.

84 J. B. Phillips. R. J. Molyneux, E. Sturm & V. Boekelheide (1967) *J. Am. Chem. Soc.*, 89 : 1704.

85 R. J. Abraham & W. A. Thomas (1966) *J. Chem. Soc.* (B), p. 127.

86 J. I. Musher (1965) *J. Chem. Phys.*, 43 : 4081.

87 J. A. Pople (1964) *ibid.*, 41 : 2559.

88 A. F. Ferguson & J. A. Pople (1965) *ibid.*, 42 : 1560.

89 J. M. Gaidis & R. West (1967) *ibid.*, 46 : 1218.

90 J. I. Musher, *ibid.*, p. 1219.

91 H. J. Dauben Jr, J. D. Wilson & J. L. Laity (1968) *J. Am. Chem. Soc.*, 90 : 811.

92 *Ibid.* (1969) 91 : 1991.

93 J. F. Labarre & F. Crasnier (1967) *J. Chim. Phys.*, 64 : 1664.

94 J. F. Labarre, M. Graffeuil & F. Gallais (1968) *ibid.*, 65 : 638.

95 A. S. Kende (1966) *Trans. N.Y. Acad. Sci.*, 28 : 981.

96 D. J. Bertelli, T. G. Andrews & P. O. Crews (1969) *J. Am. Chem. Soc.*, 91 : 5286.

97 P. Yates (1968) in : *Advances in Alicyclic Chemistry*, II (eds. H. Hart & G. J. Karabatsos), Academic Press, New York, p. 59.

98 A. S. Kende, P. T. Izzo & W. Fulmor (1966) *Tetrahedron Lett.*, p. 3697.

99 A. P. Downing, W. D. Ollis & I. O. Sutherland (1969) *J. Chem. Soc.* (B), p. 111.

100 M. Ueno, I. Murata & Y. Kitahara (1965) *Tetrahedron Lett.*, p. 2967.

101 E. D. Bergmann & I. Agranat (1966) *Tetrahedron*, p. 1275.

102 O. Kennard, D. G. Watson, J. K. Fawcett, K. A. Kerr & C. Romers (1967) *Tetrahedron Lett.*, p. 3885.

103 H. Shimanouchi, T. Ashida, Y. Sasada, M. Kakudo, I. Murata & Y. Kitahara, *ibid.*, p. 61.

104 H. Shimanouchi, Y. Sasada, T. Ashida, M. Kakudo, I. Murata & Y. Kitahara (1969) *Acta Crystallogr.*, 25 B : 1890.

105 H. L. Ammon (1969) *Tetrahedron Lett.*, p. 3305.

106 M. Dobler & J. D. Dunitz (1965) *Helv. Chim. Acta*, 48 : 1429.

107 J. Bregman, F. L. Hirshfeld, D. Rabinovich & G. M. J. Schmidt (1965) *Acta Crystallogr.*, 19 : 227.

108 F. L. Hirshfeld & D. Rabinovich, *ibid.*, p. 235.

109 E. D. Bergmann & I. Agranat (1965) *Chem. Comm.*, p. 512.

110 T. Nakajima, S. Kohda, A. Tajiri & S. Karasawa (1967) *Tetrahedron*, 23 : 2189.

111 M. J. S. Dewar (1969) *The Molecular Orbital Theory of Organic Chemistry*, McGraw-Hill, New York, pp. 181–182.

19

112 D. J. Bertelli & T. G. Andrews (1969) *J. Am. Chem. Soc.*, 91 : 5280.

113 H. Weiler-Feilchenfeld, I. Agranat & E. D. Bergmann (1966) *Trans. Faraday Soc.*, 62 : 2084.

114 M. Yamakawa, H. Watanabe, T. Mukai, T. Nozoe & M. Kubo (1960) *J. Am. Chem. Soc.*, 82 : 5665.

115 R. D. Brown, F. R. Burden & J. E. Kent (1968) *J. Chem. Phys.*, 49 : 5542.

116 R. D. Brown, F. R. Burden, A. J. Jones & J. E. Kent (1967) *Chem. Comm.*, p. 808.

117 B. A. W. Coller, M. L. Heffernan & A. J. Jones (1968) *Aust. J. Chem.*, 21 : 1807.

118 R. D. Brown & F. R. Burden (1966) *Chem. Comm.*, p. 448.

119 A. Julg & P. François (1967) *Theor. Chim. Acta* (Berlin), 7 : 249.

120 C. Trindle (1969) *J. Am. Chem. Soc.*, 91 : 219.

121 W. Kemula & T. M. Krygowski (1968) *Tetrahedron Lett.*, p. 5135.

122 G. Binsch, E. Heilbronner & J. N. Murrell (1966) *Molec. Phys.*, 11 : 305.

123 G. Binsch & E. Heilbronner (1968) in : *Structural Chemistry and Molecular Biology* (eds. A. Rich & N. Davidson), Freeman, San Francisco, p. 815.

124 Idem (1968) *Tetrahedron*, 24 : 1215.

125 G. Binsch, I. Tamir & R. D. Hill (1969) *J. Am. Chem. Soc.*, 91 : 2446.

126 G. Binsch & I. Tamir (1969) *J. Am. Chem. Soc.*, 91 : 2450.

127 D. P. Craig (1950) *Proc. Roy. Soc.*, A 200 : 390.

128 R. A. Abramovitch & K. L. McEwen (1965) *Can. J. Chem.*, 43 : 2616.

129 E. Nagel (1961) *The Structure of Science — Problems in the Logic of Scientific Explanation*, Routledge and Kegan Paul, London, p. 99.

Discussion

M. Cais:

It is very difficult to open a discussion after a review such as this one, where an attempt has been made to please everybody. However, I cannot abstain from bringing out a very serious omission. It may be that as an organometallic chemist I find myself a lone wolf in this audience, but I am rather surprised that the speaker did not make any mention of the additional complication to the concept of aromaticity which has been produced by the discovery of metallocenes. I would be very interested in hearing from some of the distinguished guests in this audience if they would like to comment on this particular point. Do these new molecules, which some people call super-aromatic in order to extend the vocabulary presented by our speaker, fit anywhere in that series of definitions that have been presented in this opening talk?

I. Agranat:

Apparently, I didn't please everybody. Mr Cais is correct—the limitations of this brief review made a certain selection of topics necessary. Of course, one can argue that if one talks even briefly about the cyclopentadienide anion as one of the aromatic structures, the case of the ferrocenes will not be more than a special case of this structure. However, we will undoubtedly have a discussion on this point after Mr Cais' paper at this Symposium.

E. Heilbronner:

I do not know exactly how I am going to say what I want to say, but somebody has to be the devil's advocate. To begin with, I think we should all realize that we are united here in a symposium on a non-existent subject. It must also be stated quite clearly at the beginning that aromaticity is not an observable property, i. e., it is not a quantity that can be measured and it is not even a concept which, in my experience, has proved very useful. Indeed, my experience as a teacher is that the amount of confusion caused by the term 'aromaticity' in the student's mind is not compensated for by a gain in the understanding of the chemistry and physics of the molecules so classified. I think we all agree that the aim of chemical theory is to explain and predict the chemical and physical properties of molecules. We also know that modern quantum chemistry has finally succeeded in providing us with a pretty good picture of the molecules we deal with and that it yields reasonably good predictions concerning the behaviour of strange molecules, even before they have been synthesized. Nobody can claim that the vague concept of 'aromaticity' must be introduced at any stage to make these theories work.

'Aromaticity', if to be used at all, should be a purely structural concept. We should define certain molecules as 'aromatic', depending on the formal aspects of their structure. These formal aspects will depend on certain conventions, such as drawing bonds between certain atoms, etc.

One thing is quite certain: 'aromaticity', let alone the 'degree' of 'aromaticity', should not depend on the chemical and physical properties of a given compound. From Mr Agranat's review it is obvious that, depending on the physical method which you are using, you will come to a different grading, to a different classification of 'aromaticities' for the same set of compounds. Even for a given type of physical measurement, according to which a compound is going to be classified, it becomes quite arbitrary where you want to draw the line that discriminates between 'aromaticity' and 'non-aromaticity'. As an unescapable consequence, you can have—and Mr Agranat has shown this—many compounds that according to one method will be classified as 'aromatic' and according to another as 'non-aromatic'. It might be a good idea to introduce the term 'schizo-aromatic' for molecules that may be one or the other, depending upon the type of measurements made.

That 'aromaticity' cannot be measured, it not being a physical quantity, is perhaps best shown by the well-known but elusive 'resonance energy'. Mr Agranat said in the beginning that a low enthalpy of formation is an indication of 'aromaticity'. On an absolute scale (i.e., relative to the elements in their standard state) this is not true. The 'lowness' of the enthalpy of formation has to be judged relative to a hypothetical molecule whose non-observable enthalpy of formation depends on many arbitrary assumptions about the geometry and the electronic structure of this fictitious molecule. As a consequence, resonance energies are defined quantities and not observed ones. And this is the reason for the widely diverging values quoted by different authors for the same compound.

This was a general remark. I now have one question about which I am not quite clear: Craig's rules, as given in his original papers, do include the statement that they can only be applied if you can write more than one unexcited valence-bond structure. Consequently, Craig's summary, given in the book on non-benzenoid aromatic compounds—which has been edited by D. Ginsburg—seems incomplete, insofar as this particular rule is missing. Am I wrong in this?

I. Agranat:

Well, if you read Craig's review in Ginsburg's book, he gives the example of fulvene and uses it to show that, according to Craig's rules, fulvene is aromatic. It was not an accidental omission that he did not mention that Craig's rules do not apply to fulvene. It is obvious, if you read the review, that at that time Craig meant that his rules assume, or predict, that fulvene should be aromatic. I think that it is clearly realized now that they should not be applied to classical conjugated molecules.

G. Berthier:

From a theoretical point of view it is possible (if not too easy) to define aromaticity in the frame of pure π-electron theories. The interpretation of all-valence or all-electron calculations in terms of aromaticity is much more difficult. This is why in works of that sort the relative stability of conjugated systems is deduced from the values of total binding energies rather than of resonance energies.

G. Del Re:

The remark that aromaticity is not an observable is perfectly valid, as long as we do not know what aromaticity really is. But if we had a clear-cut definition, it could very well be an observable. Many physical quantities (including energy) can be measured only by elaborating the results of experiment. In the case of aromaticity, one could 'observe' it by dividing appropriate experimental results into contributions associated with various effects like hybridization, conjugation, etc. Comparison of molecules, where such contributions have different weights (or, in the limit, all but one are the same), should then provide an unambiguous derivation of any one of them from experiment.

R. H. Martin:

Before Mr Heilbronner's remarks I was going to ask the following question: How are we to define aromaticity in excited molecules? Now, if we regard 'aromaticity' as a 'structural concept', the answer to my question is probably obvious.

Discussion

E. Heilbronner:

Well, I am not quite sure whether I am going to make myself very popular, but I would like first to comment upon what Mr Del Re has said. I am primarily a chemist, and my view of theoretical chemistry is terribly pragmatic. I want to know what my molecules will do when subjected to certain experiments; I want to know what the answer will be if I am going to record, for example, a photo-electron spectrum; what the answer will be when I record a UV spectrum or when I let two molecules react with each other. If I know how to deal with these theoretical problems, then maybe I can make fewer experiments or better ones.

Now, let us assume that I want to make such a prediction. There is absolutely no instance, at any time, where I have to bring in aromaticity. In other words, from the point of view of the chemist who wants to know the behaviour of his molecules, the question of whether a molecule is aromatic, pseudo-aromatic, anti-aromatic or whatever aromatic is completely irrelevant. We want to know the properties of molecules, and we don't want to know whether a molecule falls into a certain abstract category that we have invented. Looked at from this point of view, an argument concerning 'aromaticity' must lead to the same sort of discussion that theologians had in the Middle Ages when they wanted to determine how many angels could sit on the head of a pin.

The second point is the one Mr Martin talked about. As I said before, if we want to keep the word 'aromatic' at all, it must refer to a structural concept. In this case the question will be: Will the molecule in the electronically excited state have a structure which agrees with our definition, or not? This will then answer the question.

B. Pullman:

I must say that I do not agree entirely with what Mr Heilbronner just said. Of course, we are interested essentially in properties of molecules, but if we speak only in terms of properties, we are going into a jungle just the same, because, if only for practical reasons, we have to classify molecules anyhow. Of course, aromaticity is a difficult problem, and I do not think that we shall come out of this meeting with a clear-cut definition of aromaticity. Nevertheless, I think it is useful to try to divide molecules into some types; otherwise, how are you going to write encyclopaedias or books or anything? All molecules have properties. You do have to try to have some kind of organization when you want to find things you look for easily, even from the purely pragmatic point of view. When we speak about aromatic molecules, we do not have too many problems in under-standing what they are. But when we start to speak of pseudo- or anti-aromatic molecules, then we have problems. Even if no clear definition is available, I still think that some kind of a classification might be extremely useful.

When I started to work on this problem, I was concentrating on benzenoid polycyclics and found that they have very 'usual' properties. Then I came across fulvenes and found that they have very unusual properties. Well, to some extent, in a very simplified way, of course, aromatics are, for me, conjugated compounds which have 'usual' properties. It is not a definition, of course, but it helps me very nicely in classifying them.

E. Heilbronner:

I entirely agree with Mr Pullman about the last point. However, if you want to write encyclo-paedias, there is still the Geneva Convention to rely on. On the other hand, I must say again that I cannot see that it is very interesting to know whether a molecule is 'aromatic' or not. This is a classification into only two boxes. And if you think of the hundreds of thousands of compounds we know, this is a very poor classification indeed.

B. Pullman:

We do not have a hundred thousand compounds that can be classified as aromatic or non-aromatic; we may have a few hundred of them. And this division is useful, even if not clear-cut.

23

Discussion

E. D. Bergmann:

Allow me to add to your remark. I think Mr Heilbronner underestimates his own approach to the problem, because, although he said he only wants to know the properties of compounds — which I think every chemist really wants to know — he also uses theoretical approaches, and if you look at his papers, I think he has developed quite a number of theories and approaches based on concepts that are necessary if one wants to have any classification at all of compounds. We all want classification, but we do not want to be in a position where every compound has to be considered as something separate.

I would like to make a second point in order to protect Mr Agranat and myself. If you look at the heading of this Symposium, we deliberately put 'Aromaticity, Pseudo-aromaticity, Anti-aromaticity' in inverted commas; this is not accidental. We did not want to say that we can define what we are talking about. What we hope is that, as we leave at the end of the Symposium, we will know whether we can define it or not.

Double-Bond Fixation, Ring Currents and Aromaticity

G. BINSCH

Department of Chemistry and the Radiation Laboratory, University of Notre Dame, Indiana

I. INTRODUCTION

I SHOULD LIKE to begin this lecture with a few remarks of a philosophical nature. Why is it that the problem of aromaticity, though seemingly being a rather narrow aspect of chemical structure and reactivity, has intrigued chemists for a century, and yet is as much alive today as ever? I offer two answers to that question. The first is that we are dealing with a problem that is insoluble in principle, simply because aromaticity is not an observable phenomenon. This may sound like an escape-hatch for theorists and has, in fact, occasionally served in that capacity when certain concepts came under fire by the experimentalists, but it is nevertheless true. More importantly perhaps, this area is uniquely characterized by one of the most fruitful interplays of theory and experiment in the history of chemistry and has therefore succeeded in preserving its appeal to a wide spectrum of investigators. Even if the problem itself should eventually evaporate, as it probably will owing to its artificiality, it will have served an important function in the development of chemistry toward an exact science.

It is in this spirit that we have recently proposed a novel criterion of aromaticity [1], based on the theory of double-bond fixation [2–5]:

> A conjugated π-electron system is called aromatic if it shows neither strong first-order nor second-order double-bond fixation [1].

In suggesting this criterion we were not so much concerned with the question whether this definition encompasses all aspects of what we intuitively mean by aromatic character — in fact, we are certain that it does not — but rather, we were interested in gaining an insight into a genuine physical phenomenon, namely, the tendency of π electrons to cluster in certain bonds.

I intend first to give a brief summary of the theory of double-bond fixation, concentrating on the main ideas rather than on the computational details, then present some new results, and, finally, analyse the relationship of the double-bond fixation criterion to the popular ring-current criterion of aromaticity [G. Binsch & R. D. Hill, unpublished results].

The Radiation Laboratory is operated by the University of Notre Dame under contract with the US Atomic Energy Commission. This is AEC Document No. COO–38–726.

25

II. Double-Bond Fixation

Consider a collection of sp^2-hybridized carbon atoms joined together in a planar *sigma* framework of the desired topology, with all carbon–carbon bond distances equal to 1.50 Å by definition and with an empty set of p orbitals perpendicular to the plane. We now ask how this zero-order geometry is distorted if we fill in the π electrons, to successive degrees of approximation. To do this, we develop the total energy into a Taylor series in terms of the bond distortions ΔR:

$$E = E_\sigma^0 + E_\pi^0 + \sum_{\mu < \nu} \left[\left(\frac{\partial E_\sigma}{\partial R_{\mu\nu}} \right)_0 + \left(\frac{\partial E_\pi}{\partial R_{\mu\nu}} \right)_0 \right] \Delta R_{\mu\nu} + \frac{1}{2} \sum_{\mu < \nu} \left(\frac{\partial^2 E_\sigma}{\partial R_{\mu\nu}^2} \right)_0 (\Delta R_{\mu\nu})^2 +$$

$$+ \frac{1}{2} \sum_{\mu < \nu} \sum_{\kappa < \lambda} \left(\frac{\partial^2 E_\pi}{\partial R_{\mu\nu} \partial R_{\kappa\lambda}} \right)_0 \Delta R_{\mu\nu} \, \Delta R_{\kappa\lambda} + \cdots \text{(higher terms)}, \tag{1}$$

where we assume complete σ–π separation and independent σ bonds, but allow for interactions between the π electrons.

By breaking off the series of Equation (1) after the fourth term, it is easy to show [2] that one obtains a linear relationship between the 'first-order bond lengths' $R_{\mu\nu}^{(1)}$ and the bond orders $P_{\mu\nu}$ (see also Fig. 1)

$$R_{\mu\nu}^{(1)} = -mP_{\mu\nu} + b, \tag{2}$$

large bond orders corresponding to short bonds and small bond orders to long bonds.

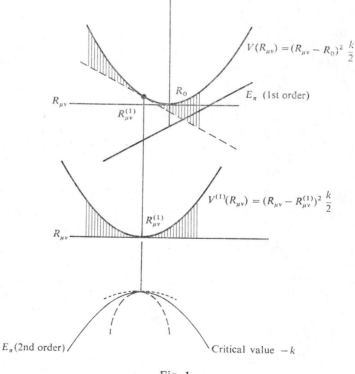

Fig. 1

Components E_π and E_σ of the total energy of a π-electron system

The differences in such lengths between adjacent bonds are then a measure of first-order double-bond fixations.

For investigating the importance of second-order distortions, one has to compare the second derivatives of E_σ and E_π in Equation (1). Because of the cross terms in the π contribution, this can only be done after subjecting them to a normal coordinate analysis. In other words, one has to diagonalize the matrix

$$\left(\frac{\partial^2 E_\pi}{\partial R_{\mu\nu} \partial R_{\kappa\lambda}}\right)_0 . \tag{3}$$

The most negative eigenvalue, λ_{max}, then corresponds to the energetically most favourable second-order bond distortion, and the normalized components of the corresponding eigenvector $\mathbf{D}_{max}$ determine the distortion type. The matrix elements (3) may be calculated by second-order perturbation theory, either in an independent-electron [2–4] or an SCF model [1, 5]. All numbers to be reported here were obtained by self-consistent perturbation theory.

Two limiting cases can now be distinguished. If λ_{max} is smaller in magnitude (dotted line in Fig. 1) than a critical value, λ_{crit}, equal to the negative curvature of the σ energy, we only expect a flattening of the potential. However, if λ_{max} exceeds λ_{crit} (dashed line in Fig. 1), the minimum in the potential will become a maximum, and the molecule will suffer second-order bond distortions, which will in general result in the destruction of the symmetry of the σ framework. The critical value applicable to the subsequent numbers was derived to be $\lambda_{crit} = 1.22$ (β_{core}^{-1}), expressed in units of the inverse core resonance integral.

Figs. 2–3 show typical results. In benzene, all bond orders are equal by symmetry, and the components of $\mathbf{D}_{max}$ alternate in sign around the ring. But, since $\lambda_{max} < \lambda_{crit}$, one expects only a flattening of the potential, which shows up in a low force constant for the B_{2u} stretching mode [6]. The situation is different for the non-alternant systems pentalene and heptalene. Here the bond orders are also more or less equalized, similar to the situation in azulene, but λ_{max} exceeds λ_{crit}. Pentalene and heptalene are therefore predicted to suffer second-order double-bond fixation, resulting in 'dynamic' [1, 3–5] bond alternation around the ring.

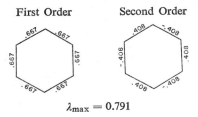

First Order   Second Order

$\lambda_{max} = 0.791$

Fig. 2
First-order and second-order double-bond fixations in benzene

What is the relevance of this phenomenon to the problem of aromaticity? Fig. 4 shows the situation for cyclobutadiene. We know that the lowest singlet state of cyclobutadiene should not be represented by a 'resonance arrow' between two quadratic canonical forms, corresponding to a single minimum in the potential, but by an equilibrating mixture of two equivalent rectangular structures and a double minimum in the potential, a theoretical

27

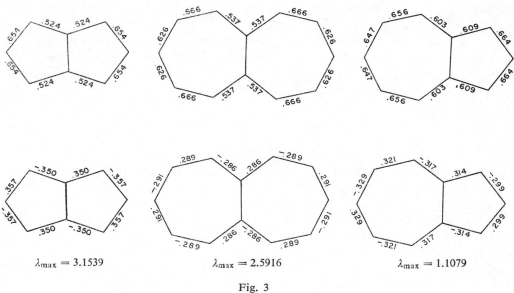

$\lambda_{\max} = 3.1539$ $\lambda_{\max} = 2.5916$ $\lambda_{\max} = 1.1079$

Fig. 3
First-order (top) and second-order (bottom) double-bond fixations
in pentalene, heptalene and azulene

conclusion for which experimental evidence has recently become available [7]. The pseudo-Jahn-Teller effect responsible for this distortion is nothing else but a special case [5] of our general theory of second-order double-bond fixation. If one insists on calling cyclobutadiene 'anti-aromatic', pentalene and heptalene would have to be called anti-aromatic as well.

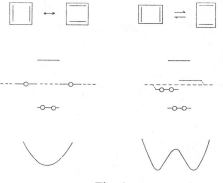

Fig. 4
Pseudo-Jahn-Teller distortions, energy-level
diagrams and potentials in cyclobutadiene

III. DOUBLE-BOND FIXATION AND RING CURRENTS

Stimulated by Jung's work [8] on the ring currents of certain fused π systems, we have performed [G. Binsch & R. D. Hill, unpublished results] self-consistent perturbation calculations for the molecules I–XI of Fig. 5. First-order double-bond fixations were

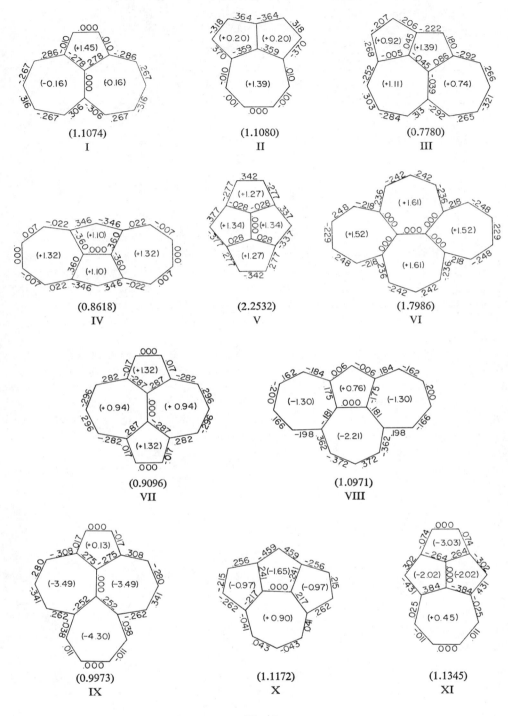

Fig. 5

Largest eigenvalues, λ_{max} (in parentheses under formulae), and normalized components of the corresponding eigenvectors D_{max} of the SCF bond-bond polarizability matrices; Jung's [8] magnetic values are shown inside the rings

found to be unimportant in IV–VII, moderate in I–III, VIII and X, and strong in certain segments of IX and XI, as indicated by the heavy lines in Fig. 6.

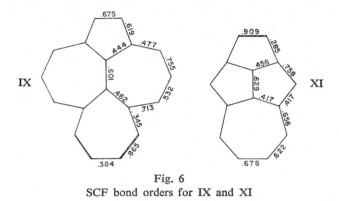

Fig. 6
SCF bond orders for IX and XI

We first note that the double-bond fixation criterion does not give rise to the ambiguities [8] of the ring-current criterion. More importantly, however, the systems V and VI are clearly predicted to be aromatic by the ring-current criterion, but anti-aromatic by virtue of their second-order double-bond fixations. What is the reason for this discrepancy?

Both the paramagnetic contributions to the ring-current and the second-order double-bond fixations are calculated by second-order perturbation theory, and can thus be attributed to the mixing of certain excited states into the ground state. But paramagnetic ring currents can only arise from those excited states that have the same transformation properties as the magnetic-dipole operator $\mathbf{M}$, perpendicular to the plane of the π system [9], whereas all excited states, in principle, contribute to second-order double-bond fixation. However, since the various second-order distortion patterns are orthogonal and therefore mutually exclusive [3], we only need to consider the excited states transforming as $\mathbf{D}_{max}$. If we confine ourselves to the three most important single-determinant singlet SCF wave functions, whose construction is indicated schematically in Fig. 7, we obtain the results collected in the table (System III has been omitted, since it lacks essential symmetry).

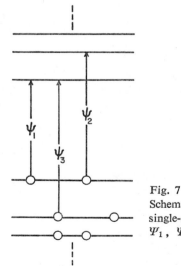

Fig. 7
Schematic representation of the
single-determinant wave functions
Ψ_1, Ψ_2 and Ψ_3

Transformation Properties of the Magnetic-Dipole Operator **M**,
the Eigenvector $\mathbf{D}_{max}$ and the Singly-Excited Singlet Configurations
Ψ_1, Ψ_2 and Ψ_3 of Fig. 7

System	Symmetry	Transformation Properties				
		Ψ_1	Ψ_2	Ψ_3	M	$\mathbf{D}_{max}$
I	C_{2v}	B_1	B_1	A_1	B_1	B_1
II	C_{2v}	B_1	A_1	A_1	B_1	B_1
IV	D_{2h}	B_{2u}	B_{3u}	B_{1g}	B_{1g}	B_{1g}
V	D_{2h}	B_{2u}	B_{3u}	B_{3u}	B_{1g}	B_{2u}
VI	D_{2h}	B_{2u}	B_{3u}	B_{3u}	B_{1g}	B_{2u}
VII	D_{2h}	B_{1g}	B_{2u}	A_g	B_{1g}	B_{1g}
VIII	C_{2v}	B_1	A_1	A_1	B_1	B_1
IX	C_{2v}	B_1	B_1	A_1	B_1	B_1
X	C_{2v}	B_1	A_1	B_1	B_1	B_1
XI	C_{2v}	B_1	A_1	B_1	B_1	B_1

It can be seen that, with the exception of IV, $\mathbf{D}_{max}$ always transforms as Ψ_1. This was to be expected, since Ψ_1 will, in general, be the singly-excited singlet state of lowest energy, and therefore dominates the second-order double-bond fixation. In IV the effective one-electron SCF level differences of Ψ_1, Ψ_2 and Ψ_3 are almost identical, and our SCF calculations show that electron-repulsion effects actually cause Ψ_3 to become the single-determinant singlet excited state of lowest energy. In most cases, the magnetic-dipole operator **M** also transforms as $\mathbf{D}_{max}$ and as the lowest excited singlet configuration, but not for V and VI. In these two molecules the very low-lying Ψ_1 configurations are inaccessible by magnetic-dipole transitions, with a component perpendicular to the plane of the π system, and the ring-current criterion is therefore incapable of detecting the important effects these states are expected to exert on the chemical structure of V and VI.

IV. CONCLUSION

In the past, the chemists' thinking about the problem of aromaticity was strongly influenced by 'magic rules', and this attitude is even reflected in the title of the present Symposium. I personally believe we ought to free ourselves from that and start to think in terms of physical phenomena.

The ring-current and double-bond fixation criteria both refer to physical phenomena. It is easy to show [10–11] that both criteria lead to the same qualitative conclusions for monocyclic conjugated hydrocarbons, but not necessarily for polycyclic systems, as we have seen. There can be no question that the phenomenon of double-bond fixation has a direct bearing on chemical structure, but to what extent it reflects the salient features of the so-called 'aromatic character' is entirely a pragmatic issue and will have to be decided by the experimentalists.

REFERENCES

1 G. Binsch & I. Tamir (1969) *J. Am. Chem. Soc.*, 91 : 2450.

2 G. Binsch, E. Heilbronner & J. N. Murrell (1966) *Molec. Phys.*, 11 : 305.

3 G. Binsch & E. Heilbronner (1968) in : *Structural Chemistry and Molecular Biology* (eds. A. Rich & N. Davidson), Freeman, San Francisco, p. 815.

4 Idem (1968) *Tetrahedron*, 24 : 1215.

5 G. Binsch, I. Tamir & R. D. Hill (1969) *J. Am. Chem. Soc.*, 91 : 2446.

6 R. D. Mair & D. F. Hornig (1949) *J. Chem. Phys.*, 17 : 1236.

7 P. Reeves, T. Devon & R. Pettit (1969) *J. Am. Chem. Soc.*, 91 : 5890.

8 D. E. Jung (1969) *Tetrahedron*, 25 : 129.

9 J. H. van Vleck (1932) *Electric and Magnetic Susceptibilities*, Oxford University Press.

10 H. C. Longuet-Higgins (1967) in : *Aromaticity (Special Publication, No. 21)*, The Chemical Society, London, p. 109.

11 J. A. Pople & K. G. Untch (1966) *J. Am. Chem. Soc.*, 88 : 4811.

Discussion

E. Heilbronner:

On Mr Binsch's slide the following definition has been proposed: 'A conjugated π-electron system is called aromatic if it shows neither strong first-order nor second-order double bond fixation.' Now, could you point out a molecule, except benzene, which classifies as 'aromatic'?

B. Binsch:

Benzene is a perfect example!

E. Heilbronner:

Name a second one.

B. Binsch:

It is, of course, a question of degree.

E. Heilbronner:

I have a more serious question. If you do your self-consistent treatment, you get λ values for the distortions that belong to the different irreducible representations. Now, for naphthalene and anthracene you have reasonably good infra-red spectral data. Is there any indication that there is a correlation between the dependence of frequencies of, say, B_{2u} vibrations or A_g vibrations with the corresponding λ values?

B. Binsch:

I must confess that we have not looked at that. We have used the B_{2u} stretching in benzene as a calibration for the calculation of this critical value — this critical value is just calibrated on experiments. But we have not looked at normal modes in other kinds.

G. Berthier:

Three years ago, the concept of ring currents was strongly attacked by J. I. Musher, because it should be an artefact of the London calculation method. I should like to ask Mr Binsch whether this controversy is now definitely settled.

B. Binsch:

Let me first say that we were not concerned with this question at all; we regarded the ring current concept as an empirical manifestation of experiments. If you want my honest opinion about Musher's theory, I can only say that in his reply to the objection raised by Gaidis and West he claims that his theory is an 'anti-theory' and cannot be disproved by experiments, no matter how devastating.

J. F. Labarre:

I should like to add something to this discussion. I have met Musher recently, and he says now 'ring currents, yes, but not properties of π electrons'. Secondly, as Mr Binsch has said so well, ring currents may be compared within a given symmetry group family only.

33

Permutation Symmetry Control in Concerted Reactions

by J.J.C. MULDER and L.J. OOSTERHOFF

Department of Theoretical Organic Chemistry, University of Leiden

CONCERTED ORGANIC REACTIONS have found a generally utilized theoretical interpretation in the so-called Woodward-Hoffmann rules [1].* Usually, these rules are derived from the idea that symmetry of molecular orbitals — and of the electronic configurations obtained from them — is conserved during all stages of a reaction [2]. A reinvestigation from a different viewpoint seemed worthwhile for two reasons:

1. Recently, van der Lugt and Oosterhoff [3–5] introduced a description of electrocyclic processes in terms of valence-bond structures. From their results it was concluded that orbital symmetry is irrelevant for processes taking place in the photochemical reactions; the question as to its importance for the interpretation of thermal reactions was left open.

2. In two recent symposia and in an extensive review [6–8], Woodward and Hoffmann have given a new formulation of the rules. The most significant changes are the abolishment of molecular orbitals and symmetry considerations and, to a certain extent, the restriction to thermal reactions. According to Woodward [7–8], 'a thermal pericyclic reaction is symmetry-allowed (proceeds smoothly) if the total number of $[4r]_a$ and $[4q + 2]_s$ components is odd.'

Using the fact that a suprafacially interacting component is to be identified with an even number of negative overlap integrals, and that an antarafacially interacting component is to be identified with an odd number of negative overlap integrals, this rule summarizes in a concise way the combined influence of the number of electron pairs and the number of negative overlap integrals in the course of the reaction.** In this communication we report how, in a valence-bond treatment, without invoking spatial symmetry, these two factors appear to determine the energy of the transition states for the two modes of a concerted reaction.

If we take the conversion of *cis*-hexatriene to 1,3-cyclohexadiene (Fig. 1), and *vice versa*, as a typical example, we may say that hexatriene is well represented by its classical formula,

---

* The Woodward-Hoffmann selection rules are obtained from permutation symmetry by means of the valence-bond method.

** These two factors have been recognized by others as well. They are present in Zimmerman's [9] elegant treatment, using the valuable concepts of normal and Möbius ring; they were also used by Dewar [10] in his discussions of aromatic and anti-aromatic transition states in concerted reactions.

which corresponds in quantum chemical terms to a structure wave function Ψ_A. Likewise, we may associate a structure wave function Ψ_B with cyclohexadiene.

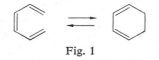

Fig. 1

A structure wave function describes an allocation of electron pairs to bonds; we will take into account only the electrons and bonds that are relevant to the reaction — in this case 6 electrons, the three π bonds in hexatriene, and the two π bonds and one σ bond in cyclohexadiene. In valence-bond theory the ground-state wave function for hexatriene should not be represented by Ψ_A only, but will contain a small contribution of Ψ_B and of other possible structure functions as well.

In the same way, the ground state wave function of cyclohexadiene will be mainly Ψ_B, with small amounts of Ψ_A and other structure functions mixed in. This may be formulated as follows:

$$\Psi_g^{\text{hex}} = \lambda_h \Psi_A + \mu_h \Psi_B + \ldots (\lambda_h^2 \gg \mu_h^2), \tag{1}$$

$$\Psi_g^{\text{cyc}} = \lambda_c \Psi_A + \mu_c \Psi_B + \ldots (\lambda_c^2 \ll \mu_c^2). \tag{2}$$

In the usual sign convention for structure wave functions [11], λ and μ will both be positive. In a concerted process the wave functions (1) and (2) will be interconverted in a continuous way by changing λ and μ monotonically. On the reaction pathway a point will be passed where λ^2 and μ^2 are equal; presumably this point will be close to the cyclic transition state. Here we will have

$$\Psi^+ = (\Psi_A + \Psi_B + \ldots)/N^+ \tag{3}$$

(N^+ takes care of normalization).

From general considerations one may infer the existence of an excited-state wave function belonging to the transition state configuration of the form

$$\Psi^- = (\Psi_A - \Psi_B + \ldots)/N^-. \tag{4}$$

The state corresponding to this wave function will appear to control the course of photochemical reactions. The selection rules in operation in the reaction will depend on the energies corresponding to the transition state wave functions (3) and (4). Evidently, the line of reasoning given for the specific example (Fig. 1) is valid for any concerted reaction of this type, and the analysis has accordingly been performed for the general case of a $2n$-electron transition state. The derivation, which is described in a separate communication [12], yields the following energies:

$$U^+ = \frac{U_i^+ - (-1)^{n+\nu} \cdot 2U_c^+}{S_i^+ - (-1)^{n+\nu} \cdot 2S_c^+}, \tag{5}$$

$$U^- = \frac{U_i^- + (-1)^{n+\nu} \cdot 2U_c^-}{S_i^- + (-1)^{n+\nu} \cdot 2S_c^-} \tag{6}$$

In these energy expressions the subscripts i and c refer to 'independent' and 'cyclic', re-

35

spectively. The quantities U_i and S_i are those parts of the energy and normalization, respectively, that are independent of the number v of negative overlap integrals. The number of electron pairs relevant to the transition state is denoted by n. Energy quantities (U) are negative; normalization elements (S) are positive. The terms U_c and S_c stand for the contributions to the energy and the normalization, respectively, resulting from two cyclic permutations of the electron numbers of order $2n$, one clockwise, the other counter-clockwise; in cycle notation: $(1, 2, 3 \ldots 2n)$ and $(2n, 2n-1, 2n-2 \ldots 1)$. Because the permutations play an essential role in the argument, their influence will be discussed briefly. Any permutation, in a semi-classical picture, can be visualized as a simultaneous jump of electrons from one orbital to another. Precisely this picture caused the introduction of the term 'exchange integral' ('Austausch Integral') in Heitler and London's treatment of the hydrogen molecule and in the subsequent development of the valence-bond theory. Within the Mulliken approximation, this exchange integral is proportional to S^2, the square corresponding to two electrons making the same jump in opposite directions. In the present problem, if the usual approximation of neglecting non-neighbour interactions is made, all the permutations — with one exception — give rise to terms in the total energy expression, with signs that are independent of the occurrence of negative overlap integrals. The one exception is formed by the cyclic permutations mentioned before. In the picture used, these cyclic permutations represent the simultaneous jumps of all electrons to their next position in the cycle of atomic orbitals. The energy terms due to these permutations are proportional to the product of all the overlap integrals, each occurring to the first power; the sign of those terms will be dependent on the number of negative overlap integrals. This gives the factor $(-1)^v$ in Equations (5) and (6). The analysis also shows that the two cyclic permutations applied to the spin functions forming part of Ψ^+ and Ψ^- yield the factors $(-1)^n$ and $-(-1)^n$. Finally, the fact that a cyclic permutation of order $2n$ has parity equal to -1 completes the sign-fixing procedure.

Formulae (5) and (6) demonstrate the existence of four energy levels at the transition state configuration in a concerted reaction. In the very schematic energy diagram (Fig. 2)

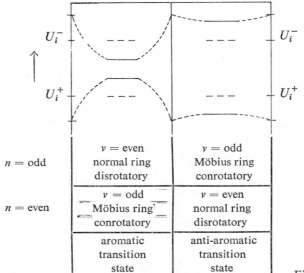

Fig. 2

the levels $U_i{}^+$ and $U_i{}^-$ have been drawn at the same height in the left and right part of the figure, and other quantities in (5) and (6) have likewise been taken equal.

From the scheme given in Fig. 2, the rule follows immediately:

Thermal reactions require either an odd number of electron pairs and an even number of negative overlap integrals, or an odd number of negative overlap integrals combined with an even number of electron pairs.

In addition, however, this alternative to Woodward's formulation is of necessity supplemented by its reversal in photochemical processes, provided Ψ^- describes the state which determines the course of the photochemical reaction. The relevance of Ψ^- in this respect follows from the observation that the present transition states can be regarded as monocyclic polyenes, for which it is known from theory, as well as experiment, that Ψ^- corresponds to the lowest excited singlet state. It should be pointed out that this state may be intrinsically different from the spectroscopic state into which the reacting system was initially excited [3–5].

REFERENCES

1 R. Hoffmann & R. B. Woodward (1968) *Acc. Chem. Res.*, 1 : 17.

2 H. C. Longuet-Higgins & E. W. Abrahamson (1965) *J. Am. Chem. Soc.*, 87 : 2045.

3 W. T. A .M. van der Lugt & L. J. Oosterhoff (1968) *Chem. Comm.*, p. 1235.

4 Idem (1969) *J. Am. Chem. Soc.*, 91 : 6042.

5 W. T. A. M. van der Lugt (1968) Thesis, University of Leiden.

6 R. B. Woodward & R. Hoffmann, in : *Symposium on Valence Isomerization, Karlsruhe, 1968.*

7 Idem, in : *Symposium on Orbital Symmetry Correlation, Cambridge, 1969.*

8 Idem (1969) *Angew. Chem.*, 81 : 797.

9 H. E. Zimmerman (1966) *J. Am. Chem. Soc.*, 88 : 1564.

10 M. J. S. Dewar (1966) *Tetrahedron*, Suppl. 8, Part 1, p. 75.

11 L. Pauling (1933) *J. Chem. Phys.*, 1 : 280.

12 J. J. C. Mulder & L. J. Oosterhoff, *Chem. Comm.* [in press]

Discussion

B. Binsch:

I should like to express an opinion and ask a short question. I wonder what is the point of re-deriving something that is already as well known as the Woodward-Hoffman rule. The point now is to go beyond that rule. My question then is: Do you arrive at any new conclusion or new approaches that go beyond the Woodward-Hoffman rule?

J. J. C. Mulder:

Our treatment shows that you may expect exceptions, especially in the kinds of cases Mr Gompper told us about, and it is only this method that treats the excited-state reaction on the same footing as the ground-state reaction. The fact is that, if you look at it, you always see that the excited state that we have here has a heavy contribution from the doubly-excited configuration (in molecular orbital language), in which the lowest unoccupied orbital is doubly occupied. This is, in my opinion, an important phenomenon. Moreover, our derivation does not use symmetry.

If I may make a comment of my own, there is an obvious question about the magnitude of the effect, because our treatment gives only the sign of the effect, and this is the right sign. The magnitude of the effect depends, of course, on the type of orbitals that you use, and, in this case, it turns out that if you use ordinary atomic orbitals, the magnitude of the effect dies down very quickly in larger systems.

Moreover, the neglect of ionic structures and the neglect of the non-neighbour terms is serious. Now, if you make a transformation of the normal atomic orbitals to a set of atomic orbitals that are more strongly non-orthogonal between neighbours, but less non-orthogonal between non-neighbours — and this is possible — you will get large overlap integrals. This also immediately gives you the inclusion of the ionic structures in the covalent ones. This transformation to a new set of orbitals is, as it were, an $S^{+\frac{1}{2}}$ transformation.

G. Wagnière:

In your treatment you neglect overlap between non-neighbours. Does that not somewhat limit the applicability of it? The neighbour–non-neighbour relationship is defined only in situations that can be described by a clear-cut chemical formula. Woodward-Hoffman's treatment is supported by extended Hückel calculations, which do not distinguish between neighbours and non-neighbours.

J. J. C. Mulder:

This is true, and I can only say that, by making the transformation that I have mentioned, the quotient of the neighbour overlap integral and the non-neighbour overlap integral, which has a certain value for normal atomic orbitals, gets a larger value for the transformed orbitals. But, of course, you are perfectly right that if the geometry of the system is such that you no longer have a clear cycle, this does not work. That is perfectly true.

However, regarding the extended Hückel calculations, I would like to say that you cannot use them at all for photochemical reactions.

38

Aromaticité et Polarisibilité

par A. PACAULT

Centre de Recherches Paul Pascal, Centre National de la Recherche Scientifique, Bordeaux

INTRODUCTION

LES POLARISIBILITÉS diamagnétique et électrique des molécules sont directement liées à la répartition spatiale des électrons autour des noyaux constituant la molécule et à leur distribution sur les niveaux d'énergie.

Un modèle simple et fructueux de répartition spatiale particulière est celui de la délocalisation des électrons π sur des orbites ayant sensiblement la dimension du cycle aromatique considéré. Il a été exploité pour expliquer l'anisotropie diamagnétique [1–2] et pour interpréter les déplacements chimiques vus par résonance magnétique nucléaire [3]. Un tel modèle fait immédiatement comprendre pourquoi une délocalisation électronique dans le plan aromatique entraîne une anisotropie des propriétés magnétique et électrique. Celle-ci cependant n'est pas seulement due aux électrons π mais encore aux électrons σ dont l'influence est surtout sensible sur l'anisotropie optique.

Si 'l'aromaticité' est la conséquence de ces répartitions électroniques particulières, les polarisibilités principales qui en sont le reflet, peuvent donc servir à la mesurer.

RAPPELS

Les polarisibilités magnétiques χ ou K et optiques* α sont des tenseurs symétriques de second rang. Elles ont donc six composantes. Si, cependant, la molécule est repérée par un référentiel *principal* choisi grâce aux conditions de symétrie, le tenseur des polarisibilités à trois composantes seulement, dites polarisibilités principales.

Les expressions suivantes définissent l'anisotropie:

$$\text{magnétique} - \gamma_M^2 = \tfrac{1}{2}[(K_{11} - K_{22})^2 + (K_{22} - K_{33})^2 + (K_{33} - K_{11})^2]**$$

$$\text{optique} - \gamma_O^2 = \tfrac{1}{2}[(\alpha_{11} - \alpha_{22})^2 + (\alpha_{22} - \alpha_{33})^2 + (\alpha_{33} - \alpha_{11})^2],$$

[6]

K_{ii} et α_{ii} étant les polarisibilités magnétiques et optiques principales moléculaires.

---

* Dans ce qui suit, la polarisibilité électrique sera appelée polarisibilité optique car ne sera envisagée que celle résultant du champ électrique d'une onde lumineuse.

** L'anisotropie magnétique est souvent définie par l'expression $K_{ii} - K_{jj}$ [4–5]. Si $K_{11} = K_{22}$, l'anisotropie est alors $K_{33} - K_{11}$ identique à γ_M.

1. *Mesure des anisotropies magnétiques*

a. *Mesure directe.* — La molécule orientée n'est accessible que dans le monocristal. On détermine donc par l'une des deux méthodes de Krishnan [7–8] ou par une méthode directe [9–17] les susceptibilités principales du monocristal. Une structure faite aux rayons X permet de déterminer le système de cosinus directeurs qui repère le référentiel lié au cristal par rapport au référentiel lié à la molécule. Une transformation tensorielle conduit des susceptibilités principales cristallines aux susceptibilités principales moléculaires ou à leur différence et de là à l'anisotropie magnétique moléculaire.

Cette méthode est rigoureuse, précise mais longue et souvent difficile à mettre en œuvre. Il faut en effet fabriquer des monocristaux, faire une structure complète aux rayons X et des mesures magnétiques assez délicates. Le nombre de molécules dont les susceptibilités magnétiques principales moléculaires ont été ainsi mesurées est de l'ordre de 50. La liste exhaustive en est donnée dans les tables de Landolt et Börnstein [18].

b. *Méthode indirecte.* — Il est parfois commode de disposer d'une méthode rapide permettant d'apprécier l'ordre de grandeur de l'anisotropie magnétique. Comme on sait calculer avec une bonne approximation la susceptibilité magnétique moyenne d'une structure moléculaire à l'aide d'une systématique magnéto-chimique [4–5, 19–33], on peut connaître la susceptibilité magnétique moyenne de la structure kekuléenne de la molécule aromatique considérée.

De la mesure généralement facile de la susceptibilité magnétique moyenne et de la valeur calculée empiriquement de la susceptibilité d'une certaine structure de la molécule (de Kekulé par exemple), on peut déduire de manière plus ou moins heureuse l'écart entre la structure vraie et la structure supposée. L'interprétation de cet écart est parfois délicate, mais, de toute manière, le caractère aromatique est toujours lié à une exaltation du diamagnétisme qui correspond à un notable accroissement de la susceptibilité magnétique principale perpendiculaire aux plans aromatiques [34–35].

2. *Mesure des anisotropies optiques*

L'anisotropie optique est obtenue par une technique appelée diffusion Rayleigh dépolarisée (DRD) [36].

Lorsqu'un faisceau lumineux polarisé rectilignement frappe des molécules à l'état liquide ou dissous, le faisceau diffusé à 90° du faisceau incident a un vecteur électrique décomposable en une composante parallèle au vecteur électrique incident et une composante très faible i perpendiculaire à ce dernier.

On démontre que l'anisotropie optique γ_0^2 est proportionnelle à cette composante i mesurable à l'aide d'un appareil récemment mis au point [37–38]. L'anisotropie optique d'un grand nombre de molécules aromatiques ou non a été mesurée par cette méthode [39–56].

RÉSULTATS MAGNÉTIQUES

Le tableau 1 réunit un ensemble de résultats expérimentaux rassemblés par nos soins [18]. La première colonne contient des composés aromatiques à v noyaux benzéniques (1ère partie du tableau) et quelques composés non aromatiques (2ème partie du tableau). Les autres colonnes contiennent successivement la susceptibilité magnétique moyenne $\bar{\chi}$, les trois susceptibilités principales moléculaires K_{11}, K_{22} et K_{33} (on aura avantage à se

Tableau 1

Anisotropies magnétiques moléculaires de composés aromatiques et aliphatiques

	$\bar{\chi}$	K_{11}	K_{22}	K_{33}	γ_m^2	γ_m	$\dfrac{K_{11}+K_{22}}{2}$
Benzène	54,8	34,9	34,9	94,6	3.564	59,7	34,9
Dichloro-1,4 benzène	82,93	78,3	50,3	120,2	3.713	60,9	64,3
Bromo-1 chloro-4 benzène	92,16	87,6	59,9	129,0	3.628	60,2	73,7
Dibromo-1,4 benzène	101,4	97,1	70,5	136,7	3.329	57,7	83,8
Dinitro-1,4 benzène	69,3	64	38	106	3.532	59,4	51,0
Tribromo-1,3,5 benzène	138,0	123	122	170	2.257	47,5	122,5
Tetraméthyl-1,2,4,5 benzène	101,2	82,4	77,3	143,9	4.122	64,2	79,8
Hexaméthyl benzène	122,5	101,1	102,7	163,8	3.833	62	101,9
p-Nitraniline	66,6	52,0	43,0	104,8	3.344	57,8	47,5
Acétanilide	72,24	55,8	44,3	116,6	4.528	67,3	50,0
Acide anthranilique	79,0	57,7	58,8	120,5	3.876	62,2	58,2
Biphényle	104,4	67,7	61,7	183,8	14.212	59,6×2	64,7
Benzidine	$\begin{cases} 111,7 * \\ 112 \end{cases}$	76	76	184	11.664	54 ×2	76
Dibenzyle	$\begin{cases} 126,5 * \\ 125,6 \end{cases}$	87,4	81	208,5	15.481	62,2×2	84,2
Stilbène	115,2	85,8	50,1	209,6	21.020	72,5×2	67,9
Naphtalène	93,6	54,7	52,6	173,5	14.367	60 ×2	53,6
β-Naphtol	97,0	46,6	50,2	194,4	21.326	73 ×2	48,4
Acide *α*-naphtoïque	107,3	70,5	58,95	192,5	16.433	64 ×2	64,7
Acénaphtène	109,3	72,0	70,5	185,5	13.055	57,1×2	71,2
Diphényl-1,4 benzène	152,0	96,8	88,1	271,3	32.044	59,7×3	92,4
Anthracène	$\begin{cases} 130,1 * \\ 134,2 \end{cases}$	76,0	72,3	254,2	32.428	60 ×3	74,1
Acridine	$\begin{cases} 122,8 * \\ 123,3 \end{cases}$	61,3	70,7	237,9	29.616	57,3×3	66,0
Quaterphényle	201,3	122	110	372	65.644	64 ×4	116
Pyrène	149,4	78,71	64,71	304,6	54.389	58,3×4	71,7
Chrysène	160,7	88,0	83,3	310,8	50.709	56,3×4	85,6
Iodoforme	117,3	128,31	128,31	95,28	1.089	33,03	128,31
Hexachloréthane	111,8	101	116,92	117,48	262,5	16,2	109
Acide oxalique	56,1	53,13	52,73	62,40	89,8	9,5	52,93
Chloracétamide	51,27	51,70	48,74	53,37	17	4,1	50,22
N-Bromosuccinimide	74,96	76,03	61,57	87,31	497,8	22,3	68,8
N-Chlorosuccinimide	64,38	64,50	51,76	76,96	472,5	21,7	58,13
Succinimide	47,3	54,5	42,1	45,6	122,6	11,1	48,3
Acide succinique	58,36	52,70	57,12	65,26	122,7	11,1	54,91
Acide adipique	81,41	72,58	81,65	90,00	227	15,1	77,11
Tetraiodoéthylène	158,0	170,26	160,65	143,09	572,8	23,9	165,4
Acide maléique	47,8	42,39	40,38	60,63	371,6	19,3	41,38

Les valeurs des susceptibilités magnétiques sont données au facteur -10^{-6} près

* La susceptibilité est donnée dans les tables; cependant, la valeur située en-dessous correspond à la moyenne des susceptibilités principales, nombre qui devrait être le même, pour des raisons évidentes de cohérence

A. Pacault

reporter à la référence [18] pour voir comment furent choisis les axes magnétiques principaux), l'anisotropie magnétique γ_m^2, γ_m et enfin l'expression

$$\frac{K_{11} + K_{22}}{2}.$$

On constate que

1. L'anisotropie γ_m des composés benzénoïdes est nettement supérieure à celle des composés aliphatiques. Cette dernière est due à la présence de substituants polarisables ou de doubles liaisons.

2. Quel que soit le type de substituant l'anisotropie magnétique des composés benzénoïdes est approximativement de $-60 \cdot 10^{-6}$ cgs par noyau benzénique [57–66].*
La présence de substituants variés modifie les valeurs de K_{11} et K_{22} mais laisse pratiquement inchangée la valeur de γ_m par cycle.

On notera cependant, qu'à la précision des mesures, des écarts entre la valeur expérimentale et $-60 \cdot 10^{-6}$ sont significatifs, mais leur amplitude maximale de 10% environ est un phénomène de deuxième ordre qui trouve tout naturellement son explication dans la nature des substituants et la variété des substitutions. A ce niveau de constatation on peut enregistrer des différences mais il est impossible de tirer des lois générales.

Il serait intéressant d'avoir des renseignements sur des molécules dont on ne connait que les susceptibilités magnétiques moyennes. Malheureusement les données expérimentales complètes comme celles du tableau 1 sont absentes, et les résultats obtenus par la méthode indirecte qui permet de calculer γ_m à partir des susceptibilités magnétiques moyennes mesurées et d'une systématique doivent faire l'objet d'une étude critique d'ensemble actuellement en cours.

RÉSULTATS OPTIQUES

Les mesures magnétiques permettent comme on vient de le voir de déterminer le tenseur des susceptibilités magnétiques de molécules à l'état solide $-\gamma_m^2$ est calculé à partir de K_{11}, K_{22}, K_{33}, $\bar{\chi}$ mesurés.

La DRD en revanche donne directement la valeur de l'anisotropie optique γ_0^2 de molécules dans l'état liquide, soit que les molécules soient de même type (étude du liquide pur) soit mélangées à d'autres (étude des solutions). On s'arrange naturellement, lorsque cela est possible, pour que l'anisotropie optique du solvant soit faible devant celle du soluté mais de toute manière on devra noter que l'anisotropie optique ainsi déterminée est une anisotropie optique moléculaire apparente puisque la polarisabilité de la molécule dont on voudrait connaître l'anisotropie optique est perturbée par la présence de ses voisines.

L'effet est très sensible comme on peut le voir sur le tableau 2, qui donne les anisotropies optiques d'alcanes normaux C_nH_{2n+2} ($5 < n < 17$) liquides ou en solution dans le cyclohexane et le tetrachlorure de carbone.

* a. Le tribromo-1, 3, 5 benzène et le β-naphtol s'écartent de cette règle. Refaire les mesures serait peut-être intéressant.
 b. A l'anisotropie des noyaux benzéniques du stilbène s'ajoute celle de la double liaison.

42

Tableau 2

Anisotropies optiques moléculaires des alcanes normaux (C_nH_{2n+2})

n	γ_0^2 liq. pur	γ_0^2 dans le cyclohexane	γ_0^2 dans le tetrachlorure de carbone
5	3,03	3,03	2,82
6	4,48	4,29	3,67
7	6,10	5,43	4,51
8	7,96	6,64	5,75
9	10,20	8,05	6,76
10	13,22	9,73	7,82
11	15,31	10,86	9,03
12	18,06	12,65	10,21
13	21,24	13,75	10,89
14	24,87	15,07	12,15
15	28,76	16,49	13,25
16	33,03	17,99	14,21
17	37,42	18,67	15,02

Les anisotropies optiques de différents corps ne pourront donc être comparées que sous réserve qu'elles aient été déterminées dans les mêmes conditions et en particulier dans le même solvant.

Le tableau 4 réunit les anisotropies optiques de quelques molécules aromatiques substituées à l'état pur et en solution dans le cyclohexane et le tetrachlorure de carbone.

On constate que

1. L'anisotropie optique γ_0^2 des composés benzenoïdes est nettement supérieure à celles des composés aliphatiques ayant le même nombre d'atomes de carbone.

2. La nature et la position du substituant modifient profondément la valeur de γ_0^2.

Même en l'absence de substituant on trouve que, dans le même solvant, le tetrachlorure de carbone par exemple, chaque cycle des composés polynucléaires a une anisotropie optique $[\gamma_0^2]$ différente.

On trouve en effet

Tableau 3

	γ_0^2		ν^2		$[\gamma_0^2]$	$[\gamma_0]$
Benzène	36	=	1	$\times$	36	6
Naphtalène	219	=	4	$\times$	54,7	7,4
Phénanthrène	563	=	9	$\times$	62,5	7,9
Anthracène	760	=	9	$\times$	84	9,16
Acridine	850	=	9	$\times$	94	9,7
Benz [a] anthracène	1.966	=	16	$\times$	128	11,3
Benz [a] acridine	1.770	=	16	$\times$	110	10,5

Tableau 4

Anisotropies optiques de composés aromatiques

Composés	$\gamma_0^2(\text{Å}^6)$ $\lambda = 5460\,\text{Å}$
Benzène	36
Toluène	43,5
ortho-Xylène	
méta-Xylène	51,5
para-Xylène	
Triméthyl-1,3,5 benzène	58,7
Triméthyl-1,2,3 benzène	58
Hexaméthylbenzène	85
Ethylbenzène	45
Propylbenzène	46,5
Isopropylbenzène	45
n-Butylbenzène	47,5
t-Butylbenzène	46,8
Fluorobenzène	46,8
Chlorobenzène	70
ortho-Dichlorobenzène	
méta-Dichlorobenzène	101
para-Dichlorobenzène	120
Trichloro-1,2,3 benzène	137
Hexachlorobenzène	280
Bromobenzène	84,5
ortho-Dibromobenzène	
méta-Dibromobenzène	122,5
para-Dibromobenzène	173
Tribromo-1,3,5 benzène	165
Tetrabromo-1,2,4,5 benzène	290
Iodobenzène	119
ortho-Diiodobenzène	162
méta-Diiodobenzène	191
para-Diiodobenzène	292
Phénol	45,72
ortho-Crésol	
méta-Crésol	54
para-Crésol	54
Di-*t*-butyl-2,6 méthyl-4 phénol	73,8
Chlorure de benzyle	49,7
Phényl-1 chloro-2 éthane	61
Phényl-1 chloro-2 propane	55,5
Bromure de benzyle	57
Phényl-1 bromo-2 éthane	66
Phényl-1 bromo-3 propane	59,4
Phényl-1 iodo-2 éthane	90

On notera cependant que les mêmes substituants perturbent de la même manière les molécules qui les contiennent. C'est ainsi que les substituants méthyl qui furent très étudiés [39–56] apportent la même contribution à l'anisotropie optique totale quelle que soit la molécule qui les contient, c'est-à-dire qu'il y a additivité tensorielle des anisotropies optiques.

Conclusion

De cet exposé rapide de synthèse qui suppose, de la part du lecteur intéressé, qu'il se reporte aux mémoires, il ressort que

1. 'L'aromaticité' engendre des polarisabilités magnétique et optique beaucoup plus grandes que celle des composés aliphatiques correspondants.

2. Cependant, les renseignements fournis par ces polarisabilités ne sont pas identiques mais complémentaires. En effet, le cycle benzénique, par exemple, a une polarisabilité magnétique spécifique de $-60 \cdot 10^{-6}$ environ alors que sa polarisabilité optique dépend de la molécule dans laquelle il se trouve.

Le calcul théorique des deux polarisabilités mené de façon parallèle et comparative n'a pas été fait jusqu'à présent, d'une façon telle qu'il puisse expliquer ce dernier fait. Une recherche dans cette voie serait intéressante.

Références

1 L. Pauling (1936) *J. Chem. Phys.*, 4 : 673.
2 F. London (1937) *J. Phys. Rad.*, 8 : 397.
3 J. A. Pople (1964) *J. Chem. Phys.*, 41 : 2559.
4 A. Pacault (1949) *Bull. Soc. Chim. Fr.*, p. D 40.
5 Idem (1954) *Experientia*, 10 : 41.
6 M. Kerker (1969) *The Scattering of Light*, Academic Press, New York.
7 K. S. Krishnan (1933) *Phil. Trans.* (A), p. 231.
8 *Ibid.* (1935) p. 234.
9 A. Pacault, B. Lemanceau & J. Joussot-Dubien (1953) *C.r. Acad. Sci.* (Paris), 237 : 1156.
10 J. Joussot-Dubien, B. Lemanceau & A. Pacault (1956) *J. Chim. Phys.*, p. 198.
11 J. Joussot-Dubien & B. Lemanceau (1956) *C.r. Acad. Sci.* (Paris), 242 : 1170.
12 A. Pacault, B. Lemanceau & J. Joussot-Dubien, *ibid.*, p. 1305.
13 J. Hoarau, N. Lumbroso & A. Pacault, *ibid.*, p. 1702.
14 A. Pacault, B. Lemanceau & J. Joussot-Dubien (1957) *Actas do Congresso*, I, Lisbon, pp. 1–8.
15 A. Pacault, J. Hoarau, J. Joussot-Dubien, B. Lemanceau & N. Lumbroso (1957) in : *Proceedings of 3rd Conference on Carbon*, Pergamon, p. 43.
16 J. Joussot-Dubien (1959) *C.r. Acad. Sci.* (Paris), 248 : 3165.
17 E. Poquet, A. Pacault, J. Hoarau, N. Lumbroso & J. V. Zanchetta (1960) *ibid.*, 250 : 706.
18 J. Favède, J. Hoarau & A. Pacault (1967) in: *Landolt & Börnstein*, Vol. II, Part 10, Springer Verlag, p. 141.

19 P. Pascal (1910) *Ann. Chim.*, p. 5.
20 *Ibid.* (1912) p. 289.
21 *Ibid.* (1913) p. 218.
22 A. Pacault & N. Pacault (1944) *C.r. Acad. Sci.* (Paris), 218 : 671.
23 P. Pascal & A. Pacault, *ibid.*, 219 : 599.
24 *Ibid.*, p. 657.
25 A. Pacault (1944) *Rev. Sci.*, pp. 465–479.
26 P. Pascal & A. Pacault (1946) *C.r. Acad. Sci.* (Paris), 222 : 619.
27 N. Pacault, *ibid.*, p. 1089.
28 A. Pacault (1946) *Ann. Chim.*, p. 527.
29 Idem (1948) *Rev. Sci.*, 3288 : 38.
30 P. Pascal, A. Pacault & J. Hoarau (1951) *C.r. Acad. Sci.* (Paris), 233 : 1078.
31 E. D. Bergmann, H. Hoarau, A. Pacault, A. Pullman & B. Pullman (1952) *J. Chim. Phys.*, 49 : 474.
32 J. Hoarau, A. Pacault & P. Pascal (1956) *Cah. de Phys.*, 74 : 30.
33 H. Shiba & G. Hazato (1949) *Bull. Chem. Soc. Japan*, 22 : 92.
34 J. Hoarau (1956) *Ann. Chim.*, 1 : 544.
35 P. Bothorel (1968) *J. Colloid Sci.*, 27 : 529.
36 J. J. Piaud (1962) *J. Chim. Phys.*, 59 : 215.
37 P. Bothorel & J. J. Piaud (1967) Brevet France du 12.4.67, No. 102348.
38 A. Rousset & A. Pacault (1954) *C.r. Acad. Sci.* (Paris), 238 : 1705.
39 P. Bothorel, A. Pacault & A. Rousset (1956) *Cah. de Phys.*, 71/72 : 66.

40 A. Unanue & P. Bothorel (1964) *Bull. Soc. Chim. Fr.*, p. 573.

41 C. Clement & P. Bothorel (1964) *J. Chim. Phys.*, p. 878.

42 *Ibid.*, p. 1262.

43 A. Unanue & P. Bothorel (1965) *Bull. Soc. Chim. Fr.*, p. 2827.

44 Luong the Man, C. Clement & P. Bothorel (1965) *C.r. Acad. Sci.* (Paris), 260 : 1405.

45 A. Unanue & P. Bothorel (1966) *Bull. Soc. Chim. Fr.*, p. 1640.

46 P. Bothorel, A. Unanue, C. Gardere, N. P. Buu-Hoi, P. Jacquignon & F. Perin, *ibid.*, p. 2920.

47 G. Fourche, A. Pacault, P. Bothorel & J. Hoarau (1966) *C.r. Acad. Sci.* (Paris), 262 : 1813.

48 R. Lapouyaoe & P. Bothorel (1967) *ibid.*, 265 : 707.

49 H. R. Craig & P. Bothorel, *ibid.*, p. 1384.

50 P. Bothorel & A. Unanue (1968) *Bull. Soc. Chim Fr.*, p. 754.

51 P. Bothorel (1968) in : *Mém. Soc. Sci. Phys. Nat. Bordeau* (*Vol. Spécial, Congrès AFAS*), p. 33.

52 G. Fourche (1968) *J. Chim. Phys.*, 65 : 1500.

53 A. Caristan, P. Bothorel & H. Bodot (1969) *ibid.*, 66 : 1009.

54 A. Caristan, H. Bodot & P. Bothorel (1969) *Bull. Soc. Chim. Fr.*, p. 2589.

55 P. Foulani & C. Clement (1969) *Bull. Soc. Chim. Fr.*, p. 3462.

56 K. Lonsdale & K. S. Krishnan (1936) *Proc. Roy. Soc.*, 156 : 597.

57 K. Lonsdale (1936) *Nature*, 137 : 826.

58 Idem (1936) *Z. Kristallogr.*, 95 : 471.

59 Idem (1937) *Proc. Roy. Soc.*, 159 : 149.

60 Idem (1937) *Z. Kristallogr.*, 97 : 91.

61 Idem (1938) *Rep. Prog. Phys.*, 4 : 368.

62 Idem (1939) *Proc. Roy. Soc.*, 171 : 541.

63 Idem (1940) *Nature*, 145 : 148.

64 *Ibid.*, p. 57.

65 Idem (1947) *Sci. Prog.*, p. 137.

66 C. J. B. Clews & K. Lonsdale (1937) *Proc. Roy. Soc.*, A 161 : 493.

Discussion

R. West:

I would like to give some of our results on direct measurement of diamagnetic anisotropy, which may be of interest to the audience. After we made this measurement, thus increasing the number of compounds studied from 50 to 51, I can understand why so few compounds have been studied by this technique.

The compound investigated was diammonium croconate, a salt of

an anion of the oxocarbon family. This salt shows a diamagnetic anisotropy of -55×10^{-6} CGS units. This number is of interest because it is, we believe, the highest diamagnetic anisotropy ever measured for a non-benzenoid delocalized system.

E. D. Bergmann:

1. Can one derive from your data on aromatic chlorine compounds a 'standard' atomic anisotropy for chlorine, as you have done for the benzene ring?
2. Obviously, the optical anisotropy must give an indication of the conformation of the molecule. Could one, e.g., decide between a planar and a twisted biphenyl, thus perhaps developing a method which would parallel the Kerr effect?

A. Pacault

1. On considère généralement que le chlore a une susceptibilité isotrope et si, comme il est probable, l'atome en combinaison présentait une légère anisotropie, elle se trouverait, par le jeu du calcul d'additivité, reportée sur le cycle.
2. C'est, en effet, une des possibilités de la DRD que de pouvoir différencier les molécules planes de celles qui ne le sont pas. La méthode est généralement plus sensible que l'effet Kerr.

Faraday Effect, Ring Current Concept and Aromaticity

by J. F. LABARRE *and* F. GALLAIS

Department of Inorganic Chemistry, Faculty of Science, Toulouse

IT IS AT PRESENT a well-established fact that magnetic rotatory power, or the Faraday effect, is what is commonly called an additive property of matter. This assertion rests on the fact that there exists [1–4] a set of additive bond magnetic rotations by means of which the experimental molecular rotation of a compound may be calculated *a priori*, often with surprising precision. The method is applicable to compounds of carbon, boron, sulphur, phosphorus, nitrogen, etc., with the single stipulation that they contain only normal covalent, diamagnetic and localized bonds. The existence of this set has been demonstrated on several occasions by utilizing a semi-empirical approach [5–6] or by adjusting the Loges theory of Daudel to the Faraday effect [7].

It may be concluded from these calculations that any digression from such a law of additivity necessarily implies the presence of certain structural peculiarities in the molecule. This is true, for instance, for molecules with a delocalized system of π electrons. For such molecules, it is noted that the experimental value (A) of the magnetic rotation is always definitely higher than the theoretical value (B), i.e., the figure calculated by neglecting the delocalization of electrons. The difference $E = (A) - (B)$ — which we will call the 'magnetic conjugation excess' — appears [8] to be constant for a given unsubstituted conjugated system, but varies appreciably from one system to another.

In the domain of aliphatic compounds we have shown [9] that E can be linked linearly to the sum, ΣI_r, of the free-valence indices of the corresponding system by the equation

$$D_{al} / E_{al} = 111 \ (\Sigma I_r - 1.464).$$

This equation has recently appeared also in the framework of a simplified theory of the Faraday effect [10].

The magneto-optical study of a great number of unsubstituted arenes [11] has enabled us to construct an equation for the aromatic hydrocarbons that is analogous to that for the aliphatic compounds:
$$D_{ar} / E_{ar} = 200 \ (\Sigma I_r - 1.464).$$

It may be noted that the straight lines D_{ar} and D_{al} originally have the same abcissa, and that the slope of the former is practically double that of the latter. This means that for a given conjugated system, characterized by a value $(\Sigma I_r)_0$, the magnetic rotation excess, E_{ar}, which is observed if the molecule is cyclic and planar, is very much higher than E_{al}, which would be observed if the molecule were alicyclic (and, therefore, non-planar).

48

$\Delta E = (E_{ar} - E_{al})$ expresses, in a way, the passage from open chain conjugation to a cyclic or 'aromatic' conjugation. The existence of ΔE seems to be due to the fact that an aromatic molecule offers to the delocalized π electrons the possibility of having their mobility increased by the action of the applied magnetic field. We therefore believe that ΔE expresses the existence, in an aromatic molecule, of the Pauling-Pople ring current. It is not surprising that such currents may be detected by means of the Faraday effect. This effect is a magnetic property of matter to the same extent, for instance, as diamagnetism or nuclear magnetic resonance, in connection with which the concept of ring current had been mentioned for the first time [12–13].

It might therefore be expected from the preceding discussion that any molecule characterized by a given ΣI_r value should have the same conjugation excess. If one studies, with the aim of verifying the above, a great number of mono- and polysubstituted benzenes, it immediately becomes clear that this is not so. Table 1 shows some of the E values we have obtained [14]. It may be seen that these are sometimes slightly higher and sometimes definitely lower than the characteristic benzene value ($+182\ \mu r$), even though it is evident that for all these molecules ΣI_r remains practically constant, at least as long as this quantity is calculated only along the benzene ring.

What, then, can the electronic or geometrical factor be on which the conjugation excess of a molecule, and therefore its aromaticity, depends so strongly? The answer may be supplied by an examination of the electronic structures that we have calculated for about sixty diversely substituted benzenes, utilizing the LCAO–UVC method perfected for our purpose in collaboration with Julg [15], and the so-called 'bond-by-bond iteration method' [16]. It is noted that if the bond characters p_{rs} remain very close to that observed for benzene (0.667), the gradient

$$G = \frac{1}{6} \sum_{i=1}^{6} \left| q_i - q_{i+1} \right| \qquad \text{(with } q_7 \equiv q_1)$$

of the $(\sigma + \pi)$ electronic charges, localized on the six carbon atoms of the benzene ring, diverges more or less definitely (Table 1) from the zero value it displays in benzene. This charge gradient is then responsible for the potential barriers that appear at the peaks of the ring, barriers which have the effect of decreasing the ability of the delocalized π electrons to generate a ring current.

It appears, therefore, that the aromaticity of a molecule, which obviously depends on the density of the delocalized π electrons, is in fact also — and, actually, most markedly — a function of the σ-electron densities of the molecule. Thus, we encounter one of the fundamental objections raised by Musher to the concept of the ring current, at least when this concept is developed in its classical form. Moreover, our magneto-optical investigations show that, at the limit, a molecule might possess a very high density of delocalized π electrons without being the seat of a ring current, if only the G gradient is considered as important. Fig. 1 justifies the reasoning that we have developed; the figure shows that there exists a very significant linear relation between E and G.

These results have therefore led us to propose the following definition for the aromaticity of a molecule:

A cyclic molecule is aromatic when it contains a distribution of delocalized π electrons susceptible to be set in ordered movement when submitted to a magnetic field, follow-

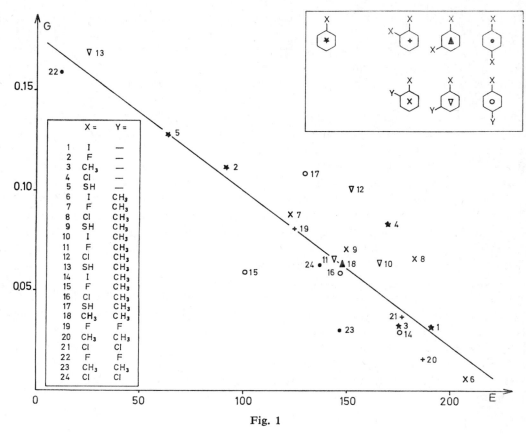

Fig. 1

ing the absence of an important gradient of localized $(\sigma + \pi)$ charges on the atoms of the ring, to give rise to a Pauling-Pople current.

This aromaticity will then either be measured experimentally by means of the Faraday effect (measure of E) or calculated theoretically from the $(\sigma + \pi)$ electronic structure (determination of G at equal p_{rs}). We therefore have two equivalent means to define the aromaticity of a compound relative to other compounds on the express condition that they all belong to the same family, that is, on condition that the distribution of the Coulomb integrals along the ring belongs to the same group.

This important result is a consequence of the following observation: If one seeks to represent E as a function of G for compounds belonging to the D_{3h} group (borazines and boroxines) [17–18] or to the C_{2v} group (furan, pyrrole, thiophene) [19], one also obtains in each case a linear relation. The slope of the representative straight line is similar to that shown in Fig. 1, but under no circumstances may it be confused with the latter. There seems, therefore, to exist a straight line, $E = f(G)$, for each of the groups having a distribution of $(\sigma + \pi)$ charges localized on the ring peaks. Here, again, we meet the second objection raised by Musher to the concept of the ring current.

From the preceding discussion it seems that a comparison of the aromaticity of benzene with that of a molecule not belonging to the D_{6h} 'group' should be strictly avoided. If we have any right to assign values for the aromaticity of various substituted benzenes studied

Table 1

E and *G* Values

for Some Substituted Benzenes

No.	$E(\mu r)$	G
6	208	0.006
1	191	0.031
20	187	0.016
8	183	0.066
21	177	0.037
14	176	0.029
3	175	0.032
4	170	0.083
10	166	0.064
12	152	0.101
9	150	0.071
18	148	0.063
23	147	0.030
16	147	0.059
24	137	0.063
17	130	0.109
19	125	0.081
7	123	0.088
2	91	0.112
11	66	0.144
5	63	0.138
15	59	0.101
13	25	0.169
22	12	0.159

in relation to benzene itself, it is because the UV consistent Coulomb integrals of carbon atoms bearing a substituent X diverge little, if at all, from the α value, at least when X = R, F, Cl, Br, I or SH. This does not hold for the aminobenzenes, for which the consistency with UV compels one to assume the relation $\alpha_{C(NH_2)} = \alpha - 0.6\beta$ [20], which, in this case, confers on the benzene ring a symmetry which is more C_{2v} than D_{6h}.

There is another magneto-optical criterion of aromaticity which in 1967 seemed very promising, in view of its simplicity of use. This criterion had been suggested by the NMR experiments carried out by Zimmerman and Foster [21]. These authors showed effectively that the signal of the benzene proton is displaced towards a lower field when benzene is diluted with an inert chemical solvent. Pople thought that this deshielding could be related to the evolution of a ring current during dilution.

An analogous experiment was attempted using the Faraday effect. We noted that the molecular magnetic rotation of benzene decreased — but in a perfectly linear manner — with increasing dilution in a solvent such as *n*-hexane. This phenomenon, we thought, afforded us a simple means for measuring the 'intensity' of the ring current of a molecule. We thought it probable — and we believed that we had verified this in a previous study [22] — that the decrease in magnetic rotation would be proportional to the intensity of the ring current.

The first measurements, carried out on fluorobenzene, on *para*-difluorobenzene and on chloro- and bromobenzene [22] in solution in *n*-hexane, allowed us to obtain a scale of relative aromaticity identical to that deduced from the measurements of E, as well as from calculations of G. But this agreement disappeared slowly as soon as the number of molecules studied increased. Moreover, we were led to note that the slope of the straight line obtained for a given aromatic compound — iodobenzene, for instance [23] — varied from one solvent to another in such a pronounced way that chemical solute-solvent interactions did not suffice to account for it.

It is for this reason that we decided to investigate anew the 'criterion for aromaticity in solution'. We measured the evolution of the molecular rotation of benzene in about fifteen usual solvents. The results obtained are presented in Fig. 2 and in Table 2; they call for the following conclusions:

1. The variation of $(\rho)_M$ is linear for a very large series of concentrations, as may be seen from the values of the correlation coefficient r.

2. The decrease observed in the case of *n*-hexane, which we had hoped would enable us to establish a criterion for aromaticity in solution, is not at all general. It is seen that in certain cases an increase, rather than a decrease, in $(\rho)_M$ is observed.

3. As to the sequence in which the investigated solvents could be arranged, it is not clear

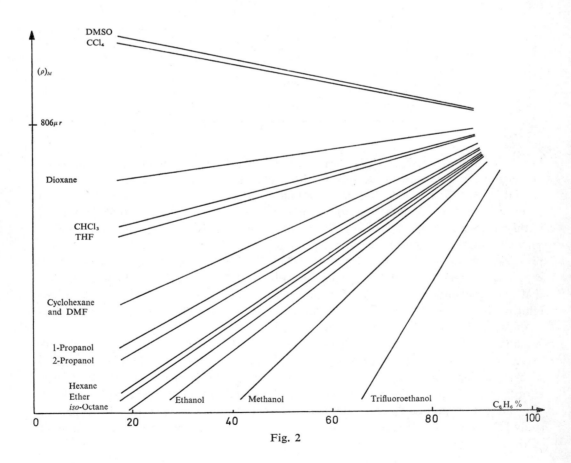

Fig. 2

Table 2

Linear Relations between y (molecular magnetic rotation of benzene)
and x (concentration in some usual solvents)

Solvent	Equation	r	n_D^{20}	P
Trifluoroethanol	$y = 0.856x + 780.3$	0.99	1.2910	0.182
Methanol	$y = 0.498x + 762.8$	0.99	1.3288	0.203
Ethyl ether	$y = 0.346x + 773.8$	0.99	1.3526	0.217
Ethanol	$y = 0.399x + 771.5$	0.99	1.3611	0.221
n-Hexane	$y = 0.338x + 773.0$	0.99	1.3749	0.229
2-Propanol	$y = 0.294x + 779.2$	0.98	1.3776	0.230
1-Propanol	$y = 0.287x + 781.0$	0.99	1.3850	0.234
iso-Octane	$y = 0.358x + 775.0$	0.99	1.3915	0.238
THF	$y = 0.140x + 794.0$	0.96	1.4076	0.247
Dioxane	$y = 0.072x + 801.9$	0.95	1.4224	0.254
Cyclohexane	$y = 0.229x + 787.2$	0.99	1.4266	0.257
DMF	$y = 0.231x + 788.9$	0.97	1.4269	0.257
Chloroform	$y = 0.130x + 798.3$	0.99	1.4433	0.265
CCl$_4$	$y = -0.103x + 818.5$	0.97	1.4601	0.274
DMSO	$y = -0.111x + 819.6$	0.99	1.4780	0.284

at first sight whether the macroscopic polarity of the solvents is the determining factor. It is, in fact, surprising that CCl$_4$ and DMSO have almost identical slopes.

It is customary to consider solute-solvent interaction phenomena as directly dependent upon two factors: (i) upon the polarity of the solvent; and (ii) upon the polarizability of the constituents of the interacting system. In view of this, it seemed to us that the next logical step would be to investigate whether the variation in $(\rho)_M$ was, in fact, a function of polarizability, rather than solvent polarity.

We calculated for each solvent the molecular polarizability, P, utilizing the Lorentz approximation:

$$P = \frac{(n_D^{20})^2 - 1}{(n_D^{20})^2 + 2}.$$

We observed (Fig. 3) that there exists an almost linear relation between the value of P and the slope of the lines obtained in the Faraday effect experiments [24].

It seems clear, therefore, that the decrease in $(\rho)_M$ is not directly dependent on the variation in the ring current intensity of a compound dissolved in a given solvent. This decrease appears to be due rather to solute-solvent interactions that modify, to a greater or lesser extent, the internal electric field of the medium that is submitted to an external magnetic field.

One is compelled to conclude that the study of the magneto-optical behaviour of aromatic molecules in solution does not seem to be utilizable as a criterion of aromaticity. This conclusion also appears to be applicable to the analogous results obtained in NMR studies; it is noted that it has been simultaneously proposed by our group [24] and by that of Falk et al. [25].

To sum up, it appears preferable to retain the criteria of '($\sigma + \pi$) charge gradient' and 'conjugation magneto-optical excess' for the estimation of the relative aromaticity of a cyclic compound, whether organic or inorganic. The 'dilution' criterion should be abandoned, at least in our present state of knowledge.

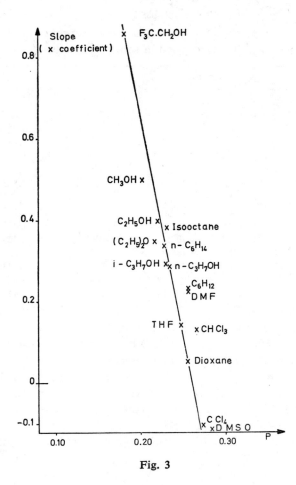

Fig. 3

Appendix

Some Considerations on the Term 'Aromaticity'

by Jean-François Labarre

We have seen that the measurements of the Faraday effect and the knowledge of the $(\sigma + \pi)$ electronic structures enable us to define the relative aromaticity of a molecule in a manner which is very similar to that utilized in NMR studies, that is, based on the ring current concept. However, it is evident that the term 'aromaticity' is not universally synonymous with 'ring current'. Thus, for the organic chemist, for instance, aromaticity may imply the ease with which a molecule may undergo nitration or sulphonation, whereas for others aromaticity means equality of bond lengths, high stability, high diamagnetic anisotropy, UV wavelengths close to the visible, or strong spin delocalizability (in ions and radicals). It seems, therefore, that the term 'aromaticity' has lost all simple meaning, and, in fact, one finds oneself incapable of defining it. A further difficulty arises from the fact that it is practically impossible to measure the aromaticity of a molecule, and that even the relative scales proposed by means of the different chemical or physical phenomena mentioned above do not generally agree.

We believe that the term 'aromaticity' should be eliminated from the scientific vocabulary in order to avoid the ambiguity that consists in postulating that 'whoever says "aromatic" says "benzenic" and *vice versa*'. We have seen that certain substituted benzenes, characterized by very weak E values and by a significant G gradient, no longer deserve to be called aromatic. Moreover, the considerable number of prefixes joined to the term 'aromaticity' (non-, anti-, quasi-, pseudo-, homo-, etc.) indicates sufficiently that this term is outdated.

The solution to this problem is not, as Coulson has remarked [private communication], to look for a new word to replace aromaticity. It is preferable to propose several (as meaningful as possible) new words, so that each one of them refers to one of the multiple chemical or physical properties that were intended to be expressed by the term 'aromaticity'.

We have considered introducing the concept of potential strobilism to describe the phenomenon measured by the Faraday effect and by NMR in the case of cyclic molecules that are the seat of a π-electronic delocalization, that is, of a ring current. This new term is not phonetically pleasant; however, it has been suggested to us by C.K. Jorgensen, on the basis of the Greek phrase ξύλινα ἄλογα στροβιλιζόμενα ('children's merry-go-round'). Strobilism would then potentially exist in the molecule, awaiting to be rendered observable by a developer (in the 'photographic' sense of the word), which is, in our case, the external magnetic field. If strobilism is a potential property of the molecule at rest, it may be conceived that it is directly dependent upon the electronic and geometrical characteristics of the molecule in its fundamental state — e.g., electron distribution (π or σ), symmetry group, etc.

The concept of potential strobilism gives rise to a number of questions which, even if they cannot be answered in the near future, deserve serious consideration:

1. One is justified in asking if the terrestrial magnetic field is not sufficient to reveal the

potential strobilism of a molecule (as it suffices to reveal the NQR of the nuclei), and whether it is not responsible for the possibilities of nitration and sulphonation of molecules. The answer would be supplied by an experiment of the type achieved in an 'anti-H' room; the violent reaction that would probably be observed would cause a certain sensation.

2. It may be that potential strobilism may only be revealed in its totality by a magnetic field whose intensity is higher than a certain saturation threshold. If this were so, it might be expected that the yield of a substitution reaction would be increased by performing the reaction in an intense magnetic field. Here, too, experiments in magneto-catalysis remain to be done. However, we have already a certain number of arguments that allow us to believe in the validity of our reasoning: the molecular magnetic rotation of benzene (and thus its strobilism) decreases in a nearly exponential manner with the value of the external magnetic field. This phenomenon seems to occur only for molecules that are the seat of a ring current, a current which we shall call a strobilic current.

To conclude this thought, we would like to present a visual image of the situation in which we find ourselves. Chemists and physicians are at present in the middle of a cavern which Plato would not have disavowed. They observe on the walls of the cavern 'certain shadows': an agreeable odour, an aptitude to nitration and sulphonation, a ring current, a magneto-optical excess, a diamagnetic anisotropy, a resonance energy and even, for some, a mathematical term. Do these shadows all belong to the same invisible reality? No one is able, at present, to answer such a question.

It is desirable, therefore, that at the present stage one should try to define and to measure whatever is measurable with the existing techniques. It is hoped that the data thus accumulated will eventually fit into a unified theory.

REFERENCES

1 F. Gallais & D. Voigt (1960) *Bull. Soc. Chim. Fr.*, p. 70.
2 F. Gallais, D. Voigt & J. F. Labarre (1960) *ibid.*, p. 2157.
3 R. Daudel & F. Gallais (1969) *Rev. Chim. Min. Fr.*, 6: 61.
4 F. Gallais, *ibid.*, p. 71.
5 F. Gallais, J. F. Labarre, D. Voigt & P. de Loth (1966) *J. Chim. Phys.*, 63: 1175.
6 J. F. Labarre & M. C. Labarre (1967) *ibid.*, 64: 1670.
7 R. Daudel, F. Gallais & P. Smet (1967) *Int. J. Quant. Chem.*, 1: 873.
8 J. F. Labarre (1963) *Ann. Chim. Fr.*, 8: 45.
9 F. Gallais & J. F. Labarre (1964) *J. Chim. Phys.*, 61: 717.
10 J. F. Labarre (1969) *ibid.*, 66: 1155.
11 J. F. Labarre, P. de Loth & M. Graffeuil (1966) *ibid.*, 63: 460.
12 L. Pauling (1936) *J. Chem. Phys.*, 4: 673.
13 J. A. Pople, W. G. Schneider & H. J. Bernstein (1959) *High Resolution Nuclear Magnetic Resonance*, McGraw-Hill, New York, p. 180.
14 J. F. Labarre, F. Crasnier & J. P. Faucher (1966) *J. Chim. Phys.*, 63: 1088.
15 J. F. Labarre, A. Julg & F. Crasnier (1965) *C. r. Acad. Sci.* (Paris), 261: 4419.
16 F. Gallais, D. Voigt & J. F. Labarre (1965) *J. Chim. Phys.*, 62: 761.
17 J. F. Labarre, F. Gallais & M. Graffeuil (1968) *ibid.*, 65: 638.
18 J. F. Labarre, M. Graffeuil, J. P. Faucher, M. Pasdeloup & J. P. Laurent (1968) *Theor. Chim. Acta* (Berlin), 11: 423.
19 J. Devanneaux & J. F. Labarre (1969) *J. Chim. Phys.*, 66: 1780.
20 B. Graziana (1969) Ph. D. Thesis, University of Toulouse.
21 J. R. Zimmerman & M. R. Foster (1957) *J. Phys. Chem.*, 61: 282.
22 J. F. Labarre & F. Crasnier (1967) *J. Chim. Phys.*, 64: 1664.
23 M. Bonnafous, B. Graziana, F. Crasnier & J. F. Labarre (1969) *ibid.*, 66: 462.
24 J. F. Labarre, R. Moezi & J. F. Keruzore (1969) *ibid.*, 66: 2010.
25 K. Bauer, H. Eberhardt, H. Falk, G. Haller & H. Lehner (1970) *Monatsh. Chem.*, 101: 469

Discussion

G. Berthier:

Could you give us some details concerning your definition of the charge gradient with respect to the usual charge distributions?

J. F. Labarre:

My charge gradient, for benzene or for any substituted molecule, is equal to

$$\frac{1}{6} \sum_{i=1}^{6} \left| q_i - q_{i+1} \right|$$

(with q_7 identical to q_1, of course). I know that this notion is theoretically not very convenient, but it allows us to explain all our experiments.

I want to say one thing more: the 'strobilism' I have proposed is dependent on the group symmetry of the molecule (as defined in the paper) and it can be measured.

Some Comments on Homoconjugation
in Unsaturated Compounds and in Bridged Annulenes

by E. HEILBRONNER

Physikalisch-Chemisches Institut der Universität Basel

THE INTERACTION between n basis orbitals ϕ_j of orbital energy A_j leads to n orbitals $\psi_J = \Sigma_j C_{Jj}\phi_j$ of energy ε_J. If the system is occupied by $2n$ electrons, i.e., if all orbitals are filled, then

$$E = 2 \sum_j A_j = 2 \sum_J \varepsilon_J.$$

Thus, the total energy E is independent of the size and sign of integrals $\beta_{ij} = \langle \phi_i | H | \phi_j \rangle$ [1]. Second-order interaction with the anti-bonding basis orbitals $\phi_j{}^*$ (orbital energy $A_j{}^*$) will result in a second-order depression of E. However, this effect will be small, as the following particular example of three interacting double bonds shows:

We assume that the three double bonds are arranged in a C_{3v} pattern, as for example in *cis-cis-cis*-1,4,7-cyclononatriene (I).

I C_{3v}

The parameters are defined as follows:

$$A_1 = A_2 = A_3 = \alpha + B,$$
$$A_1{}^* = A_2{}^* = A_3{}^* = \alpha - B,$$
$$\beta_{12} = \beta_{23} = \beta_{31} = \tfrac{1}{2}mB.$$

These yield the results [2–4]:

$$E(m) = 6\alpha + 2B[1 + m + 2(1 - m + m^2)^{1/2}],$$
$$\varepsilon_2(e) - \varepsilon_1(a_1) = -B[1 + m - (1 - m + m^2)^{1/2}].$$

As seen from Table 1, for $m < 0.3$ the percentage depression $\{[E(m) - E(0)]/[E(1) - E(0)]\} \cdot 100\%$ is small, while the orbital split $\varepsilon_2(e) - \varepsilon_1(a_1)$ is essentially a linear function of m with slope $\sqrt{2}$.

Table 1

ZDO-Model for Three Interacting Double Bonds

m	$\dfrac{E - 6\alpha}{B}$	%-Depression	$\dfrac{\varepsilon_2(e) - \varepsilon_1(a_1)}{-B}$
0.0	6.000	0.0	0.000
0.1	6.016	0.8	0.146
0.2	6.066	3.3	0.284
0.3	6.155	7.8	0.4112
0.4	6.287	14.4	0.5282
.	.	.	.
.	.	.	.
.	.	.	.
1.0	8.000	100.0	1.000

We conclude that thermodynamic data of closed-shell systems (e.g., ΔHf) are poor criteria for the assessment of weak interactions between filled orbitals ($|\beta_{ij}| \ll |B_k|$), but that photo-electron spectroscopy data [5–6] yield direct information concerning their size.

Table 2 lists a selection of interaction constants, x, which are formally defined as $x = \frac{1}{2}[I_2(\pi) - I_1(\pi)]$, i.e., as half the difference between the vertical ionization potentials $I_1(\pi)$ and $I_2(\pi)$ of the two π peaks in the photo-electron spectrum of the compounds. This assumes implicitly the validity of Koopmans' theorem [7–8] (for comparison, the interaction constants x of $C\equiv C$ π bonds in diacetylene, of $C\equiv N$ π bonds and N-lone pairs in cyanogen and of the axial lone pairs in dihaloacetylenes have been included in Table 2).

The constant x is the resultant of at least two contributions [16]: (a) a through-space interaction (first order) roughly proportional to overlap ($\rightarrow$ homoconjugation); and (b) a through-bond interaction (second order; $\rightarrow$ hyperconjugation). The contribution of effect (a) towards x has a negative sign and will thus tend to place the anti-symmetric linear combination (A) of the interacting π orbitals above the symmetric (S) one. In contrast, (b) may lead to a positive or negative contribution to x, depending on the number and the relative conformation of the σ bonds interacting with the π orbitals. Thus, in cyclo-hexadiene-1,4, the hyperconjugation of the π orbitals with the two methylene bridges dominates the through-space interaction and places the S combination above the A combination. In the case of cyclooctadiene-1,5, the two effects cancel each other.

One of the most intriguing ideas concerning the delocalization of π electrons among weakly interacting double bonds is that covered by the catchword 'homo-aromaticity' [17]. This concept is due to Winstein [18]. The classical example for a neutral 'homo-aromatic' hydrocarbon is I [2–4, 19]. In agreement with the results derived above, neither the heat of hydrogenation of I [20] nor the NMR data [2–4] or the structure determination [20–21] are indicative of any significant departure from a model which assumes three non-interacting double bonds. Such a behaviour would demand that $m \leq 0.3$.

In contrast, the photo-electron spectrum of I allows a reliable determination of the parameter m [22]. Two π bands are observed at 8.8 eV and at 9.80 eV. The first of the two bands shows a Jahn-Teller split. This is consistent with the prediction derived from our molecular orbital model, namely, that the top occupied orbitals of I are $e(\pi)$ and $a_1(\pi)$, respectively. From the difference $\varepsilon[e(\pi)] - \varepsilon[a_1(\pi)] = 0.90$ to 0.97 eV, a value $mB = 2\beta_{ij} = -0.68$ eV obtains, if $B = -2.5$ eV is used as a reference. This leads to $m = 0.27$. This interaction

Table 2

Interaction Constants (x) in eV for the Interaction of
π- and n-Orbitals from Photoelectron Measurements

(all values are derived from vertical ionization potentials)

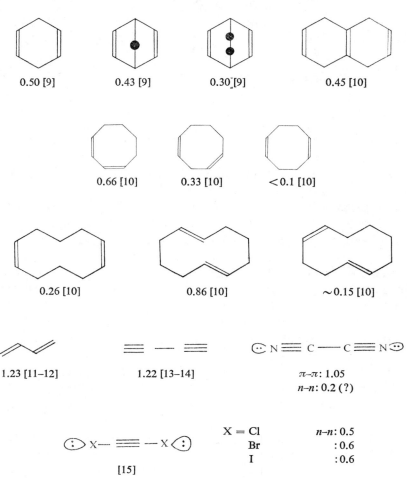

0.50 [9]   0.43 [9]   0.30 [9]   0.45 [10]

0.66 [10]   0.33 [10]   <0.1 [10]

0.26 [10]   0.86 [10]   ~0.15 [10]

1.23 [11–12]   1.22 [13–14]   π–π: 1.05
                                             n–n: 0.2 (?)

$\odot$ X— ≡ —X $\odot$

	X = Cl	n–n: 0.5
	Br	: 0.6
	I	: 0.6

[15]

represents almost exclusively through-space interaction between the π orbitals. Through-bond interaction (hyperconjugation) is only a minor effect in I.

Homoconjugative or hyperconjugative interactions between π-centres of unsaturated or aromatic compounds do manifest themselves in the electronic spectra [23–24]. However, even in those cases where a naive orbital picture can be used as a basis of discussion, two orbitals are involved in one transition, which makes it difficult to calibrate interaction parameters other than by trial and error. The fact that in the majority of cases one has to consider configurations and their interaction even in a qualitative treatment makes such a determination of the size and sign of interaction parameters almost impossible.

A special case of homoconjugation is spiroconjugation [25–26; see also 27]. Fig. 1 shows the electronic spectrum of the spirotetrene derivative II and that of the dienone III [28].

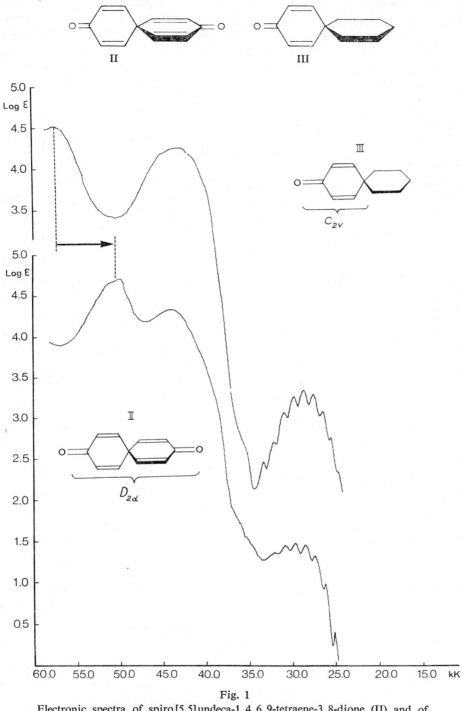

Fig. 1

Electronic spectra of spiro[5.5]undeca-1,4,6,9-tetraene-3,8-dione (II) and of spiro[5.5]undeca-1,4-dien-3-one (III)

In this particular case, spiroconjugation leads to a bathochromic shift of -7400 cm^{-1} of the high energy $\pi^* \leftarrow \pi$ transition $[B_2 \leftarrow A_1\ (D_{2d})]$. In contrast, the long-wavelength bands ($\pi^* \leftarrow n$ at 29 kK, $\pi^* \leftarrow \pi$ at 43 kK) are insensitive to spiroconjugation, in agreement with the predictions derived from symmetry considerations. This shows that the influence of homoconjugation on the electronic spectrum is strongly dependent on local symmetry. Contrary to what is usually observed for normal conjugation, the behaviour of the low-energy transitions is not always a safe indication for the presence or absence of homo-conjugative interactions.

From the above example the size of the interaction parameter between two spiroconnected double bonds was found to be approximately 4000 cm^{-1}, or 0.5 eV.

It should also be noted that homo- or spiroconjugation can introduce non-alternant character into an otherwise alternant system. This, as well as hetero-atoms participating in the π system, may lead to natural hypsochromic shifts as a consequence of spiroconnection. An example is shown in Fig. 2.

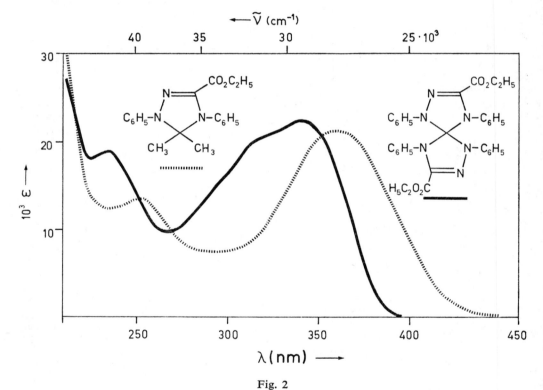

Fig. 2

Example for a natural hypsochromic shift as consequence of spiroconjugation
[29; E. Heilbronner & G. Hohlneicher, unpublished results]

The phenomenon of homoconjugation and/or hyperconjugation is also of importance in aromatic compounds with non-planar π systems and/or parts of their σ frame in optimal conformations for σ–π interaction. An example, taken from the benzenoid series, is provided by paracyclophane, where the interaction of the two benzene π systems with the two C–C σ orbitals between the methylene groups yields the orbital scheme shown in Fig. 3 [30].

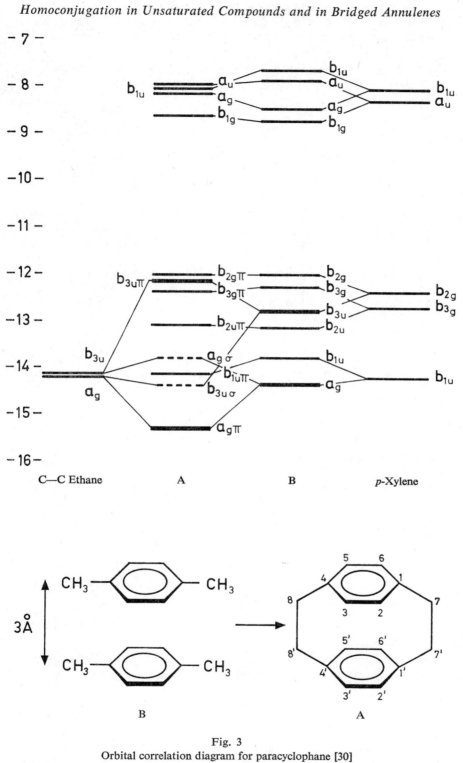

C—C Ethane   A   B   *p*-Xylene

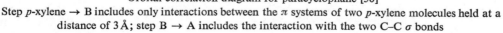

B   A

Fig. 3

Orbital correlation diagram for paracyclophane [30]

Step *p*-xylene → B includes only interactions between the π systems of two *p*-xylene molecules held at a distance of 3 Å; step B → A includes the interaction with the two C–C σ bonds

It explains the structure of the electronic spectrum, the ESR spectrum and the photo-chemical behaviour of paracyclophane.

Tables 3 and 4 contain the experimental assignment of Platt-labels [31] to the bands in the electronic spectra of the bridged [10]annulenes (IV) and [14]annulenes (V) [32].

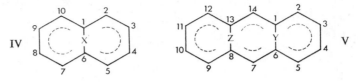

Table 3

Assignment of Bands in the Electronic Spectra of Bridged [10]annulenes (IV)

(band positions in kK = 10^3 cm^{-1}; all values refer to the band maxima)

1, 6-X-[10] annulene	^1L_b	^1L_a	?	^1B_a	^1B_b
– CH$_2$–	25.0	32.0	33.5	40.0	39.0
– CF$_2$–	24.4	32.5	33.5	39.8	39.4
– O –	24.0	31.5	33.2	37.4	39.0

Table 4

Assignment of Bands in the Electronic Spectra of Bridged [14]annulenes (V)

(band positions in kK = 10^3 cm^{-1}; all values refer to the band maxima)

1, 6-Y-8, 13-Z-[14] annulene		^1L_b	^1L_a	?	^1B_a	^1B_b
Y	Z					
CH—CH$_2$CH$_2$—CH		20.5	25.3	27.5	31.3	32.8
CH — CH$_2$ — CH		20.3	25.0	27.5	31.2	32.8
CH$_2$	O	19.0	24.7	26.8	29.8	32.8
O	O	17.8	24.7	26.0	29.0	32.6

The assignment is based on the experimental determination of the direction of polarization of the electronic transitions by two independent methods: (a) the measurement of the fluorescence polarization (example: Fig. 4) [33]; and (b) the measurement of the dichroism of the molecules dissolved in stretched foils (example: Fig. 5) [34–35]. Both methods show that for the molecules IV and V the assignment of the lowest singlet states is the following:

S_0	S_1	S_2	S_3	S_4	S_5
^1A	^1L_b	^1L_a	?	^1B_a	^1B_b
Ground state	(∥)	⊥	∥	⊥	∥

Increasing energy →

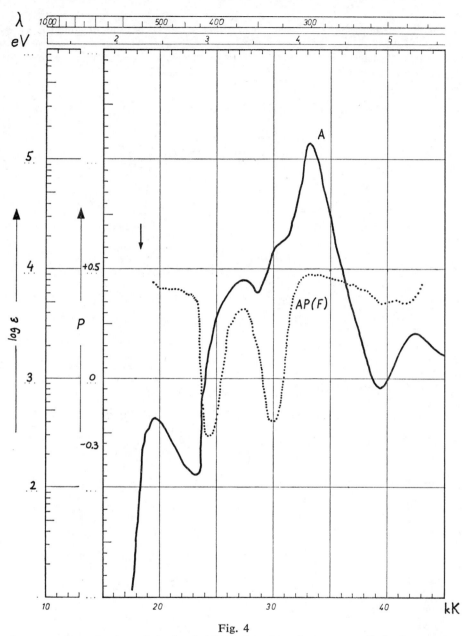

Fig. 4

Example for the determination of the band polarizations from fluorescence polarization measurements
A = Electronic absorption spectrum; AP(F) = Fluorescence polarization curve; the arrow indicates the wavelength at which the fluorescence has been measured; the data refer to the bridged [14]annulene (V) with Y = CH$_2$ and Z = O

Fig. 5

Example for the determination of the band polarizations from stretched film dichroism measurements
A = Optical densities measured parallel (EP) and perpendicular (ES) to the stretch direction
B and C = Computer reduction according to Eggers for the parallel bands (B) and the perpendicular
bands (C); the data refer to the bridged [10]annulene (IV) with X = CF_2

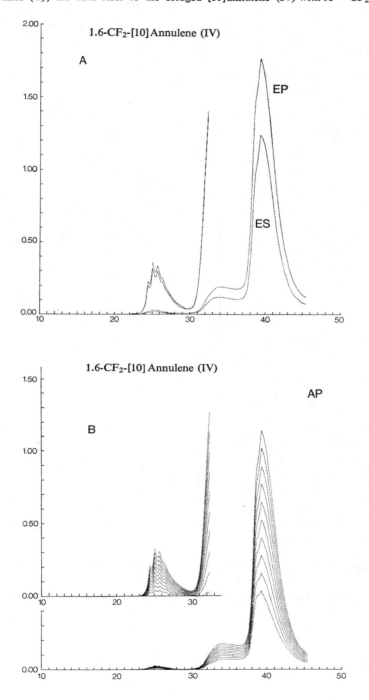

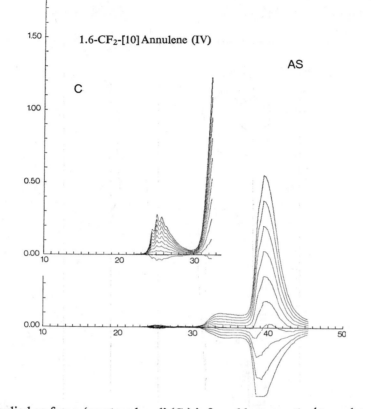

In all cases studied so far, a 'mystery band' (S_3) is found between the $^1L_b \leftarrow {}^1A$ and $^1B_a \leftarrow {}^1A$ transitions. It is polarized parallel to the first band ($^1L_b \leftarrow {}^1A$). The origin of this additional band is unknown. It is not accounted for by π-electron calculations — e. g., Pariser-Parr-Pople (PPP) — and may be due to the interaction of the σ bonds present in IV and V with the non-planar π systems. Inclusion of homoconjugation between the centres 1,6 in IV, or 1,6 and 8,13 in V, in the theoretical models will not yield improved predictions for the positions and intensities of the L and B bands, let alone the prediction of an additional low-lying state of appropriate symmetry. However, it is known from ESR experiments that homoconjugation plays an important role in these compounds [36–37]. An estimate for the interaction between the pairs of centres 1,6 and 8,13, derived from ESR coupling constants, is again of the order of 0.5 eV.

A special kind of homoconjugation, similar to that encountered in the paracyclophane series, is found in [14]annulene derivatives, such as dimethyl-dihydropyrene (VI) [38–39].

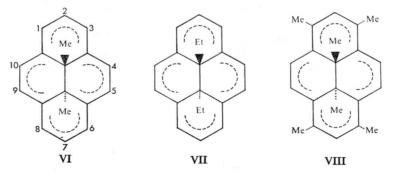

The electronic spectrum of VI, shown in Fig. 6 and, similarly, those of diethyl-dihydro-pyrene (VII) and of hexamethyl-dihydropyrene (VIII), consist of four main bands, which

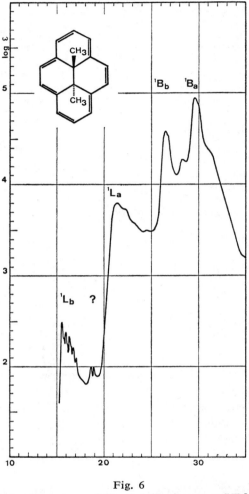

Fig. 6
Electronic spectrum of dimethyl-dihydropyrene (VI) [40]

have been shown by fluorescence polarization measurements [40] to correspond to the following sequence of states (cf. Table 5):

S_0	S_1	S_2	S_3	S_4	S_5
^1A	^1L_b	?	^1L_a	^1B_b	^1B_a
Ground state	($\parallel$)		$\perp$	$\parallel$	$\perp$

Increasing energy $\longrightarrow$

Table 5

Electronic Spectra of Dimethyl-dihydropyrene (VI), Diethyl-dihydropyrene (VII), Hexamethyl-dihydropyrene (VIII) and Some of Their Derivatives

(all values in kK = 10^3 cm^{-1}; solvent: ethanol)

Bond	^1L_b	^1L_a	^1B_b	^1B_a
VI	15.6	21.4	26.6	29.7
2-Nitro-	14.9	18.6	24.4	28.6
2-Formyl-	15.0	19.4	25.0	29.0
2-Benzoyl-	15.0	19.4	24.9	28.7
2-Acetamido-	15.3	20.6	26.0	29.2
2-Acetamido-7-formyl-	15.2	18.1	24.0	28.6
2, 7-Diacetoxy-	15.6	21.5	26.9	29.6
4-Formyl-	14.6	22.1	26.8	32.9
4-Carboxy-	15.3	21.2	26.1	29.2
4-Bromo-	—	21.2	26.2	29.3
VII	15.0	20.8	25.6	28.8
2-Nitro-	14.6	17.8	23.6	17.9
VIII	15.3	21.0	25.7	28.1
2-Nitro-	15.2	21.0	25.9	28.0
2-Formyl-	14.2	19.3	24.3	27.2
2-Cyano-	14.7	19.8	24.9	27.5
2-Acetyl-	15.2	21.0	25.6	27.9
2-Methyl-	—	21.0	25.6	28.1
2-Formyl-oxim-	15.0	20.7	25.4	27.8
4-Nitro-	15.0	20.8	25.6	28.4
4-Formyl-	14.6	19.9	22.2	25.3
2, 7-Diacetyl-	15.1	20.8	25.4	27.7
2, 7-Diacetoxy-	15.4	21.4	26.1	28.2

The spectra contain again a weak additional band (S_2) of unknown degree of polarization. Low temperature spectra indicate that this band cannot be part of either the $^1L_b \leftarrow {}^1A$ or the $^1L_a \leftarrow {}^1A$ transition [H. R. Blattmann & E. Heilbronner, unpublished results]. It is an open question whether or not it is related to the 'mystery band' found in the electronic spectra of the systems IV and V, or if it has its origin in the interaction of the planar peripheral π system, with the central σ chain of four or six sp^3 carbon atoms present in VI, VII, VIII and in their derivatives.

That there exists such an interaction is again suggested by ESR measurements [41], and especially by the remarkable phototropic rearrangement of the type VI → IX [42].

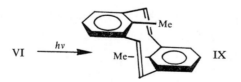

The parallelism between the quantum yields for the photoreaction VI → IX and the thermal back-reaction IX → VI (see Table 6 for some examples) can be explained in terms of the Frosch-Robinson mechanism [H. R. Blattmann & W. Schmidt, unpublished results].

Table 6

Comparison of Quantum Yields ϕ for
the Phototropic Rearrangement (VI) → (IX) with the Rates for
the Thermal Back-Reaction (IX) → (VI) as a Function of Substitution

Rate constants k_d in min^{-1} (30°C); E_s is the energy of the lowest singlet state
in eV and $\Delta G^{\neq}$ the free enthalpy of activation for the back-reaction

	$k_d 10^3$	ϕ	E_s	$\Delta G^{\neq}$
VI	1	0.02	1.93	1.05
2-Nitro-(VI)	70	0.37	1.85	0.95
2-Formyl-(VI)	50	0.26	1.86	0.95
VII	6	0.01	1.86	1.00
VIII	1	0.008	1.90	1.05
2-Nitro-(VIII)	2	0.005	1.88	1.05
2-Formyl-(VIII)	12	0.03	1.81	1.00
2-Acetyl-(VII)	1	0.013	1.88	1.00

To conclude, we wish to draw attention to the fact that the long-wave absorption maximum of [18]annulene (X), shown in Fig. 7 [43], has recently been claimed not to be of the $\pi^* \leftarrow \pi$, but rather of the $\pi^* \leftarrow \sigma$ type [44]. This interpretation would imply a non-planar structure of X in solution.

Low temperature electronic spectra of X and the study of the influence of substituents on band positions and band intensities [E. P. Woo & F. Sondheimer, to be published]

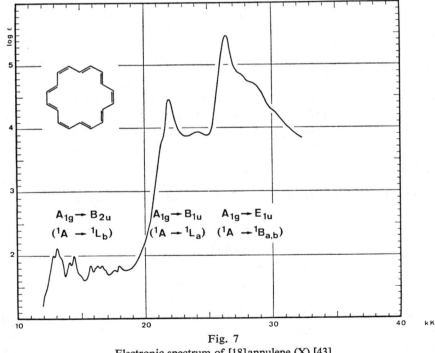

Fig. 7
Electronic spectrum of [18]annulene (X) [43]

are difficult to reconcile with such an assignment. On the other hand, π calculations of the PPP type (including bond localization effects) do not account properly for the observed band locations.

REFERENCES

1 M.J.S. Dewar (1969) *The Molecular Orbital Theory of Organic Chemistry*, McGraw-Hill, New York, p. 138.

2 P. Radlick & S. Winstein (1963) *J. Am. Chem. Soc.*, 85 : 344.

3 K. G. Untch, *ibid.*, p. 345.

4 K. G. Untch & R. J. Kurland (1964) *J. Molec. Spectrosc.*, 14 : 156.

5 D. W. Turner (1969) 'Molecular Photoelectron Spectroscopy', in : *Molecular Spectroscopy*, The Institute of Petroleum, p. 209.

6 A. D. Baker (1970) *Acc. Chem. Res.*, 3 : 17.

7 T. Koopmans (1934) *Physica*, 1 : 104.

8 W. G. Richards (1969) *J. Mass. Spectrosc. Ion Phys.*, 2 : 419.

9 P. Bischof, J. A. Hashmall, E. Heilbronner & V. Hornung (1969) *Helv. Chim. Acta*, 52 : 1745.

10 P. Bischof, J. A. Hashmall, E. Heilbronner & V. Hornung, *Tetrahedron Lett.* [in press].

11 M. I. Al-Joboury & D. W. Turner (1964) *J. Chem. Soc.*, p. 4434.

12 J. H. D. Eland (1969) *J. Mass. Spectrosc. Ion Phys.*, 2 : 471.

13 C. Baker & D. W. Turner (1967) *Chem. Comm.*, p. 797.

14 Idem (1968) *Proc. Roy. Soc.*, A 308 : 19.

15 E. Heilbronner, V. Hornung & E. Kloster-Jensen (1970) *Helv. Chim. Acta*, 53 : 331.

16 R. Hoffmann, E. Heilbronner & R. Gleiter (1970) *J. Am. Chem. Soc.*, 92 : 706.

17 S. Winstein (1969) *Q. Rev.*, 23 : 141.

18 Idem (1959) *J. Am. Chem. Soc.*, 81 : 6524.

19 W. R. Roth (1964) *Liebigs Ann. Chem.*, 671 : 10.

20 W. R. Roth, W. P. Bang, P. Göbel, R. L. Sass, R. B. Turner & A. P. Yü (1964) *J. Am. Chem. Soc.*, 86 : 3178.

21 R. B. Jackson & W. E. Streib (1967) *ibid.*, 89 : 2539.

22 P. Bischof, R. Gleiter & E. Heilbronner, *Helv. Chim. Acta* [in press].

23 H. Labhart & G. Wagnière (1959) *Helv.*, 42 : 2219.

24 S. Winstein (1969) *Q. Rev.*, 23 : 141.

25 H. E. Simmons & T. Fukunaga (1967) *J. Am. Chem. Soc.*, 89 : 5208.

26 R. Hoffmann, A. Imamura & G. D. Zeiss (1967) *ibid.*, p. 5215.

27 G. Hohlneicher (1967) Habilitationsschrift, Technische Hochschule, Munich.

28 R. Boschi, A. S. Dreiding & E. Heilbronner (1970) *J. Am. Chem. Soc.*, 92 : 123.

29 R. Huisgen, R. Grashley, R. Kunz, G. Walbillich & E. Aufderhaar (1965) *Chem. Ber.*, 98 : 2174.

30 R. Gleiter (1969) *Tetrahedron Lett.*, p. 4453.

31 J. R. Platt (1949) *J. Chem. Phys.*, 17 : 484.

32 E. Vogel (1968) *Chimia*, 22 : 21.

33 F. Dörr (1966) *Angew. Chem.*, 78 : 457.

34 J. H. Eggers & E. W. Thulstrup, *8th European Congress of Molecular Spectroscopy, Copenhagen, 1965.*

35 E. W. Thulstrup & J. H. Eggers (1968) *Chem. Phys. Lett.*, 1 : 690.

36 F. Gerson, E. Heilbronner, W. A. Böll & E. Vogel (1965) *Helv. Chim. Acta*, 48 : 1494.

37 F. Gerson, J. Heinzer & E. Vogel (1970) *ibid.*, 53 : 95, 103.

38 V. Boekelheide & J. B. Phillips (1963) *J. Am. Chem. Soc.*, 85 : 1545.

39 Idem (1964) *Proc. Natn. Acad. Sci. USA*, 51 : 550.

40 H. R. Blattmann, V. Boekelheide, E. Heilbronner & J. P. Weber (1967) *Helv. Chim. Acta*, 50 : 68.

41 F. Gerson, E. Heilbronner & V. Boekelheide (1964) *ibid.*, 47 : 1123.

42 H. R. Blattmann, D. Meuche, E. Heilbronner, R. J. Molyneux & V. Boekelheide (1965) *J. Am. Chem. Soc.*, 87 : 130.

43 H. R. Blattmann, E. Heilbronner & G. Wagnière, (1968) *J. Am. Chem. Soc.*, 90 : 4786.

44 F. A. Van-Catledge & N. L. Allinger (1969) *ibid.*, 91 : 2582.

Discussion

J. F. Labarre:

Using the Faraday effect, I find exactly the same order as you for interactions in cyclooctatetraene, 1,3-cyclooctadiene, norbornadiene and 1,5-cyclooctadiene, the effect being almost nil in the last molecule.

E. Heilbronner:

We have not yet recorded the PE spectrum of cyclooctatetraene. The splitting pattern will be more complicated in view of the D_{2d} symmetry of the molecule. As a consequence of this symmetry, one of the π orbitals will be degenerate, and the corresponding PE band must show a Jahn-Teller split. It is very pleasing to see that two different experimental techniques predict the same orbital sequence.

Mrs A. Pullman:

The question of the 'mystery band' of your bridged annulene compound is very interesting. One may, of course, wonder if it is not a $\pi^* \leftarrow \sigma$ transition, and this could be detected by making an all-valence calculation by the Del Bene-Jaffe CNDO procedure, which is rather good for this sort of investigation. In fact, this would give you at the same time the role of the bridge atom or group of atoms.

E. Heilbronner:

Such a calculation is now being carried out in Basel by W. Schmidt. We are quite convinced that σ–π interactions cannot be neglected in this case. However, only a detailed calculation of the type suggested by you will show how much '$\pi^* \leftarrow \sigma$ character' has to be assigned to the transition responsible for the 'mystery band'.

We have tried to account for the contribution of the bridge by using a theoretical model based on locally excited states of the peripheral π system and on charge-transfer states involving the bridging groups, i. e., with CH_2, CF_2, NH, $N(CH_3)$ and O as donors. This model predicts that the transition energy from ground to mixed excited state should depend markedly on the nature of the group bridging positions 1 and 6. However, the experiment tells us that no such dependence exists for those bands that dominate the electronic spectrum of these compounds; at least, not to the extent expected on the basis of our calculations. On the other hand, it should be pointed out that extensive ESR investigations by F. Gerson have shown that the rearrangement of the radical anions obtained by populating the first anti-bonding π orbital with a single electron depends very much on the type of bridge. This could be interpreted in terms of different degrees of σ, π mixing.

G. Wagnière:

Mr Heilbronner has generously associated me with the discovery of the long-wavelength absorption in [18]annulene, which disproves bond alternation. I wish to emphasize that the discovery of this band is Mr Heilbronner's. Before our collaboration I actually tried to prove that [18]annulene does have alternation of bond lengths.

Concerning through-space *versus* through-bond interaction of double bonds in cyclic dienes, do extended Hückel calculations, for instance, support this distinction?

72

E. Heilbronner:

For all the cases that I have discussed we have carried out extended Hückel theory (EHT) calculations. These have been checked very often by more sophisticated methods (MINDO/2, CNDO/2, etc.). Contrary to what one might have expected, the results are quite reasonable. Of course, the absolute values of the predicted ionization energies, using Koopmans' theorem, are much too high, but the relative spacings of the orbital energies, and thus of the ionization potentials, are of the correct order of magnitude. Experimental and theoretical spacings may differ by a factor of 2. From our point of view, the most important observation is that the EHT model predicts the same sequence for the top 4 or 5 occupied orbitals, i.e., the same labels of irreducible representations, as do the many-electron treatments.

The great advantage of an independent-electron treatment, such as EHT, resides in the ease with which the computed orbitals can be 'read' by the chemist. Furthermore, these orbitals can be used very simply as a basis for qualitative perturbation considerations, which in turn will provide a rationalization for the observed PE spectrum and, in many cases, predictions which can be subjected to experimental proof. The success of this simple theory in photo-electron spectroscopy makes me believe that it is, for the moment, the most useful model to be applied to this technique.

Now, to come back to the original question: The EHT model does indeed allow a clear distinction to be made between through-space and through-bond interactions of π orbitals. Pure through-space interaction will, by definition, not involve any of the semilocalized σ orbitals, which will therefore be excluded from the final linear combination. On the other hand, through-bond interaction will be characterized by sizeable coefficients of those σ orbitals that mix with the interacting π orbitals.

It is perhaps worthwhile to point out that quite a few of the predictions derived by R. Hoffman before the advent of photo-electron spectroscopy (concerning $\pi-\pi$ interactions or interactions of lone pairs) have since been successfully confirmed.

M. Cais:

I would like to refer to the electronic spectrum of that spirotetraene-dione. When you extended the range of the measurement, you observed the additional band, whereas before that the impression was that there is no interaction. If you take the mono-substituted compound, do you expect that by extending the range of the measurement you might observe the same kind of effect that you have just observed here?

E. Heilbronner:

I do not know.

M. Cais:

Do you know what the conformation of this compound is? Is it planar?

E. Heilbronner:

It is difficult to see through this problem at such short notice. All one can say, in quite general terms, is that high local symmetry can constitute a considerable barrier to conjugative interactions of semi-localized orbitals.

Aromaticity and Delocalization in Five-Membered Heterocycles

by G. DEL RE

Istituto Chimico dell' Università, Napoli,
and Gruppo Chimica Teorica del CNR, Roma

I. Introduction

IN THE COURSE of several studies on five-membered 'aromatic' heterocycles (in particular furan, pyrrole, thiophene and isoxazole), we have often had to try to decide why and to what extent those compounds are really aromatic. Now, as has been illustrated in the excellent introduction by Bergmann and Agranat [1], if aromaticity is not altogether one of Croce's pseudo-concepts, it belongs certainly to those *voces metaphysicae* which, as Leibniz said, 'men use because of some need, and think they understand only because they learned to give them a name' [2]. In particular, the extension of the term 'aromatic' to the five-membered heterocycles of the pyrrole series was mainly prompted by a desire for generality, and its only support was the existence in those rings of a potential, and perhaps actual, aromatic sextet [3]. The same desire for generality has suggested further extension, so that already in 1960 it could be claimed that 'almost any structure in which the electrons are non-localized is now given the name "aromatic"' [4]. The existing confusion is witnessed by the fact that the various criteria for aromaticity give contradictory results in the series furan, thiophene, pyrrole: ring currents suggest a decreasing aromaticity in the order thiophene > pyrrole > furan [5–6]; the UV spectra suggest the order pyrrole > thiophene > furan [7]; and so on.

At the origin of this situation we find contradictory theoretical views. Some believe that aromaticity should be associated with the energy-level distribution of a given molecule; others think in terms of reactivity; still others exclude both by claiming that aromaticity is a property of the ground state of a molecule, so that its physical manifestations are to be sought in the energy content, in the bond lengths and in the importance of ring currents [8]. Therefore, further emphasis on the theoretical aspects of the question is not out of place, especially as regards the five-membered heterocycles.

As a preliminary to any discussion, one should justify the attempt to see some aromaticity in the compounds under study. But the need for criteria permitting to establish analogies and classifications in organic chemistry is such a well-known and important justification that any further consideration is superfluous. Coming then to the concept itself, it may be assumed to be an abbreviation for the set of peculiar qualities arising from the existence of cyclic conjugation in a given compound.

This definition is probably acceptable for everybody, but leaves three main questions

74

open: What are the requirements for aromaticity? How much aromaticity is present in any given case? What 'peculiar qualities' are associated with aromaticity?

The answer to the first question is immediate: The requirements for aromaticity are, of course, that the molecule be a planar ring with a conjugated π system. The other two questions are much more complicated.

II. Aromaticity and Delocalization Energies

It has been sometimes overlooked that the closure of an open chain to form a ring, like, say, the cyclization of the hypothetical imino-alcohol I to form the isoxazole II, involves, in addition to changes associated with the change in the σ system, a number of changes in π properties which have nothing to do with increase in delocalization in the Hückel sense. This may be one of the most important reasons for the disagreements found in

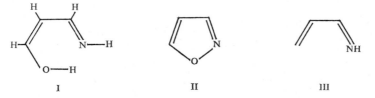

| I | II | III |

experiment. To mention the most important point, we recall that the so-called aromatic five-membered heterocycles are, first of all, dienes like III blocked in a *cis* geometry by the hetero-atom bridge. Moreover, this geometry is distorted with respect to that of the true *cis*-diene. These effects are quite distinct from aromaticity. For instance, 1,2-dimethylenecyclopentane (IV) has an absorption peak at 248 nm (5.00 eV), as do many other compounds in which a 1,3-butadiene chain is held in a *cis* position by an appropriate saturated bridge; *trans*-butadiene has a band of the same intensity around 217 nm (5.71 eV) [9–11]. In short, a red shift with respect to *trans*-butadiene, or the appropriate *trans*-diene system, should be expected for the 'aromatic' five-membered heterocycles even if the function of one of the hetero-atoms were solely that of blocking the diene system in its regular *cis* position. Further effects could be ascribed to distortion.

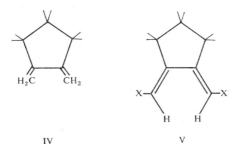

| IV | V |

The above considerations strongly suggest that an approach to the problem of aromaticity should start in the frame of a π-electron method, not taking into account differences in geometry and long-range interactions; with some reservations and alternatives, which we shall mention later, such a method is the old simple Hückel method.

An examination of various sophisticated calculations, including those of Clementi [12], suggests that a general discussion can be largely (but not completely!) based on a very simple model of the Hückel type. In fact, a σ–π separation is always possible as a very

75

good approximation [13], and an MO scheme is essential for understanding results [12]. Granted a convenient choice of the AO basis [14], it may then be concluded that we may confine ourselves to what is formally a Hückel problem, provided we do not draw conclusions based on specific values of matrix elements. It is reasonable to include overlap in such a model, because it seriously affects the distribution of energy levels, but this can be done by a very simple linear transformation of energies [15]. Of course, this extremely simplified approach allows only an incomplete qualitative discussion, useful mainly for analysing definitions. To become complete, such a discussion would have to be supplemented by quantitative considerations extending beyond the MO scheme.

A particularly useful feature of our 'Hückel model' (as we shall call it to distinguish it from the Hückel method) is its ability to give ideal cases, as those required according to the preceding section, where differences in geometry and long-range interactions are neglected. It is true that the idea that formally equivalent bonds have the same bond parameter, or that 'no bond' is the same as 'zero bond parameter', is not strictly correct, but this does not introduce essential differences into the scheme of analysis.

Examples of Hückel Hamiltonians obtained from more refined calculations and discussions of them have been given many times. For the particular cases at hand, a justification of the various approximations and an interpretation of the parameters are suggested by Carpentieri et al. [16].

The first measure of aromaticity is certainly the change in the delocalization energy of a diene system brought about by the hetero-atom bridge. In accordance with previous remarks, we can analyse this quantity—which we may call 'extra delocalization energy' (EDE)—by assuming that the diene system is an ideal conjugated system (all bonds equivalent) and by varying both the atomic parameter of the hetero-atom and the bond integral associated with the bridge. Table 1 reports the results obtained when the parameters in question are used with inductive corrections on the terminal carbon atoms of the diene system equal to 1/10 of the atomic parameter of the hetero-atom (see Table 1 for details). The orbital energies E_i' obtained from the standard Hückel calculation have been transformed to account for overlap by the linear transformation [15]

$$E_i = 4E_i'/(4 + E_i'),\qquad(1)$$

which is very important because it alters the distribution of the energy levels. In calculating the EDE's, the appropriate transformations have been carried out also on the atomic and bond parameters [17].

The use of EDE as a measure of aromaticity is satisfactory only at first sight. In fact, the increased values of the atomic parameters of the carbon atoms may and do change the delocalization energies of the diene system. Is this an effect to be included in aromaticity? This question is strictly similar to that which arises in the definition of concepts like localization and conjugation energies in the frame of more sophisticated MO methods [18], which is not surprising, because the inclusion of inductive effects is already a refinement of the 'pure' Hückel method. Therefore, a study of it is of general importance.

Following the line already suggested by Berthier and Del Re [18], we remark that, if we are defining EDE as an ideal observable—i.e., a quantity which could be measured in some ideal experiment—we must include in it the effect of the hetero-atom on the diene chain, because then EDE is essentially a π dissociation energy; but, if we do that, we are not studying a quantity associated only with the delocalization of the lone pair of the

Table 1

Extra Delocalization Energies of Cyclic Five-Membered π Systems

η_{cx} \ δ_x	0.800 (0.098)		1.333 (0.190)		1.714 (0.279)		2.000 (0.364)	
	a	b	a	b	a	b	a	b
0	0.137	—	0.275	—	0.413	—	0.551	—
0.2	0.162	**0.025**	0.300	**0.025**	0.421	**0.008**	0.551	**0.000**
0.4	0.266	**0.129**	0.345	**0.070**	0.441	**0.028**	0.552	**0.001**
0.6	0.476	**0.339**	0.430	**0.155**	0.477	**0.064**	0.563	**0.012**
0.8	0.675	**0.538**	0.539	**0.264**	0.563	0.123	0.584	**0.033**
1.0	0.869	**0.732**	0.663	**0.398**	0.621	**0.208**	0.619	**0.068**
1.2	1.062	**0.925**	0.796	**0.521**	0.736	**0.323**	0.668	**0.117**

NOTES:

1. The values correspond to parameters $\delta_x = 1, 2, 3, 4$, with inductive effect $0.1\,\delta_x$, when referred to the standard Hückel calculation giving the same charge distributions [17]. For the discussion, a linearization of the scale of atomic parameters is not necessary.

2. The energies are in units such that the total π-delocalization energy of ideal butadiene is 3.3746 γ units.

3. The LPDE values for pyrrole, furan, oxazole and isoxazole, obtained by the Hückel method [19], are 0.362, 0.160, 0.295 and 0.053, respectively (the hetero-atom considered in the latter two heterocycles is obviously oxygen).

a = with respect to butadiene plus the hetero-atom (EDE)
b = with respect to the diene with inductive effects plus the hetero-atom (LPDE)

The atomic parameter of the hetero-atom is denoted by δ_x, and the bond parameter of the CX bond by η_{cx}; the value of the inductive parameter is given in parentheses.

hetero-atom. Therefore, it is better to keep the inductive effect out, and introduce the 'lone-pair delocalization energy' (LPDE), i.e., the difference between the π-electron delocalization energy of the ring and the delocalization energy of the same system, where only the bond integral η_{cx} is equal to zero.

To see the difference between EDE and LPDE one should think of the example of a charged metal sphere in the presence of a point charge. The mutual energy is made up of two contributions: one is the same as if the metal sphere were replaced by a point charge concentrated in its centre; the other is associated with the mobility of charge carriers on the metal surface. This example also brings out the key difficulty we face when we apply such schemes to molecules: the metal sphere can be replaced very easily by what is practically a point charge, and at least one of the two contributions to the mutual energy can be measured separately. A molecule where a given bond integral is zero but the inductive effects are not zero is difficult to imagine, except in cases like biphenyl and its derivatives. Nevertheless, one can always try to build saturated substituents for the terminal hydrogen atoms of, say, a butadiene chain such that the inductive effect be there without any conjugation. An example is given by compounds of the type IV, with substituents X like in V. Such substituents could be, for instance, appropriate alkyl groups, giving a pure inductive effect. The vibrational spectra permit also an estimate of the situation for the ideal geometry, if the actual geometry is known.

The figures in Table 1 speak for themselves. We only point out that if the atomic parameter of oxygen is higher than that of nitrogen, the LPDE criterion is in full agreement with the conclusion that furan is less aromatic than pyrrole. But it is not so if we take a π-dissociation energy like EDE — Columns a of Table 1 — because for values of δ_x around 0.7 (which correspond to values around 1 for the corresponding overlap-neglecting calculation), that energy first decreases and then increases with increasing values of η_{cx}. The question then hinges on the physical meaning of the various parameters. Investigations of this have subsided for several years, mainly because of the emphasis on *ab initio* calculations; but it is now becoming clear that empirical one-electron schemes are essential in quantum chemistry, and those parameters are worth investigating. After all, they are the matrix elements of a well-defined truncated-basis one-electron Hamiltonian.

A completely different kind of difficulty derives from the problem of deciding how much aromaticity there is in any given case. At this point, another aspect of the concept under study comes out: whereas the purely qualitative definition of aromaticity as cyclic conjugation is independent of any reference, here we have to consider two extreme reference compounds — one which is completely non-aromatic, one which is fully aromatic. The former is clearly an ideal butadiene; the latter is benzene. Now, if we consider benzene split into an ethylene π bond and a butadiene π system, the extra delocalization energy is 0.892 γ units; thus, on a linear scale, the aromaticity relative to benzene (if we accept the definition given above) is given by 1.1, the measure of the LPDE in γ units. With the Hückel parameters of Reference 19, for instance, we find relative aromaticities of 6%, 33%, 18% and 40% for isoxazole, oxazole, furan and pyrrole, respectively.

III. Aromaticity and Charge Transfer

The ground-state energy criterion for defining quantitatively the aromatic character of five-membered heterocycles is thus reduced theoretically to the adoption of a clearly-founded, highly simplified method, and experimentally to the introduction of *ad hoc* experiments permitting to isolate, at least ideally, the contribution to the energy which we have chosen to associate with aromaticity.

Another way of deciding how aromatic a potentially aromatic ring is, may be found from dipole moments. Table 2 shows the transferred charges (which are just the net charges of the hetero-atom sharing the lone pair, changed in sign) in some of the cases recorded in Table 1. They evidently follow the same trend as the delocalization energies of Columns b in Table 1. If we admit that it is possible somehow to evaluate separate σ and π contributions to dipole moments, then the π dipole moment should be a good indication of aromaticity. Of course, this would require measuring the dipole moments of the appropriate *cis*-diene systems, but there is no difficulty in principle in that connection.

If we look at the results obtained by a 'PPP-variable electronegativity' calculation [7], we find that the transferred charges are as follows: furan –0.195, thiophene –0.237 and pyrrole –0.295. The order of aromaticity is thus fixed with thiophene between furan and pyrrole. That thiophene should have properties that place it either as the first or as the last term of the triad should then be explained in terms of other effects, mainly due to the fact that sulphur belongs to the second row of the periodic table.

Even though it has not been calculated with the same method as the systems mentioned above, isoxazole appears to be less aromatic than the three compounds just mentioned. This conclusion is borne out both by a sophisticated treatment [18] and by a simple Hückel calculation [19]; in both cases the transferred charges are close to –0.12 (provided one

Table 2

Net π Charges of the Hetero-Atom

(i.e., electrons lost to the diene system, in the general heterocycles of Table 1)

η_{cx} \ δ_x	0	0.8	1.333	1.714
0.6	0.317	0.161	0.100	0.080
0.8	0.370	0.223	0.152	0.111
1.0	0.380	0.277	0.201	0.153
1.2	0.422	0.313	0.241	0.194
1.4	0.438	0.342	0.276	0.236

NOTE:

For pyrrole, thiophene and furan the values are 0.275, 0.237 and 0.195, respectively [6]; for isoxazole a different but comparable method gives 0.124 [18]; the simple Hückel method of Ref. 19, when the charges are divided by 1.6 according to the recipe for calculating dipole moments with the given parameters, gives 0.112 for isoxazole and 0.201 for oxazole; this sequence is in qualitative agreement with other indications regarding the relative character of the heterocycles in question, but places oxazole near furan, contrary to the indications of LPDE

applies the recipe of dividing the Hückel charges by 1.6, as is required by the set of parameters used in Reference 19).

The transferred-charge criterion is especially good because it does not take into account rearrangements of charges upon ring closure. Unfortunately, there is no one-to-one correspondence between transferred charges and the delocalization energies defined as in Columns *b* of Table 1, although the trends are strictly similar. This is by no means surprising, and is understood if one thinks of a non-symmetric (double) potential well as a model for the potential energy associated with the CX bond; but it bears out again the question whether aromaticity should be explained in terms of wave functions or in terms of energies.

IV. INTERACTION ENERGIES AND AROMATICITY

There are many other ways of looking at aromaticity in an MO scheme. One which is very close to resonance theory consists in studying the 'interaction energies'. These are the changes in delocalization energies when two conjugated systems are connected. Even by a simple Hückel scheme one finds results like those of Table 3. It is evident that head-to-tail coupling of open chains gives the highest interaction energy, while coupling of benzene π systems gives the lowest interaction energy. Therefore, the π-interaction energies of phenyl-substituted five-membered heterocycles should give an indication of how aromatic the π systems of the latter are. This approach presents the advantage of connecting a measure of aromaticity directly to the sensitivity of a π system closed to external effect; for the rest, it has the same difficulties and drawbacks as the LPDE approach.

V. CONCLUSION

The preceding considerations show some aspects of a possible treatment of aromaticity in the frame of the MO-LCAO scheme. In a wider context, based on molecular orbitals (like any MO-LCAO calculation), an extension of those considerations is relatively straightforward. Taking pyrrole as an example, the following quantities have to be calculated:

a. The total energy of the ground state of the system, in the real geometry, E_{true}.

Table 3

Interaction Energies in Some π Systems

(units γ; interacting units separated by broken lines)

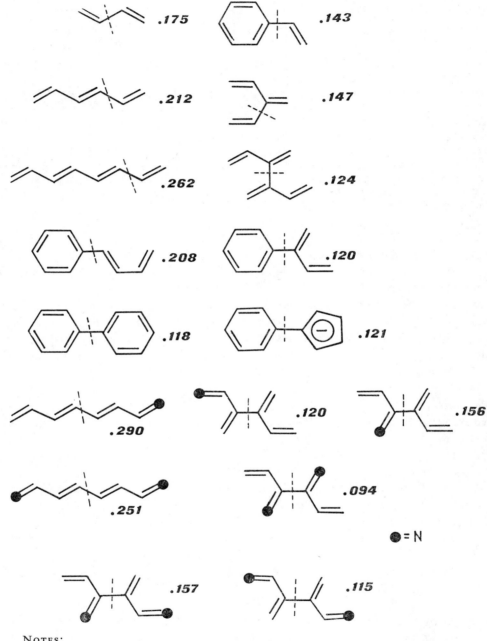

NOTES:

1. The interaction energy is defined as the difference between the energy of the whole system and the sum of the energies of the separated units

2. For nitrogonic compounds, the parameters used are those of Del Re [19]

b. The energy E_{na} of the same system when the π orbitals lose one degree of freedom, and have necessarily either all the coefficients of the carbon atom orbitals or the coefficients of the hetero-atom orbitals equal to zero.

c. The energy E_{id} of the pyrrole ring for an ideal geometry of the diene system, viz., the geometry corresponding to the isolated *cis*-diene.

d. The energy E_{naid}, which corresponds to b for the geometry considered in c.

e. The net π charges Q_{π}.

All these quantities are perfectly well defined, and can be used for repeating the above discussion. It is encouraging that they can be defined in an all-electron scheme, even though aromaticity should be taken purely as a π property.

It is also possible to go out of the MO scheme completely in analysing aromaticity, either by extending the above treatment or by looking directly at the energy level distribution. As is well known, this distribution is important both for the spectroscopic properties and for the effect of external agents, owing to the fact that the energy differences play a role in perturbation effects.

In the case of the five-membered heterocycles, a simple procedure consists in comparing the energy level distribution with that of the cyclopentadienyl anion, on the one hand,

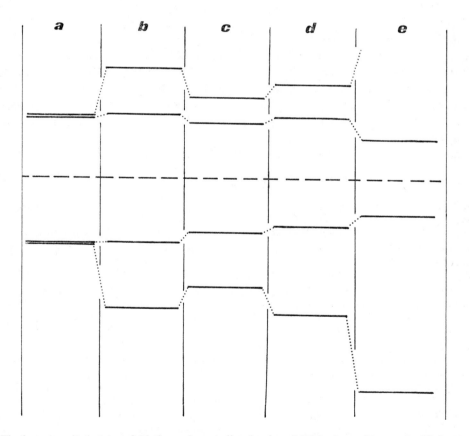

The lowest excited states of (a) the cyclopentadienyl anion, (b) thiophene, (c) pyrrole, (d) furan, (e) cyclo-pentadiene, according to my MO-LCAO-CI calculations [16]

The zero point for the energy has been chosen for each compound so as to facilitate comparison

81

G. Del Re

and with that of cyclopentadiene, on the other hand (see figure). One sees that the degeneracies of the former are gradually removed in passing to pyrrole, thiophene, furan and cyclopentadiene; this gives an order of decreasing aromaticity, which can be directly confirmed by spectroscopic investigations [7], again with some reservations for thiophene.

The separation of levels that are degenerate in benzene is 0.01 eV in pyridine, while the separation of levels corresponding to degenerate ones of the cyclopentadienyl anion in pyrrole is 0.25 eV. This justifies saying that pyridine is aromatic, whereas pyrrole is only slightly so.

Although reference to energy levels is physically very significant, the difficulties arising in using them as an indication of aromaticity are largely the same as those encountered with other definitions; they reduce to the question of how many of the observed differences, with respect to the reference hydrocarbons, are really due to reduced cyclic conjugation. We all know, for instance, that degeneracy can be removed by simple distortion. Therefore, one cannot avoid the above discussion altogether. The additional question is that, from this standpoint, aromaticity is not taken as a ground-state property, as some would like to have it [8].

The latter remark throws the ball back to experimentalists, and thus may serve as a conclusion to the present paper. It seems that it is by no means impossible to give a clear-cut definition of aromaticity even for non-benzenoid cyclic systems, at least in the framework of a given theoretical scheme; but it remains to be decided what use we have ready for the concept itself, once it has been translated into a mathematical formalism. This is a difficult problem, as has been seen, because different definitions place emphasis on different aspects of experimental evidence. It is to be hoped that experimentalists choose one definition and work on the basis of it, accepting the risk that the aromaticity thus specified may no longer lend itself to explaining away anything that is vaguely related with cyclic conjugation and cannot be easily attributed to some other specific effect.

REFERENCES

1 E. D. Bergmann & I. Agranat (1970) These Proceedings, pp. 9 ff.
2 W. Leibniz (ed. 1875–1890) *Phil. Schriften,* IV, Berlin, p. 468.
3 R. Robinson (1958) *Tetrahedron,* 3 : 323.
4 J. M. Tedder (1960) *Ann. Rep. Chem. Soc.* (London), p. 223.
5 J. A. Elvidge (1965) *Chem. Comm.,* p. 160.
6 D. W. Davies (1965) *ibid.,* p. 258.
7 F. Momicchioli & G. Del Re (1969) *J. Chem. Soc.* (London) (B), p. 674.
8 W. Pfleiderer (1967) *Topics in Heterocyclic Chemistry* (ed. R. N. Castle), Wiley-Interscience, New York, p. 56.
9 A. T. Blomquist, J. Wolinsky, Y. C. Meinwald & D. T. Longone (1956) *J. Am. Chem. Soc.,* 78 : 6057.
10 K. Alder & H. H. Möllss (1956) *Chem. Ber.,* 89 : 1960.
11 W. J. Baley & J. C. Goossens (1956) *J. Am. Chem. Soc.,* 78 : 2804.
12 E. Clementi (1968) *Chem. Rev.,* 68 : 341.
13 W. Kutzelnigg, G. Del Re & G. Berthier, *Fortschr. Chem. Forsch.* [in press].
14 G. Del Re (1969) *Hung. Phys. Acta,* 27 : 477.
15 Idem (1960) *Nuovo Cimento,* 17 : 644.
16 M. Carpentieri, L. Porro & G. Del Re (1968) *Int. J. Quant. Chem.,* 2 : 807.
17 G. Del Re, O. Mårtensson & J. Nordling (1959) *Tech. Note,* No. 29, Quantum Chemistry Group, Uppsala University.
18 G. Berthier & G. Del Re (1965) *J. Chem. Soc.* (London), p. 3109.
19 G. Del Re (1965) *ibid.,* p. 3324.

Discussion

J. F. Labarre:

Can you give me the relative aromaticity scale that you get for the thiophene, pyrrole, furan series?

I think that the order you get is due to your 'only π' model. I agree with you that aromaticity depends on the charge transfer induced in the ring by the X atom, but only under the express condition that this transfer be a $\sigma + \pi$ one! (cf. J. Devanneaux & J. F. Labarre (1969) *J. Chim. Phys.*, p. 11).

Moreover, you cannot compare the aromaticity of C_{2v} molecules to that of D_{6h} benzene.

G. Del Re:

The relative scale of aromaticity that we get from different kinds of calculation is: furan, thiophene, pyrrole. I should like to add that we also calculated the σ charges, but we did not find much change; the spectra appeared to be explained essentially in terms of π electrons, even though we did try to allow for σ effects. But my main point is this: If one vaguely associates aromaticity with cyclization, one includes in it almost everything. If we want to find a clear-cut definition of aromaticity, we must try to be more specific, even at the cost of changing its name. Now, I proposed that we define it as a pure π property and work on it in that way. With this premise, my answer to you is that if you do not find the same order, other effects of cyclization are playing a role.

Mrs A. Pullman:

I am a little puzzled about the inclusion of σ charges in the discussion of aromaticity. Nor do I like the idea of associating aromaticity with charge transfer, because, from a theoretical point of view, the charges in this case are not well defined, and one can find whatever one wants, according to the definition and the way one interprets the calculations.

I think that Mr Del Re's ideas are better. For instance, I particularly like his idea of looking at the splitting of the orbital energies. Regarding the charges in heterocycles, especially pentaheterocycles, in a pure π calculation, when you introduce your δ parameter on the hetero-atom, you include implicitly a certain amount of σ polarization.

M. Cais:

May I introduce just one additional complication to this problem. Take the benzene molecule, and let us take butadiene in the *cis* form. Now, you tried to introduce a hetero-atom. I also introduce a hetero-atom, only I will make a π complex with the benzene, say, the chromium tricarbonyl derivative. It is very well known that butadiene forms another kind of π complex, a butadiene iron tricarbonyl derivative. If you take 1-phenylbutadiene or 1,4-diphenylbutadiene and you try to make these complexes, you can get very stable and defined structures where the chromium will be at one point and the iron at another, and, if you want, you can add another chromium tricarbonyl. Now, what do you do about the delocalization of the system here and your problem of aromaticity of the π system?

Discussion

G. Del Re:

When there are well-defined cyclic π systems, it should be possible to carry out the analysis suggested. Otherwise, it is better not to speak of aromaticity at all. Now, in the complexes you mentioned, complex formation seems to be a property of both open and closed π systems; moreover, the latter cease to be strictly π in the complex, because the metal atom lies on one side of what should be the nodal plane. On the other hand, it is possible that the reactivity of either butadiene or benzene, or of other π systems, when bonded in those complexes, is of the 'aromatic' type; then, however, we are using a definition of aromaticity that is based purely on chemical analogy with benzene.

S. Sarel (Hebrew University, Jerusalem):

I do not want to complicate the situation even more, but I have the impression that we did depart from experimental evidence. One of the arguments for the delocalization could come from measuring the resonance energy of the systems. Lately, a series of compounds, including furan, has been measured, and furan turned out to be non-aromatic. The resonance energy (1.59 Kcal/mole) was much too low; see M. J. S. Dewar, A. J. Harget & N. Trinajstić (1969) *J. Amer. Chem. Soc.*, 91:6324.

G. Del Re:

Our conclusions were in full agreement with that. I believe that in the case of furan the role of oxygen is something like hyperconjugation; at any rate, it is of minor importance. Of course, it all depends on the values of the parameters you have. Therefore, in all rigour, I can only say that with a reasonable choice of parameters and with our own way of analysing the situation, even taking into account the UV absorption spectra, we reached the same conclusion. This is due to the fact that spectra often show effects related not to delocalization, but to other effects, like change in shape.

E. D. Bergmann:

I do not remember whether the electronic spectra of 2- and 3-phenylthiophene have been measured. They should be quite different in your theory.
Whilst the 1,2-dimethylenecyclopentane derivative that you suggest is certainly of interest, there already exist some data on the *cis*-form of butadiene. Its presence is indicated in the infra-red spectrum of butadiene: its concentration is a function of the temperature. From this temperature dependence one might be able to calculate some energy data.

B. Binsch:

I should like to sound a general warning in trying to base a definition of aromaticity on energetic criteria. This can work only if you know the correct model for your conjugated system. Let me make myself clear by citing one particular example. If you do a calculation on pentalene, using the symmetry of the σ framework, you find a substantial resonance energy, and this result is not changed even though you may use an iterative scheme for adjusting the bond parameters in a PPP calculation.
The conclusion, which was reached in the literature by Dewar in doing such a calculation, was that pentalene should be an aromatic system, though strongly reactive, whatever that means. It turns out now that the essential feature of the electronic structure of pentalene is that you have a geometry that has a lower symmetry than the σ framework. If you did not know this beforehand in doing such a calculation, you would miss out on the most essential feature of the electronic structure of pentalene, and then the energetic criterion might not give you the correct conclusion. I think that Mr Nakajima would entirely agree with my point of view.

84

Quasi-Aromatic Compounds

Examples of Regenerative, or Meneidic, Systems

by D. LLOYD* and D. R. MARSHALL**

Department of Chemistry, University of St. Andrews, Fife, Scotland,* and
Department of Chemistry, University College of North Wales, Bangor, Caernarvonshire**

WE DID NOT INTRODUCE the term 'quasi-aromatic', which has been used to describe the chemical behaviour of some non-benzenoid compounds, but in 1964 we attempted to make a definition of the term, as follows:

> We propose that molecules should be called quasi-aromatic only if they contain an acyclic conjugated π-electron system and show chemical properties typical of aromatic compounds, especially reaction by substitution with retention of type. A significant mesomeric stability is implied [1].

We think that we should possibly have added the word 'electrophilic' in front of substitution.

This definition has been criticized, for example, by Marchand [2], who said that there was a need to resolve the meaning of aromaticity in some suitable physical terms before introducing extra terms such as 'quasi-aromatic', and also by Daltrozzo [3], who considered that the typical chemical properties of quasi-aromatic compounds did not justify any term including 'aromatic'.

While standing by our definition, which was apposite at the time it was made, we also consider that there is substance in both of these criticisms.

The real problem, however, is that mentioned by Marchand — namely, the various meanings that have been ascribed, and are still ascribed, to the term 'aromaticity'.

Quasi-aromatic has been used most frequently as a term to describe the behaviour of the metal complexes of acetylacetone (I), which are characterized by the ease with which they undergo electrophilic substitution at the β-carbon atom [inter alia 4–8].△

It was suggested [11] that the metal atom played a part in a cyclic conjugated system (II) in these β-diketone complexes and thus contributed to their stability.

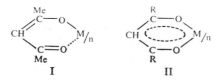

I   II

△ The term quasi-aromatic has also been used, and was probably first used, specifically, in connection with sugar phenylosazones [9–10].

A number of workers [12–13] considered that the NMR spectra of these complexes indicated the presence of a diamagnetic ring-current in the molecule, thus justifying their description as a kind of aromatic compound, but other workers [14–17] have interpreted the NMR spectra as evidence of lack of any type of aromaticity in these chelate complexes.

In contrast, 2,3-dihydro-1,4-diazepinium salts (III) cannot have a closed cyclic conjugated system, since part of the ring is made up of sp^3 atoms.

III

Therefore, in this case there is no question of these molecules possessing an aromatic system as normally understood.

Yet these compounds possess an extremely stable electronic system, which resists breakdown and shows a great 'tendency to retain the type', to use Armit and Robinson's classic characterization of the properties of aromatic compounds [18]. (This classic paper, with its statement of the concept of the aromatic sextet, marks a notable contribution from St. Andrews to the development of the concept of aromaticity.)

The stability constants of the dihydrodiazepinium salts with respect to the 1,2-diamines and β-dicarbonyl compounds from which they are prepared are mostly of the order of 10^6–10^9, although the value is somewhat lowered by neighbouring substituent groups in the conjugated system [19]. We believe that the planarity or near-planarity of the conjugated system may be important or even essential in determining its character, and that adjacent substituent groups may partially interfere with this planarity.

A simple demonstration of the stability of this system is afforded by the enormous pH range over which the monocation is the predominant species. The pK values for the equilibria beween III and IV are about 13–14 [20–21], while spectra of solutions indicate the absence of any noticeable concentration of the dication (V) in 40% sulphuric acid; only in > 70% sulphuric acid does the dication predominate over the monocation [21–22].

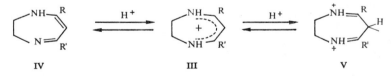

IV   III   V

The chemical similarity of the dihydrodiazepinium salts to benzenoid compounds is clearly demonstrated by the ease with which they undergo electrophilic substitution at the 6-position. Thus, they are readily halogenated [23–25], and the kinetics of these halogenation reactions closely resemble those for activated benzene derivatives, such as phenols and amines [26–27]. The dihydrodiazepinium salts are also nitrated under conditions used for benzenoid compounds [28–31].

The mechanism for these electrophilic substitutions obviously involves formation of a dication intermediate (VI), which, as in the case of benzenoid substitution reactions, loses a proton and reverts to the original stable system. It is noteworthy that the initial reaction in, for example, nitration involves attack by a cation on another cationic species.

The resultant 6-nitrodihydrodiazepinium salts can be reduced to amines, which in turn closely resemble benzenoid amines [31–32]. In particular, they form diazonium salts which

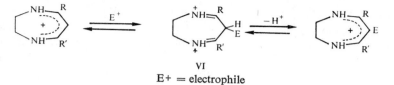

VI

E+ = electrophile

undergo Sandmeyer reactions [31–32]. These diazonium salts are, incidentally, abnormally stable, presumably due to contributions from alternative electronic structures, as depicted in VII.

VII

All of these chemical properties and reactions are just those which have always been firmly associated with what have been called aromatic compounds. For this reason, we originally described these dihydrodiazepinium salts as quasi-aromatic compounds [1]. Yet no one is likely to regard such a cyclic system, part conjugated, part saturated, as an aromatic system. This thus provides an excellent illustration of the pitfalls of trying to use chemical reactivity as an index of aromaticity.

Therefore, we were perhaps wrong in adding to the confusion by using the term 'quasi-aromatic', but the real nub of the problem is, as Marchand pointed out in his criticism [2] of our usage, that the use of the term 'aromatic' itself leaves much to be desired. In turn, this problem arises, of course, from the fact that its usage, and hence implied meaning, has changed many times from its initial meaning of 'possessing a smell'. It has been interpreted at different times in terms of molecular structure, of reactivity and of electronic structure, and, in consequence, there has been much confusion over its precise meaning and definition.

We suggest that because of this confusion, it would be better if the use of this term were discontinued, save perhaps with its general and original connotation of 'perfumed', and that it should pass with other technical terms which have outlived their precision and usefulness into the realm of the historian of chemistry.

When, about a hundred years ago, Kekulé was describing the cyclohexatriene structure for benzene, he used 'aromatic compound' as a synonym for 'benzene derivative'. Before the nineteenth century ended, the five-membered ring heterocycles were considered to have aromatic character, and in more recent years a plethora of so-called non-benzenoid aromatic compounds has been described. Yet, in almost all textbooks 'aromatic' continues to be a synonym for 'benzenoid'. We repeat our suggestion [33–34] that 'it is preferable to denote derivatives of benzene as "benzenoid" rather than "aromatic" '. This word is unambiguous. The term 'arenes', presently defined as 'the generic name of monocyclic and polycyclic aromatic hydrocarbons',* should be specifically restricted to benzenoid hydrocarbons. Aryl derivatives should be similarly defined.

We also suggest that cyclic conjugated systems should be termed either 'Hückel' or 'anti-Hückel' systems, and that these terms should replace the common present-day usage of 'aromatic' and 'anti-aromatic' [e.g. 35–36]. This replacement is necessary because of the lack of a universally recognized usage or meaning of this troublesome term 'aromatic'. Further, we

---

* IUPAC rule for the nomenclature of Organic Chemistry, A – 12, 4.

suggest the introduction of a term to mean 'the tendency to retain the type'. We suggest either 'regenerative' or 'meneidic'.* Thus, benzene is regenerative, or meneidic; ethylene is not.

Our studies on the chemistry of the 2,3-dihydro-1,4-diazepinium salts, which we have up to now described as quasi-aromatic compounds, have thus led us to see even more clearly the need to rethink the nomenclature of what have hitherto been called, all too frequently without any real definition, aromatic compounds.

We therefore propose this new nomenclature, which is based on the terms 'BENZENOID' (referring to structural features); 'HÜCKEL' and 'ANTI-HÜCKEL' (referring to ground-state properties); and either 'REGENERATIVE' or 'MENEIDIC' (referring to chemical reactivity). All of these terms refer to specific features.

In keeping with this suggestion, we therefore end with the alternative preferred description of 2,3-dihydro-1,4-diazepinium salts as compounds which, although they are not Hückel compounds, are examples of regenerative, or meneidic, systems.

---

* This term is based on a suggestion kindly made by Prof. K. J. Dover, Department of Greek, University of St. Andrews, and conveys the sense of retention of form.

REFERENCES

1 D. Lloyd & D. R. Marshall (1964) *Chemy Ind.,* p. 1760.

2 A. P. Marchand (1965) *ibid.,* p. 161.

3 E. Daltrozzo & K. Feldmann (1968) *Ber. Bunsenges. Phys. Chem.,* 72 : 1140.

4 C. Djordjevic, J. Lewis & R. S. Nyholm (1959) *Chemy Ind.,* p. 122.

5 J. P. Collman, R. A. Moss & W. S. Trahanovsky (1960) *ibid.,* p. 1213.

6 J. P. Collman, R. A. Moss, H. Maltz & C.C. Heindel (1961) *J. Am. Chem. Soc.,* 83 : 531.

7 R. W. Kluiber (1960) *ibid.,* 82 : 4839.

8 *Ibid.* (1961) 83 : 3030.

9 L. Mester (1955) *J. Am. Chem. Soc.,* 77 : 4301.

10 Idem (1965) *Angew. Chem. (Int. Edn),* 4 : 574.

11 M. Calvin & K. M. Wilson (1945) *J. Am. Chem. Soc.,* 67 : 2003.

12 J. P. Collman, R. L. Marshall & W. L. Young (1962) *Chemy Ind.,* p. 1380.

13 R. E. Hester (1963) *ibid.,* p. 1397.

14 R. H. Holm & F. A. Cotton (1958) *J. Am. Chem. Soc.,* 80 : 5658.

15 J. A. S. Smith & E. J. Wilkins (1966) *J. Chem. Soc. (A),* p. 1749.

16 R. C. Fay & N. Serpone (1968) *J. Am. Chem. Soc.,* 90 : 5701.

17 M. Kuhr & H. Musso (1969) *Angew. Chem. (Int. Edn),* 8 : 147.

18 J. W. Armit & R. Robinson (1925) *J. Chem. Soc.,* 127 : 1604.

19 C. Barnett, D. R. Marshall & D. Lloyd (1968) *J. Chem. Soc. (B),* p. 1536.

20 G. Schwarzenbach & K. Lutz (1940) *Helv. Chim. Acta,* 23 : 1162.

21 D. Lloyd, R. H. McDougall & D. R. Marshall (1966) *J. Chem. Soc. (C),* p. 780.

22 A. M. Gorringe (1968) Ph. D. Thesis, University of St. Andrews, Scotland.

23 D. Lloyd & D. R. Marshall (1958) *J. Chem. Soc.,* p. 118.

24 C. Barnett, H. P. Cleghorn, G. E. Cross, D. Lloyd & D. R. Marshall (1966) *J. Chem. Soc. (C),* p. 93.

25 A. M. Gorringe, D. Lloyd, F. I. Wasson, D. R. Marshall & P. A. Duffield (1969) *ibid.,* p. 1449.

26 R. P. Bell & D. R. Marshall (1964) *J. Chem. Soc.,* p. 2195.

27 D. Lloyd & D. R. Marshall (1970) *Paper presented at meeting of the Heterocyclic Group of the Chemical Society,* London.

28 C. Barnett (1967) *Chem. Comm.,* p. 637.

29 Idem (1967) *J. Chem. Soc. (B),* p. 2436.

30 A. M. Gorringe, D. Lloyd, D. R. Marshall & L. A. Mulligan (1968) *Chemy Ind.,* p. 130.

31 A. M. Gorringe, D. Lloyd & D. R. Marshall (1970) *J. Chem. Soc. (C),* p. 617.

32 Idem (1968) *Chemy Ind.,* p. 1160.

33 D. Lloyd (1963) *Alicyclic Compounds,* London, Arnold, p. 103.

34 Idem (1964) *The Chemistry of Simple Organic Compounds,* London, University of London Press, p. 132.

35 R. Breslow, J. Brown & J. J. Gajewski (1967) *J. Am. Chem. Soc.,* 89 : 4383.

36 M. J. S. Dewar (1967) *Chem. Soc. (London), Special Publication, No. 21,* p. 177.

Discussion

S. M. Sprecher (University of Tel-Aviv):

I wanted to ask the following question: What is the real evidence that your electrophilic substitution is taking place on the mono-cation, rather than on the small amount of neutral molecule in equilibrium with your mono-cation? Not too long ago there has been work on the nitration of aromatic amines and the like which has shown a similar phenomenon: there is a small amount of amine which is not protonated and which substitutes, of course, very much faster than the protonated species.

D. Lloyd:

The direct evidence is of a kinetic nature. Furthermore, we have, by using the correct conditions, been able to isolate compounds of the intermediate type we suggest.

S. M. Sprecher:

Since your systems are essentially vinylogous amidines, it would mean that one could substitute the non-protonated nitrogen of an amidine, and such reactions are not known to me.

D. Lloyd:

I will not use the word unique, but the reactions of this system are, let us say, quite unusual.

E. D. Bergmann:

We are still trying to find a definition of aromaticity. If you take benzene itself, to which of your three classifications would you say it belongs? To all three?

D. Lloyd:

Yes.

E. D. Bergmann:

Would you then say it is an aromatic compound?

D. Lloyd:

In Britain, at any rate, there is now a tendency to separate benzene from all other aromatic compounds and say it is arch-aromatic. I did not introduce that term!

B. Binsch:

I cannot help but object to the new term you have introduced, and the fact that it is derived from the venerated Greek does not change my opinion; I do so on philosophical grounds. I think that, in order to use this criterion, you first have to do an experiment to find out, and any criterion that is based on such an experiment is, in my opinion, unsatisfactory.

D. Lloyd:

But surely fifty per cent of us here are experimental chemists.

B. Binsch:

My answer to that is that we should at least attempt — though we may not succeed — to come up with a criterion that we can apply even if the molecule is still unknown.

D. Lloyd:

No, I disagree. I merely want this as a term to describe observed behaviour; I do not want to read more into it than that.

New Hetero-Aromatic Non-Benzenoid Compounds

by W. TREIBS

Heidelberg

IN MY PAPER I will discuss the synthesis of iodides of aromatic and hetero-aromatic non-benzenoid compounds, and of aromatic heterocyclic substances. I will also present some remarks on azapentalenes (dehydroindoles).

I. IODIDES

Olefinic organic substances are characterized by the easy addition of chlorine and bromine; iodine is added with difficulty. However, iodine can be added to certain heterocyclic compounds, such as, for instance, pyrrole. We found that iodine may be added rapidly, without heat, light or catalysts, to many polar or easily polarized organic substances. The primary addition reaction is often followed by secondary reactions:

$$RH + I_2 \rightarrow [RHI_2 \leftrightarrow RHI^{\oplus} I^{\ominus}] \xrightarrow{-HI} RI$$

In proper solvents, especially in hydrocarbons and methanol, the solid iodides separate quantitatively, and are often analytically pure. They are soluble in chlorinated hydrocarbons and, to a greater extent, in acetone.

From among the many results we have obtained, I will present a few that are particularly interesting. These are divided according to their chemical behaviour.

1. *Azulenes*

Azulenes are deeply coloured, and their chemical changes are therefore visible. As a result of their ready polarization they are more accessible to electrophilic substitution than real aromatic hydrocarbons. Till now they have been halogenated by N-halogeno-succinimides in positions one and three of the five-membered ring and, when both of these positions were occupied, in the seven-membered ring.

We found that elementary iodine substitutes azulene (vis. max. 580 mμ) almost instantaneously in positions one and three, to form the diiodide I (vis. max. 668 mμ; IR freq. 1,670 and 1,550 μ). In addition, we obtained a yellow product of low solubility, which has not yet been identified.

Guaiazulene (vis. max. 608 mμ), in the same type of reaction, is converted predominantly to a blue diiodide II (vis. max. 670 mμ; IR freq. 1,670 and 1,620 μ). Considering the large bathochromic shift in the visible maximum of the product II, the substitution must have

91

taken place on two odd carbon atoms; these can only be C_3 in the five-membered ring and C_5 in the seven-membered ring.

Both iodides are not very stable and are decomposed by metals to the original azulenes and dimers.

2. *Azulenes with Nitrogen in the Five-Membered Ring*

This class of non-benzenoid hetero-aromatic compounds was first prepared by us. They react instantaneously with iodine to form dark solid iodides, whose solutions are intensely yellow. In contrast to the bathochromic shift of the diiodides of azulenes, the iodides of these azazulenes show a strong shift to shorter wavelengths. This behaviour is also exhibited by the methiodides of the azazulenes, which are soluble in water. Since the azazulene iodides are insoluble, they are not regarded as salts.

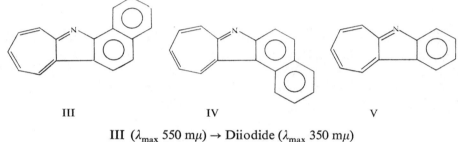

III (λ_{max} 550 mμ) → Diiodide (λ_{max} 350 mμ)
IV (λ_{max} 508 mμ) → Diiodide (λ_{max} 336 mμ)
V (λ_{max} 504 mμ) → Diiodide (λ_{max} 325 mμ)

3. *Tropilidenes*

Tropilidene (cycloheptatriene) is precipitated quantitatively by elementary iodine from solution in petroleum ether. The analytically pure diiodide (characteristic IR freq. 725) surely corresponds in its formula (VI) to the dibromide of W.v.E. Doering. Its iodine is easily transferred to other suitable molecules.

4. *Metallocenes*

Ferrocene in petroleum ether is quantitatively precipitated by elementary iodine. The analytical formula of the product corresponds to a triiodide (vis. max. 610 mμ; characteristic IR frequencies 1,000 and 880 μ). In polar solvents more iodine is taken up. Nickelocene reacts in the same manner, but not cobaltocene.

5. *Pyrrocolines (Indolizines)*

From among a large number of aromatic heterocyclic substances giving solid iodides, mention may be made of the 2-methyl- and the 2-phenylpyrrocoline. Their diiodides

(VII, R = CH$_3$, C$_6$H$_5$) are slightly soluble in methanol and methylene chloride; they are more soluble in acetone. The 2-phenyl derivative shows a visible maximum of 364 mμ.

VII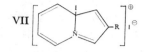

6. *Indoles*

The investigation of the dehydrogenation of indoles by bromine to azapentalenes was the starting point of this work on diiodides. The two classes chiefly examined, the pentindoles and the benzylenindoles, differ in their reactivity.

In treating pentindole (and its benzologues) with iodine, one molecule is added and one hydrogen iodide is split off, to give the monoiodide VIII (vis. max. 362 and 507 mμ). Benzyl-enindole (as well as its benzologues), however, adds iodine and gives a stable diiodide IX (vis. max. 332 mμ; IR freq. 3,100, 1,680 and 1,590 μ).

VIII  IX

7. *Other Iodides*

Iodides have been prepared, by addition of elementary iodine, from cyclopentadiene, from many hetero-aromatic products and from aromatic hydrocarbons activated by hetero-atoms. Most reactions involving iodine can also utilize bromine effectively.

8. *Reactions of Iodides*

Diiodides have hitherto been used in a variety of chemical reactions, such as dehydro-genation, hydrolysis, photolysis, reduction and treatment with cationic compounds.

II. AZAPENTALENES

The behaviour of the pentindoles differs from that of the benzylenindoles with regard to bromine, as well as to iodine.

1. *Benzazapentalenes*

Pentindole (as well as its unsymmetrical benzologues) adds bromine, even in nonpolar solvents, to form at first a colourless and then a deep blue intermediate that splits off hydro-bromic acid. The final product is the purple-red hydrobromide of the azapentalene X (vis. max. 580 mμ; IR freq. 2,850, 1,680 and 1,580 μ). X is liberated by treatment with caustics

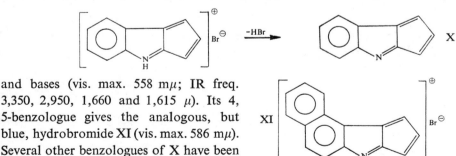

and bases (vis. max. 558 mμ; IR freq. 3,350, 2,950, 1,660 and 1,615 μ). Its 4, 5-benzologue gives the analogous, but blue, hydrobromide XI (vis. max. 586 mμ). Several other benzologues of X have been prepared.

2. Symmetrical Dibenzazapentalenes

In nonpolar solvents benzylenindole and its benzologues are substituted by bromine in the benzene ring in the *para*-position to the indole nitrogen, but are not dehydrogenated to the desired azapentalene hydrobromides. The latter are obtained in very good yield by boiling the indole with N-bromosuccinimide in polar solvents, especially in nitromethane. Benzylenindole is thus converted to the hydrogen bromide adduct XII (IR freq. 3,050, 1,700 and 1,600 μ), which, by splitting off hydrogen bromide, is converted to the dibenzazapentalene XIII (vis. max. 496 mμ; IR freq. 3,450, 3,080, 1,700 and 1,008 μ).

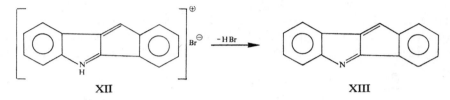

XII   XIII

3. Chemical Nature of the Acid Adducts

What is the true chemical nature of these adducts? A comparison of their chemical properties may perhaps provide an answer. Azulene is a very weak base, adds strong acids, but its salts (cation XIV) are rapidly hydrolysed by water. Benz-azazulene, however, is a strong base and gives water-soluble and water-resistant salts even with weak acids (cation XV). The acid adducts of the azapentalenes, however, are insoluble in water and wholly resistant

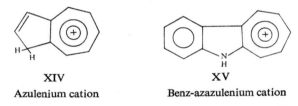

XIV   XV

Azulenium cation   Benz-azazulenium cation

to it. Contrary to our proposed formulae, these compounds probably are not at all salts. In both azapentalenes the conversion of the acid adduct to the base gives rise to new frequencies in the IR spectrum at 3,350 and 3,450 μ, probably due to the group N$\diagdown$C. The azapentalenes are indolenines, which are known to add acid chlorides to their double bond:

Perhaps the azapentalenes add hydrogen bromide and other acids in the same manner, and the bromine atom is linked to the five-membered ring.

Discussion

E. D. Bergmann:

If the compounds you have obtained by iodination of the heterocyclic systems are not salts, one might determine the dipole moments and derive from them the nature of the halogen bonding.

W. Treibs:

They have a very weak tendency to ionization.

M. Cais:

In connection with the last remark, perhaps it wouldn't be fair to include the metallocene iodides as a class of compounds We have done some work with these ferrocene iodides, and Prof. Herbstein in our Department has done an X-ray analysis of ferrocene triiodide; there is no question about it, this is a ferricenium triiodide, a salt. We have also isolated from the same mixture a ferricenium pentaiodide and a heptaiodide. In addition, we have isolated what seems to be, from analysis, the compound ferrocene-I_2, but we are not sure about this. But for the others: the triiodide, pentaiodide, etc., there is no question that they are salts.

W. Treibs:

We had not enough time to work all these things out.

Aromaticity Problems in Metallocenes

by M. CAIS

Department of Chemistry, Technion – Israel Institute of Technology, Haifa

I. Introduction

IT HAS OFTEN BEEN STATED that 'aromaticity can be expected wherever conditions of stereochemistry, availability of orbitals and numbers of electrons allow electron delocalization to occur in a cyclic system' [1]. On the basis of this definition, the cyclopentadienyl rings in ferrocene, I, have been regarded as having aromaticity conferred upon them by complex formation with the metal atom. This criterion for the aromaticity of ferrocene is certainly an oversimplification, because the cyclopentadienide anion, II, by itself, already meets the above-mentioned requirements for aromaticity. The relegation of the metal atom to an unimportant place, as implicit in this viewpoint, is necessarily in conflict with the idea that the presence of the metal atom may play a significant role in the bestowal of aromaticity on the ferrocene molecule as a whole.

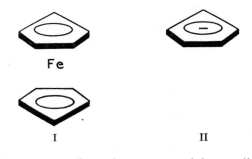

Is it justified to ignore the presence of metal atoms as such in metallocenes on the tenet that 'the kind of atom participating in the delocalized system is not important; the kind of orbital *is*' [1]? Can one draw a reasonable analogy between the concept of molecular orbitals of the π-electron sextet of the aromatic benzene (or cyclopentadienyl) ring and the pertinent molecular orbitals of the ferrocene molecule?

The problem of chemical bonding in ferrocene has been the subject of a large number of theoretical analyses, at various degrees of sophistication, and some fundamental differences have yet to be resolved [2].

The ferrocene molecule is usually treated as an 18-electron problem. All the various analyses [2; cf. 3–9; E. Heilbronner & M. Cais, unpublished results] agree that the two highest occupied orbitals (ψ_8, ψ_9) are the degenerate pair of weakly bonding E_{2g} molecular orbitals

96

which have a high degree of metal $3d$ character. Similarly, the various treatments concur in considering the metal $3d_{z^2}$ a non-bonding, occupied orbital, but they differ in the placement of this energy level relative to the bonding orbitals. Whereas in some treatments this molecular orbital, $A'_{1g}(d_{z^2})$, occupies an energy position next to the highest occupied MO level, in another analysis this position is occupied by a degenerate pair of bonding E_{1g} molecular orbitals, each of which has only about 20% metal $3d$ character (Table 1).

Table 1

Per Cent Electron Density on Metal Atom
in Each MO of Ferrocene *

	SD**	DB**	HC**
$A_{1g}(4s)$	24.0 (ψ_1)***	40.0 (ψ_1)	13.0 (ψ_1)
$A'_{1g}(3d_{z^2})$	100 (ψ_7)	100 (ψ_5)	100 (ψ_7)
$A_{2u}(4p_z)$	1.0 (ψ_2)	22.2 (ψ_2)	1.3 (ψ_2)
$E_{1u}(4p_x, 4p_y)$	34.8 (ψ_3, ψ_4)	34.9 (ψ_3, ψ_4)	22.5 (ψ_3, ψ_4)
$E_{1g}(3d_{xz}, 3d_{yz})$	13.7 (ψ_5, ψ_6)	20.6 (ψ_6, ψ_7)	27.2 (ψ_5, ψ_6)
$E_{2g}(3d_{x^2-y^2},$			
$3d_{xy})$	72.2 (ψ_8, ψ_9)	80.6 (ψ_8, ψ_9)	88.2 (ψ_8, ψ_9)
A^*_{1g}	76.0 (ψ_{10})	60.0 (ψ_{19})	87.0 (ψ_{10})
E^*_{1g}	86.3 (ψ_{18}, ψ_{19})	79.4 (ψ_{14}, ψ_{15})	72.8 (ψ_{11}, ψ_{12})
A^*_{2u}	99.0 (ψ_{11})	77.8 (ψ_{16})	98.7 (ψ_{13})
E^*_{1u}	64.2 (ψ_{12}, ψ_{13})	65.1 (ψ_{17}, ψ_{18})	77.5 (ψ_{14}, ψ_{15})
E^*_{2u}	0.0 (ψ_{14}, ψ_{15})	0.0 (ψ_{10}, ψ_{11})	0.0 (ψ_{16}, ψ_{17})
E^*_{2g}	27.8 (ψ_{16}, ψ_{17})	19.4 (ψ_{12}, ψ_{13})	11.8 (ψ_{18}, ψ_{19})

* Calculated from published data as per cent contribution of $[C_n \phi_n$ (metal orbitals)$]^2$ in each MO of the form ψ_n(MO) $= C_n \phi_n$ (Cp rings) $+ C_n \phi_n$ (metal orbitals)
** SD = Shustorovich and Dyatkina [5–6]; DB = Dahl and Ballhausen [4]
HC = Heilbronner and Cais [unpublished results]
*** ψ_n ($n = 1$ to $n = 19$) in order of increasing energy $\psi_1 < \psi_2 \ldots < \psi_{19}$

Even more significant disagreement arises in the ordering of anti-bonding orbitals. For example, a treatment by Shustorovich and Dyatkina [5–6] ($E_{2g} < A^*_{1g} < A^*_{2u}$), an independent-electron calculation by Heilbronner and Cais [unpublished results] ($E_{2g} < A^*_{1g} < E^*_{1g}$) and an all-valence electron calculation of Schachtschneider, Pruis and Ros ($E_{2g} < A^*_{1g} < E^*_{1g}$) [7] are in agreement that the lowest unoccupied orbitals have much metal $3d$ character. On the other hand, the analysis by Dahl and Ballhausen ($E_{2g} < E^*_{2u} < E^*_{2g}$) [4] produced the E^*_{2u} orbitals on the lowest anti-bonding level. These ring orbitals have no metal $3d$ character at all, nor can they have any, since no metal orbitals of E_{2u} symmetry are available for mixing. Using the ligand-field model, Scott and Becker [8] obtained good agreement between calculated and experimental ligand-field bands of ferrocene. The fitted results were then used to calculate the one-electron orbital energies. The order of the d levels was found to be $E_{2g} < A'_{1g} < E^*_{1g}$, and the energy differences were $E^*_{1g} - A'_{1g} = 3.16$ eV and $A'_{1g} - E_{2g} = 0.57$ eV.

The general result is that, with only 18 valence electrons available (ten π electrons from the rings and eight from the iron atom), any attempts to make ten equivalent bonds from the metal atom to the ten ring carbon atoms will cause ferrocene to appear to be an electron-deficient molecule. According to the MO treatments, twelve of the eighteen valence electrons are placed in six strongly-bonding molecular orbitals. The remaining six electrons, which for the purpose of the aromaticity concept have been regarded [9] as analogous to the sextet of π electrons in benzene, must 'take the blame' for the chemical features of ferrocene. This intuitive approach allows us to comply with the generalization that the kind of orbital participating in the delocalized system is important, and not the kind of atom. At the same time, we can appropriately include in our consideration of chemical properties of ferrocene the high degree of metal character mixed into the MO orbitals occupied by this 'sextet' (see Table 1). All this provided, of course, that we are not disturbed by Linnett's statement [10] that the molecular orbital approach is not well suited for describing the electronic distribution even in such a highly symmetric and relatively simple π complex as ferrocene!

II. CHEMICAL REACTION CONSIDERATIONS

For our present purpose, it may be fruitful to summarize some of the chemical properties of ferrocene from the following aspects: (1) electrophilic substitution; (2) intra- and inter-annular effects; (3) oxidation-reduction reactions; (4) substituent effects of the ferrocenyl group.

1. *Electrophilic Substitutions*

In a reaction considered typical of chemical aromatic character, the Friedel-Crafts acetylation, ferrocene was found to react faster than benzene by a factor of 10^6 (in methylene chloride at 0°) [11]. In classical aromatic systems, such as benzene, the aromatic character is attributed to the high density of polarizable electronic charge above and below the plane of the carbocyclic system. By analogy and extrapolation, the ferrocene molecule should be considered to possess an even more highly mobile and polarizable electron cloud, in order to account for its 'super-aromaticity', relative to benzene, in electrophilic reactions such as Friedel-Crafts acylations. The immediate question arising from this analogy is: Could this be due to to the high percentage of metal $3d$ character in the highest occupied molecular orbitals of ferrocene? Rosenblum and co-workers [11] attempted to provide an answer to this question by suggesting that the initial step in ring substitution involves attack by the electrophile on the E_{2g} orbital electrons to form the complex depicted by III. The latter, in a rate-determining step, rearranges to IV, followed by expulsion of H$^+$ to give the product V.

In a subsequent paper [12], Rosenblum and Abbate showed that the isomeric *exo* and *endo* acids, VIa and VIb, respectively, cyclized in the presence of trifluoroacetic acid to the homocyclic ketones, VIIa and VIIb, with a somewhat faster rate of cyclization for the *exo* acid. Since this result seems to indicate that attack by the electrophilic reagent can occur from either side of the ring, Rosenblum and Abbate conclude that the metal atom is not an essential participant in the electrophilic substitution of ferrocene. In other words, this can be taken to mean that the rearrangement of the cation III to the σ complex IV cannot be the rate-determining step, and thus agrees with the suggestion of Ware and Traylor [13] that electrophilic reactions in metallocenes can be explained on the basis of direct formation of a σ complex without metal participation.

More recently, Yakushkin and co-workers [14] studied the kinetic isotope effect in the exchange of deuterium- and tritium-labelled ferrocene with trifluoroacetic acid in dichloromethane or acetic acid. The k_D/k_T ratio for ferrocene, pentamethylbenzene and thiophene was found to be 1.2, 1.8 and 1.9, respectively. The small isotope effect for ferrocene in acid

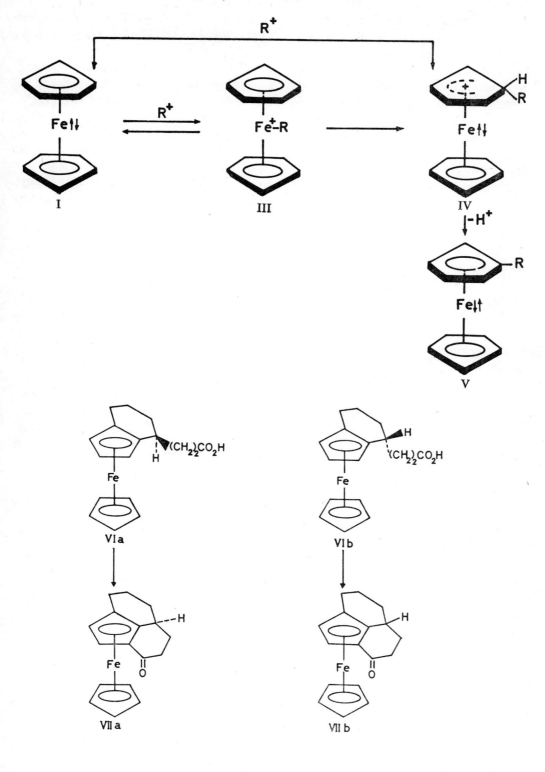

solution is said to suggest that the rate-determining step is addition of the proton to the metal, I $\rightleftarrows$ III (R$^+$ = D$^+$), followed by fast rearrangement to IV (R = D), and subsequent splitting out of a proton to give V (R = D).

In basic media, the same workers found that the kinetic isotope effects for ferrocene and benzene were substantially similar, and, consequently, it was concluded that under those conditions the mechanism involved direct proton attack on the ring, I $\rightleftarrows$ IV (R$^+$ = D$^+$), the rate being determined by cleavage of the C–H bond.

Mangravite and Traylor [15] have reported that the proton exchange in ferrocene shows an isotope effect higher ($k_H/k_D = 1.8$) than that found by Yakushkin et al. ($k_H/k_D = 1.4$). The former authors have also determined the specific rate constants for isotopic hydrogen exchange at the various positions of phenylferrocene. They have found that the ferrocenyl group accelerates exchange on benzene by a factor of 10^5, whereas the phenyl group scarcely changes the reactivity of the ferrocene rings.

This latter finding could be interpreted as showing that the phenyl substituent may have very little effect on the basicity of the E_{2g} electrons, and thus would not be expected to cause any marked change in the rate of formation of the cation complex III. On the other hand, the well-established property of the ferrocenyl group to act as an electron-donor substituent (*vide infra*) may significantly alter the π-electron density on the benzene nucleus, so as to facilitate greatly formation of the σ complex in benzenoid substitution, and thus account for the accelerated rate of exchange on the ferrocenyl-substituted benzene.

However, Mangravite and Traylor [16] have not suggested this explanation, and instead, they have been led by their results to propose a new mechanism for electrophilic substitution on metallocenes, based on the concept of σ–π conjugation. They suggest that two concurrent mechanistic pathways are operative. Path *a* consists of a two-step process in which the electrophile R$^+_{inside}$, coming from the metal side of the ring, cleaves the metal-carbon bond in an S$_E$2 reaction with retention of configuration. This is followed by rate-limiting removal of a proton in a second step, which constitutes an S$_E$2 reaction with inversion.

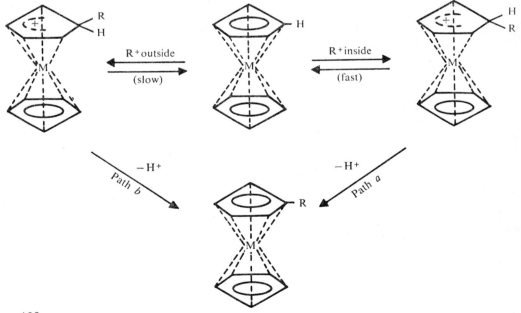

Path *b* involves rate-limiting attack of electrophile $R_{outside}^+$ from the side of the ring opposite the metal atom in an S_E2 displacement with inversion of configuration, followed by fast removal of proton from the inside with retention of configuration. In the case of proton exchange (R = H), Path *a* is the microscopic reverse of Path *b*, and the process will proceed in both senses. If R ≠ H (e.g., acylation), the operative pathway will depend on the relative rate constants for the individual steps, and thus on the nature of the electrophile and substrate. In proposing the above scheme, the authors [16] have consciously refrained from ascribing any role to the metal-complexed cation III on the grounds that they consider it to be the counterpart of the π complex in benzenoid substitution. Justification for this view does not seem to be unquestionable.

2. *Intra- and Inter-Annular Effects*

In a discussion of substitution patterns in ferrocene we have to consider electronic effects both within and across the cyclopentadienyl rings. These are usually referred to as intra- and inter-annular effects, respectively. It is rather obvious that at present any discussion of these effects must necessarily be handicapped from the start, since, as described above, we have not yet reached sufficient understanding of the electronic structure of, and thus the chemical bonding in, metallocenes. Recently, Knox, Pauson and co-workers [17] have collected and summarized the available experimental data on substituent effects in ferrocene. Therefore, we shall mention only a limited number of observations and emphasize a few special points.

Acetylation, formylation, isobutyrylation and succinylation of alkylferrocenes take place with preference for 3-substitution (Fig. 1) [17]. It is rather doubtful whether one could attribute this effect simply to steric influences, which, as pointed out by Rosenblum [18], are markedly stronger in the acetylation of toluene (the ratio of *o-*:*p-*isomers is 1:67) or cumene (no *o*-product detected by GLC analysis).

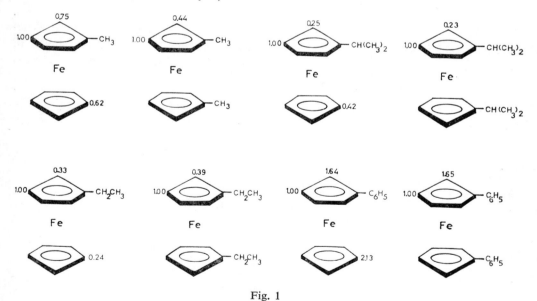

Fig. 1
Site reactivities in acetylation of alkyl- and arylferrocenes
(the 3-position has been assigned unit reactivity)

The activating effect of the alkyl group appears to be transmitted also to the unsubstituted ring, which has been found to be approximately twice as reactive as one of the rings in ferrocene itself [19]. At this point it should perhaps be noted that simple molecular orbital calculations have predicted for electrophilic substitution a higher reactivity for the 2-position in ferrocenes bearing either electron-donating or electron-withdrawing groups. However, it would be rather unwise to dwell too much on the discrepancy between these calculated results and the experimental findings mentioned above [20–21].

Friedel-Crafts acylation of acetyl-, methoxycarbonyl-, cyano-, bromo-, chloro-, acetamido-, urethano- and phthalimido-ferrocenes occurs either exclusively or mainly at the unsubstituted ring. Aryl substituents have been found to exert a smaller, but nevertheless significant, deactivating effect [17, ref. in notes 13, 17–19, 21–23]. As shown in Fig. 1 for phenyl-ferrocene and 1,1'-diphenyl-ferrocene, substitution in the 2-position is favoured over the 3-position. A similar reversal of reactivity (2- > 3-) was found for methoxy- and methyl-thio-ferrocenes [17].

The data mentioned above, as well as other results (e.g., deuterium-exchange experiments with alkylferrocenes [22]), provide rather strong evidence in support of the viewpoint that it is fruitless to attempt direct analogies between substituent effects in the benzene series and the intra-annular influences in metallocenes. In the case of inter-annular effects, such analogies must break down altogether. Virtually little is understood about the mechanism by which various effects can be transmitted from one ring 'through' (or 'across') the metal atom to the other ring. Let us review some of the facts.

The electronic absorption spectra of hetero-annularly disubstituted ferrocenes — e.g., 1,1'-diarylferrocenes — are practically identical, except for higher intensities, with those of the respective monosubstituted arylferrocenes [11, 23]. The implication is that the metal atom seems to act as if isolating electronically one aryl group from the other, that is, there appears to be little or no conjugation between the aryl substituents in 1,1'-disubstituted ferrocenes.

This is in contrast to other, chemical evidence. For instance, the acidity of ferrocene-carboxylic acid is noticeably affected by substituents in the 1'-position. As shown in Table 2, electron-releasing substituents in the second ring lower the acidity, whereas electron-withdrawing groups increase it.

Table 2
Inter-Annular Effects in Ferrocene *

	R	pK_a**
	H	6.29$^\triangledown$
	C_2H_5	6.43
	C_4H_9	6.50
	CO_2CH_3	6.08
	$COCH_3$	5.91
	CN	5.82
	SO_2NH_2	5.56

* Taken from a review by Rausch [24]
** In 68% methanol at 20°
$^\triangledown$ For comparison, pK_a for benzoic acid is 5.93 (under the same conditions)

Nesmeyanov and co-workers [25] have recently studied the protolysis of hetero-annularly substituted mercury derivatives, where the substituent is halogen, methoxy, acetoxy and methyl carboxylate. They have found that the total polar effect of substituents in one ring on the second ring is weaker by a factor of 10^2 than the effect transmitted from the *para*-position in a benzene ring.

There are also a number of studies [26–30] that provide data for linear free energy correlations of hetero-annular substituent effects in the ferrocene series. Hall, Hill and Richards [31] have recently carried out solvolysis studies on hetero-annularly substituted methyl-ferrocenylcarbinyl acetates in order to investigate the nature of the transmission of substituent effects from one cyclopentadienyl ring of ferrocene to a reaction centre on the other ring. They found a linear relationship between substituent effects in the solvolysis reaction and the reversible quarter-wave potentials of the appropriately substituted ferrocenes. From their data the authors conclude that inductive or field effects predominate, and that resonance interactions are of minor importance.

Initially, Rosenblum [11] attempted to reconcile the apparently conflicting results concerning inter-annular substituent effects in electronic spectra and those in chemical reactions by assigning a fundamental role to the metal atom (in the mechanism for electrophilic substitution). His subsequent reversal [12] on this position should be seen as a typical example of the general, as yet unsettled, controversy on whether presence of the metal atom does or does not play an important role in the chemical reactions of metallocenes.

3. *Oxidation-Reduction Reactions*

The simplest and probably the most characteristic reaction of metallocenes is their oxidation to cationic species through loss of one or more electrons. Ferrocene, for instance, can be oxidized to the ferricenium cation, VIII, by polarographic oxidation, by photolysis, by chemical oxidizing agents such as ferric chloride or ceric sulfate, or simply by bubbling oxygen through a suspension of ferrocene in an aqueous acidic solution. The ferricenium cation can be just as easily reduced back to ferrocene by chemical reagents such as stannous chloride, ascorbic acid, sodium bisulfite, sodium or potassium hydroxide solutions, and others.*

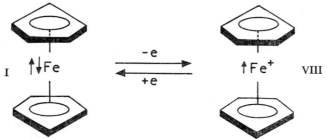

Chronopotentiometric measurements of oxidation potentials of substituted ferrocenes have shown that ease of oxidation is decreased by electron-withdrawing substituents and increased by electron-donating substituents relative to the parent ferrocene.*

Polarographic half-wave potentials and oscillopolarographic measurements have been reported for a variety of substituted ferrocenes. The effect of the substituents in hetero-annularly disubstituted ferrocenes was found to be additive and the reactions were reversible.*

* For leading references, see Rosenblum [3], pp. 48–45.

103

The various authors who obtained Hammett-type free-energy correlations are not in general agreement with regard to either the best kind of σ values to be used or to the significance of the correlations obtained. It is not at all unlikely that additional extensive efforts in the study of electron-transfer processes involving metallocenes may cause far-reaching changes in our reaction mechanism ideas and our aromaticity concepts with respect to these compounds.

4. *Substituent Effects of the Ferrocenyl Group*

Abundant experimental evidence is available [32] to show that the ferrocenyl group, $C_5H_5FeC_5H_4$-, is a very powerful electron-releasing group. This property has been best demonstrated in the remarkable stabilizing effect imparted by a ferrocenyl group substituted on an electron-deficient atom, such as in α-ferrocenylcarbonium ions IX.

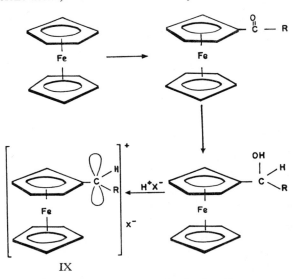

Synthesis of α-Ferrocenylcarbonium Ions

In fact, derivatives of IX have been isolated as diamagnetic, stable salts [33–35], which can be handled with relative ease under usual laboratory conditions. Diferrocenylmethyl cation (IX, R = ferrocenyl) as the fluoroborate salt has been recrystallized by us from methanol and recovered unchanged, as shown by IR and NMR spectra. This result has initiated work, currently in progress in our laboratories, towards an X-ray analysis of this stable carbonium ion.

The remarkable stability of α-ferrocenylcarbonium ions has provided a novel and convenient route for entry into the sesquifulvalene X and calicene XI systems [34].

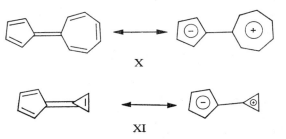

X

XI

The iron π complexes XII and XIII have been obtained in one-step syntheses by reacting ferrocene with tropylium fluoroborate and 3,3-dichloro-1,2-diphenylcyclopropene, respectively.

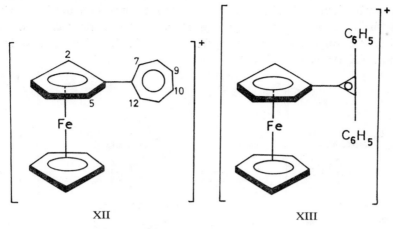

XII   XIII

It is noteworthy that a similar approach [36], aimed at preparing an iron π complex of benzopentalene, XIV, in the form of a stable α-ferrocenylcarbonium ion, was foiled because the expected cation XVI apparently prefers to exist as the cation radical XVII, which undergoes dimerization to yield the ferricenium species XVIII.

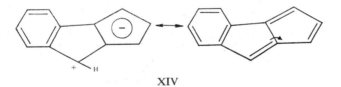

XIV

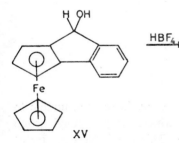

XV

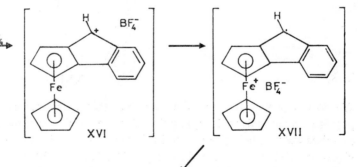

XVI   XVII

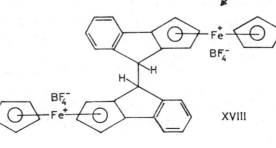

XVIII

105

The latter result provides clear-cut evidence in support of the argument that one cannot relegate to a minor position the role which the metal atom can play in the bestowal of special stability on α-ferrocenylcarbonium ions. An answer is required to the question: Why does the carbonium ion XVI prefer a diradical (or triplet) configuration, such as XVII, whereas experiment shows that the ground state of the analogous ferrocenylphenyl-carbonium ion is a singlet, pictorially represented by XIX, rather than a diradical (or triplet), such as XX (or its equivalent)?

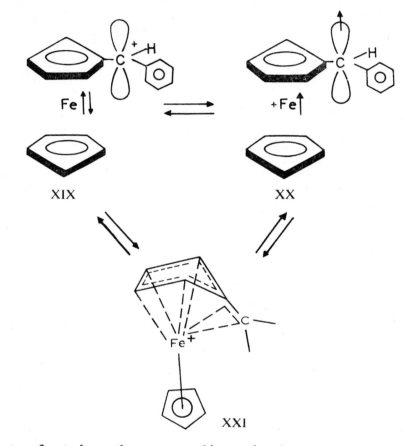

We do not profess to know the answer to this question, because we have been unable to provide an unequivocal explanation for the mechanism by which the ferrocenyl group exercises its stabilizing influence in ferrocenylcarbonium ions. This has been a matter of vigorous debate for over a decade [32–56]. No attempt will be made to discuss here the arguments advanced by the various workers in the field [32], except to say that several years ago there appeared to exist two diametrically opposed viewpoints: one group ascribing a major role in the stabilizing effect to the metal atom, i.e., the 'metal participation' group, and the second group maintaining that the metal atom contributes at best an insignificant amount of assistance to the resonance stabilization of α-ferrocenylcarbonium ions. More recently, there seems to be some kind of convergence in the views of both groups towards recognition of the importance of there being carbon-metal bonds in the molecule, though the distribution and form of these bonds continue to be a matter of controversy. Whatever the mechanism of stabilization, the ferrocenyl group is undoubtedly a better electron-releasing group than other aromatic groups, such as phenyl or phenyl-substituted moieties [32, 48, 55, 57; V. Belanic-Lipovac & M. Cais, unpublished results]. The reluctance of the ferrocenyl group to accept additional electron density has been demonstrated experimentally also in the measurement of ESR spectra of electrochemically generated radical anions of benzoyl-, *p*-toluoyl-, *p*-(methoxycarbonyl)-benzoyl-, *p*-nitrophenyl-, *p*-cyano-phenyl- and nitroferrocene [54]. Independent-electron MO calculations for the above compounds [54], as well as for a number of α-ferrocenylcarbonium ions [E. Heilbronner & M. Cais, unpublished results; 31, 37, 40; M. Cais & A. Schwarz, unpublished results; 58], indicate that the iron atom carries a significant portion of positive charge, which for the carbonium ions would be expected from structures such as XX.

The experimentally determined diamagnetic nature of the ferrocenylcarbonium salts precludes the presence of any significant amounts of XX in a potential equilibrium system, XIX ⇌ XX. To overcome this difficulty, it is possible to envisage the existence of an additional equilibrium, such as illustrated in XIX ⇌ XXI, if the differences in energy between the three species XIX, XX and XXI were rather insignificant. Our simple calculations show that the highest occupied orbital in these ferrocene derivatives is a non-bonding orbital, with high metal $3d$ character, and relative to it the lowest unoccupied orbital is only very slightly higher in energy. The idea of a very rapid interchange between these three species, XIX–XXI, might perhaps lead to a new explanation for the unusual stability of α-ferrocenylcarbonium ions and could provide the basis for a novel hypothesis in explaining many of the characteristic reactions of metallocenes.

The charge distribution in the α-ferrocenylcarbonium ion is shown in XXII, where eighteen electrons are distributed in the first nine (ψ_1–ψ_9) bonding orbitals, and in XXIII, where one electron from ψ_9 has been promoted to ψ_{10}, the first anti-bonding orbital.

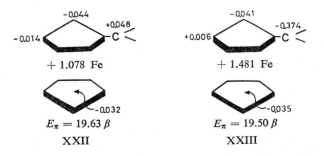

As can be seen from this simple exercise, XXII corresponds to the pictorial representation XIX, and XXIII fits the formulation of the cation radical XX, with practically no difference in the π-electron energy of the two systems.

It was stated in Sub-section 3 that the most characteristic reaction of metallocenes is probably the oxidation-reduction process. In line with this, one can regard the envisaged electron rearrangement XIX ⇌ XX as an intramolecular oxidation-reduction reaction. If, then, the α-ferrocenylcarbonium ion is reacted with a reducing agent, the first step in such a reaction could be an electron transfer to species XX (or XXI) to form the radical XXIV, which then dimerizes to yield the appropriate 1,2-diferrocenylethane derivative. This reaction, called 'reductive dimerization', has been shown [33] to be general for α-ferrocenylcarbonium ions.

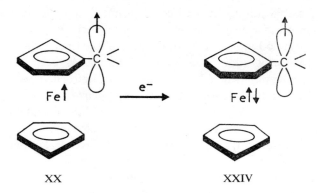

XX   XXIV

A similar electron-transfer process has been postulated in the mechanism of the arylation of ferricenium salts with aryldiazonium salts in the presence of small amounts of ferrocene [3, pp. 201–208]. The initial step involves an oxidation-reduction process, in which ferrocene reduces the diazonium salt, leading to the formation of a ferricenium cation and an arylazo-radical.

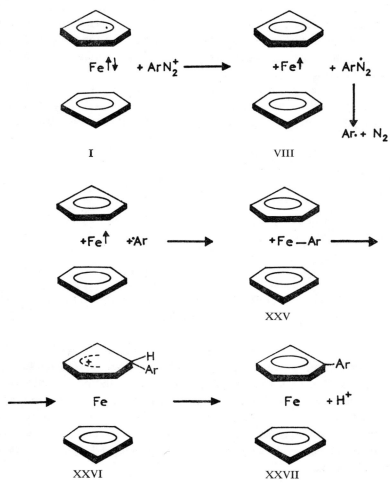

The latter decomposes as to form an aryl radical, which then couples with the ferricenium cation (radical), leading to the cation complex XXV. The latter, which is structurally identical to the intermediate III postulated in electrophilic substitutions, rearranges to the σ complex XXVI (analogous to IV), followed by expulsion of a proton to form the aryl-ferrocene XXVII. Could it not be that even in the electrophilic substitution reactions the first step involves an oxidation-reduction process like the one just described? It is of interest that a stable ferrocenylnitroxide radical has recently been reported [59].

The easy access to the highly stable, secondary α-ferrocenylcarbonium ions XXVIII has prompted us to investigate the possibility that deprotonation of XXVIII might lead to the nucleophilic carbenes XXIX.

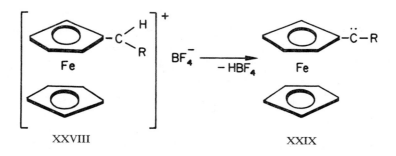

As pointed out previously, it is generally accepted that the net operative mechanism for the stabilization of α-ferrocenylcarbonium ions involves a flow of electrons from the ferrocenyl moiety towards the α-carbon atom, meaning that the newly-vacated p orbital becomes incorporated into the π-electron system of the ferrocenyl group. If a second, doubly-occupied, p orbital could be generated on the α-carbon atom by the elimination of a substituent group without its bonding electron pair, in this case a proton, the result should be formation of the α-ferrocenylcarbene XXIX.

The special interest in such a carbene would arise from the possibility that the mechanism operative in the stabilization of the α-ferrocenylcarbonium ions might be effective also in the newly generated α-ferrocenylcarbene. This might be sufficient to cause the removal of degeneracy of the two p orbitals on the divalent carbon atom, thus providing the opportunity for a stabilized singlet carbene [60].

When XXVIII (R = C_6H_5) was reacted with diisopropylethylamine (chosen for its reported [61] reluctance to enter into alkylation reactions, whilst having good proton-abstracting properties), a quantitative yield of disopropylethylammonium fluoroborate was obtained [62–63]. This seemed to indicate that deprotonation had taken place. The major metallocene-containing products were identified as either benzylferrocene XXX or 1,2-diphenyl-1,2-diferrocenylethane XXXI, depending on reaction conditions. These results could be rationalized by postulating formation of the carbene intermediate XXXII, which in turn abstracts a hydrogen atom from the reaction medium, leading to the ferrocenylphenylcarbinyl radical. This radical could either abstract another hydrogen atom to form benzylferrocene, or it could dimerize to form 1,2-diphenyl-1,2-diferrocenylethane. The same products were obtained from the photolysis of ferrocenylphenyldiazomethane XXXIV, where the postulated formation of the carbene intermediate XXXII is less likely to be questioned [62–64].

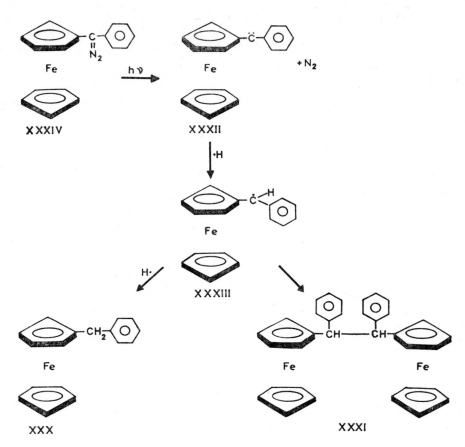

This result seemed to strengthen the idea of a carbene intermediate in the deprotonation reaction, as did the observation that the reaction between the amine and ferrocenylphenyl-deuterocarbonium ion yielded a quaternary salt, which, from NMR experiments, appeared to contain mostly deuterium, i.e., $R_3N^+D\ BF_4^-$. However, conflicting with this observation, the mass spectra of the benzylferrocene XXX and dimer XXXI, isolated in this reaction, showed the presence of deuterium in these compounds. From this result it appears obvious that if deprotonation occurs, it does so only partially, if at all, and, consequently, there must be a different path for the formation of XXX and XXXI, with retention of the deuterium present in the original carbonium ion. We have reason to believe that here too we may be faced with an oxidation-reduction process in which the amine acts as the reducing agent by transferring an electron to the cation-radical XX, itself being oxidized to a cation-radical XXXV.

The latter then abstracts a hydrogen atom from an appropriate source in the reaction medium to form the isolated quaternary ammonium salt. The radical XXXIII, containing the original deuterium atom in the case of a deuterated starting material, then proceeds to form the products XXX and/or XXXI.

We have been unable to decide as yet whether the reaction between a tertiary amine and ferrocenyltropylium fluoroborate, XII, leading to the products shown in the scheme below, is really analogous to that described for the ferrocenylphenylcarbonium ion. Independent

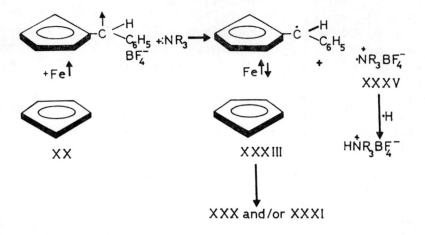

electron MO calculations for the ferrocenyltropylium ion place most of the positive change (~ 88%) on the tropylium ring [34], as opposed to the ferrocenylphenylcarbonium ion, where, as we have seen, most of the positive charge appears to reside on the Fe atom.

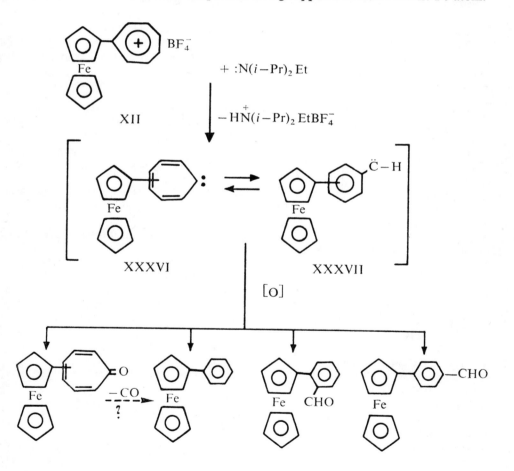

Our proposal [61] for ring contraction of the postulated ferrocenyl-substituted cyclo-heptatrienylidene XXXVI to the ferrocenylphenylcarbene XXXVII is analogous to the recently proposed [65] ring expansion of phenylcarbene to cycloheptatrienylidene.

It may well be that the ferrocenyl substituent exerts a decisive effect on the preferred direction of the rearrangement phenylcarbene $\rightleftarrows$ cycloheptatrienylidene.

We are currently investigating the postulated oxidation-reduction mechanism in the deprotonation reaction, as well as the photolytic and thermal production of ferrocenyl-carbenes from suitable precursors.

REFERENCES

1 D. P. Craig (1959) in : *Theoretical Organic Chemistry*, Butterworths, London, p. 20.

2 G. Wilkinson & F. A. Cotton (1959) in : *Progress in Inorganic Chemistry* (ed. F. A. Cotton), Interscience, New York, I, pp. 85–99.

3 M. Rosenblum (1965) *Chemistry of the Iron Group Metallocenes*, Wiley, New York, pp. 13–28.

4 J. P. Dahl & C. T. Ballhausen (1961) *K. Danske Vidensk. Selsk. (Mat.-fys. Medd.)*, Vol. XXXIII, No. 5.

5 E. M. Shustorovich & M. E. Dyatkina (1959) *Dokl. Akad. Nauk SSSR*, 128 : 1234.

6 *Ibid.* (1960) 131 : 113.

7 J. H. Schachtschneider, R. Pruis & N. Ros (1967) *Inorg. Chim. Acta*, 1 : 462.

8 D. R. Scott & R. S. Becker (1965) *J. Organometal. Chem.*, 4 : 409.

9 C. J. Ballhausen (1961) in : *Advances in the Chemistry of Coordination Compounds* (ed. S. Kirschner), MacMillan, New York, p. 13.

10 J. W. Linnett (1956) *Trans. Faraday Soc.*, 52 : 904.

11 M. Rosenblum, J. O. Santer & W. Glenn Howells (1963) *J. Am. Chem. Soc.*, 85 : 1450.

12 M. Rosenblum & F. W. Abbate (1966) *ibid.*, 88 : 4178.

13 J. C. Ware & T. G. Traylor (1965) *Tetrahedron Lett.*, p. 1295.

14 F. S. Yakushkin, V. N. Setkina, E. A. Yakovleva, A. I. Shatenshtein & D. N. Kursanov (1967) *Izv. Akad. Nauk SSSR (Ser. Khim.)*, p. 206.

15 J. A. Mangravite & T. G. Traylor (1967) *Tetrahedron Lett.*, p. 4457.

16 *Ibid.*, p. 4461.

17 G. R. Knox, I. G. Morrison, P. L. Pauson, M. A. Sandhu & W. E. Watts (1967) *J. Chem. Soc.* (C), p. 1853.

18 M. Rosenblum (1965) *Chemistry of the Iron Group Metallocenes*, Wiley, New York, p. 72.

19 R. A. Benkeser, Y. Nagai & J. Hooz (1964) *J. Am. Chem. Soc.*, 86 : 3742.

20 J. H. Richards & T. J. Curphey (1956) *Chemy Ind.*, p. 1456.

21 M. Rosenblum & W. Glenn Howells (1962) *J. Am. Chem. Soc.*, 84 : 1167.

22 D. N. Kursanov, V. N. Setkina, M. N. Nefedova & A. N. Nesmeyanov (1965) *Izv. Akad. Nauk SSSR (Ser. Khim.)*, p. 2218.

23 R. T. Lundquist & M. Cais (1962) *J. Org. Chem.*, 27 : 1167.

24 M. D. Rausch (1963) *ACS Advances in Chemistry Series*, 37 : 56.

25 A. N. Nesmeyanov, E. G. Perevalova, S. P. Gubin & A. G. Kozlovskii (1968) *Dokl. Akad. Nauk SSSR*, 178 : 616.

26 A. N. Nesmeyanov & O. A. Reutov (1959) *Izv. Akad. Nauk SSSR*, p. 926.

27 W. F. Little & R. Eisenthal (1961) *J. Org. Chem.*, 26 : 3609.

28 Idem (1961) *J. Am. Chem. Soc.*, 83 : 4936.

29 R. A. Benkeser & L. W. Hall Jr (1960) *Chem. Abst.*, 54 : 21025b.

30 L. A. Kazitsnya, B. V. Lokshin & A. N. Nesmeyanov (1959) *Dokl. Akad. Nauk SSSR (Ser. Khim.)*, 127 : 333.

31 D. W. Hall, E. A. Hill & J. H. Richards (1968) *J. Am. Chem. Soc.*, 90 : 4972.

32 M. Cais (1966) *Organomet. Chem. Rev.*, 1 : 436.

33 M. Cais & A. Eisenstadt (1965) *J. Org. Chem.*, 30 : 1148.

34 Idem (1967) *J. Am. Chem. Soc.*, 89 : 5468.

35 A. N. Nesmeyanov, V. A. Sazonova, G. I. Zudkova & L. S. Isaeva (1966) *Izv. Akad. Nauk SSSR (Ser. Khim.)*, p. 2017.

36 M. Cais, A. Modiano & A. Raveh (1965) *J. Am. Chem. Soc.*, 87 : 5607.

37 J. H. Richards & E. A. Hill (1959) *ibid.*, 81 : 3484.

38 K. L. Rinehart Jr, P. A. Kittle & A. F. Ellis (1960) *ibid.*, 82 : 2082.

39 D. S. Trifan & R. Backsai, *ibid.*, p. 5010.

40 E. A. Hill & J. H. Richards (1961) *ibid.*, 83 : 3840.

41 E. Berger, W. E. McEwen & J. Kleinberg, *ibid.*, p. 2274.

42 G. L. Hoh, W. E. McEwen & J. Kleinberg,
    ibid., p. 3949.

43 G. R. Buell, W. E. McEwen & J. Kleinberg
    (1962) *ibid.*, 84 : 40.

44 E. A. Hill (1963) *J. Org. Chem.*, 28 : 3856.

45 J. C. Ware & T. G. Traylor (1965) *Tetrahedron Lett.*, p. 1295.

46 T. G. Traylor & T. T. Tidwell (1966) *J. Am. Chem. Soc.*, 88 : 3442.

47 J. D. Fitzpatrick, L. Watts & R. Pettit (1966) *Tetrahedron Lett.*, p. 1299.

48 M. Cais, J. J. Dannenberg, A. Eisenstadt, M. J. Levenberg & J. H. Richards, *ibid.*, p. 1695.

49 T. G. Traylor & J. C. Ware (1967) *J. Am. Chem. Soc.*, 89 : 2304.

50 C. U. Pittman Jr (1967) *Tetrahedron Lett.*, p. 3619.

51 E. A. Hill & R. Wiesner (1969) *J. Am. Chem. Soc.*, 91 : 509.

52 J. Feinberg & M. Rosenblum, *ibid.*, p. 4324.

53 R. E. Hester & M. Cais (1969) *J. Organometal. Chem.*, 16 : 283.

54 C. Elschenbroich & M. Cais (1969) *ibid.*, 18 : 135.

55 M. J. A. Habib & W. E. Watts, *ibid.*, p. 361.

56 G. Gokel, P. Hoffmann, H. Klusacek, D. Marquarding, E. Ruch & I. Ugi (1970) *Angew. Chem. (Int. Edn)*, 9 : 64.

57 A. Eisenstadt (1967) D. Sc. Thesis, Technion – Israel Institute of Technology.

58 A. Schwarz (1970) M. Sc. Thesis, Technion – Israel Institute of Technology.

59 A. R. Forrester, S. P. Hepburn, R. S. Dunlop & H. M. Mills (1969) *Chem. Comm.*, p. 698.

60 R. G. Gleiter & R. Hoffmann (1968) *J. Am. Chem. Soc.*, 90 : 5457.

61 S. Hunig & M. Kiessel (1958) *Chem. Ber.*, 91 : 380.

62 P. Ashkenazi, S. Lupan, A. Schwarz & M. Cais (1969) *Tetrahedron Lett.*, p. 817.

63 P. Ashkenazi, P. M. Druce, S. Lupan, A. Schwarz & M. Cais (1969) *Proceedings of the 4th International Conference on Organometallic Chemistry, Bristol, England, August 1969*, G 1.

64 A. Sonoda, I. Moritani, T. Saraie & T. Wada (1969) *Tetrahedron Lett.*, 34 : 2943.

65 R. C. Joines, A. B. Turner & W. M. Jones (1969) *J. Am. Chem. Soc.*, 91 : 7754 .

Discussion

L. Friedman (Ohio):

I have a comment and a question. Hedaya and Jones in Florida have shown that the cyclohepta-trienylidene is formed from phenyl carbene, and not the other way around, so apparently there is no equilibration as you have indicated, although one might extrapolate. The question is this: You have indicated that you can get stable secondary carbonium ions. Now, what happens if you take, let us say, tert-butyl ferrocenyl methylenephosphorane, or something like that, and try to make either the carbonium ion or the carbene from that? You would have an internal way of differentiation between carbonium ion and carbene.

M. Cais:

A reaction was carried out with ferrocenyl tert-butyl-carbinol. First of all, I must comment that this is a remarkable example of the stability of a secondary carbonium ion; you have no rearrangement whatsoever. There is no intramolecular reaction – we are getting back the benzyl derivative. We are still looking for a way to explain exactly what happens with this compound, and if the steric effect of the tertiary butyl group is really significant in this case.

S. Sarel:

I am concerned with the question of the ability of metal to transmit electronic effects. Now, this could either be a real electronic transmission, or it could be a through-bond interaction or a through-space field effect. How can this be discerned?

M. Cais:

Well, I wish somebody would tell me. I meant to refer earlier in my paper to the concept that has been used here, through-bond and through-space effects. I think this is actually the question that is being debated. Is there a bond formed between the exocyclic carbon atom and some suitable metal orbitals, or is this simply a through-space effect? I don't think that anybody knows the answer yet. There is an attempt being made to carry out an X-ray analysis of one of these carbonium ions, and perhaps this will teach us something about the bonding question.

S. Sarel:

I asked that question because the problem arises of whether there is a difference between ferrocenes and aromatics in substitution. We know that the aromatic substitution goes from the first σ complex through a π complex to the substitution product. Does the metal take an active role in the mechanism of substitution reaction, or not? This depends on whether the metal has the ability to transmit electronic effects in the ferrocene molecule.

M. Cais:

I don't know the answer to this question. I can only mention an experimental fact: If you try to acetylate ferrocene and benzene under the same conditions, say with acetyl chloride at 0°, ferrocene acetylates faster by a factor of 10^6. Now, what is the reason for this? It comes back to the question that I asked at the beginning, and I think it is wrong to try and provide an answer by making a direct comparison between benzene and ferrocene.

Problems Raised by the Synthesis of the Helicenes

by R.H. MARTIN

Service de Chimie Organique, Faculté des Sciences, Université Libre de Bruxelles

THE PURPOSE OF THIS REVIEW is to discuss briefly the many problems raised by the synthesis of the helicenes. Some of these have already been solved, others will require more extensive research work.

The helicenes which have been synthesized so far can be divided into three main classes: hydrocarbons, azahelicenes and heterohelicenes containing two or three thiophen rings. There is no doubt that other types of helicenic structures will be prepared in the near future. The first benzologue of [5]helicene, namely, [6]helicene, was synthesized in 1956 by Newman and Lednicer [1]. The main advantage of the very elegant synthetic scheme used by these authors is that the structure of the final product is unambiguous. It is therefore most unfortunate that this particular scheme should have proved so difficult to apply to the synthesis of higher benzologues.

A second synthesis of [6]helicene was described eleven years later by Bogaert-Verhoogen and Martin [2]. The main step in this synthesis is the cyclization of I by Hewett's method (potash fusion):

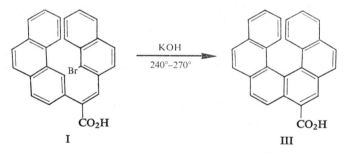

The relative simplicity of this scheme is, however, partially offset by the following limitations:

Only the *cis*-1,2-diarylethylenes are suitable for this type of cyclization.
Two hydrocarbons can be formed by the 'normal' cyclization.
Rearranged products are frequently obtained from this reaction.

There are no obvious reasons why this scheme could not be used for the synthesis of [7]helicene, but the experiment has not been attempted yet. The helicene carboxylic acids,

115

isolated from the potash fusion, could be used for the resolution of these chiral molecules.

The third synthesis published in the literature is that of Martin et al. [3]. This synthesis is based on the photo-induced cyclodehydrogenation of III, following the procedure described by Wood and Mallory [4] for the preparative-scale photoconversion of stilbene to phenanthrene:

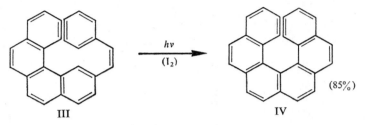

III   IV   (85%)

The same procedure was used to prepare hepta-, octa-, nona- and tridecahelicene [3, 5–6]. It has also proved very useful for the synthesis of substituted helicenes, such as 3-methoxyheptahelicene [Martin and coll., unpublished work], azahelicenes (4-azahexahelicene [7], 1- and 2-azahexahelicene [Martin and coll., unpublished work]), and hexa- and heptaheterohelicenes containing thiophen rings [8–9].

The following (preliminary) conclusions can be drawn from the work already carried out in this field:

The method is well suited for the synthesis of helicenes, heterohelicenes and a variety of substituted derivatives. The method has great potentialities for the synthesis of new helicenic structures.

The required 1,2-diarylethylenes are reasonably easy to prepare, either by the Wittig condensation or by the Siegrist reaction [10–12].

Because of the well-known cis ⇌ trans photo-isomerization of 1,2-diarylethylenes, either the cis or the trans isomer (or a mixture of both) can be used in the last stage.

The yields of helicenes follow the general trend: azahelicenes < heterohelicenes containing thiophen rings and helicenes (hydrocarbons).

The photolyses involving two cyclizations in one step are becoming more and more useful.

The main drawbacks of the photoconversion of 1,2-diarylethylenes to polycyclic aromatic hydrocarbons are the following:

The photocyclization can give rise to many isomeric structures, particularly in the case of the double cyclizations. This remark does not apply to the syntheses of heterohelicenes containing thiophen rings, carried out by Wynberg et al. [8–9].

A case of rearrangement, leading to an anthracenic structure (VII), has been described by Cohen, Mijovic and Newman [13].

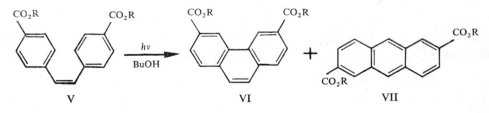

V   VI   VII

In order to avoid the dimerization of the 1,2-diarylethylenes, the photocyclizations are carried out in rather dilute solutions ($\leq 0.02\,M$).

The helicenes and azahelicenes are photosensitive. [7]helicene is 'destroyed' to the extent of 85% after three hours irradiation in benzene solution containing a trace of iodine (Hanovia 450 W medium-pressure mercury lamp, pyrex well [Martin and coll., unpublished results]).

From the synthetic point of view, it is clear that the *prediction* and the *determination* of the course of the photo-induced cyclodehydrogenations are of utmost importance. In order to be able to predict which isomer(s) is (are) most likely to be formed in the cyclization process, it would be essential to know the exact nature of the 'excited state(s)' leading to the final product(s). In spite of the large amount of work done in this field, the problem is still controversial. Thus, the photochemical ring closure of stilbenes leading to phenanthrenes is believed to procede either 'via higher vibrational levels of the ground state of the stilbenes (Lewis mechanism)' [14] or 'via excited singlets' [15].

Attempts to correlate the ease of cyclization with the sum of the free-valence numbers, in the first excited state, of the atoms involved (simple Hückel MO method) appears to be satisfactory in alternant hydrocarbons only [16–18]. We have recently confirmed Laarhoven's conclusions in the helicene series (hydrocarbons). For example, in the double photocyclization of VIII, leading to [13]helicene in 52% yield, the calculated $F_{r,s}^{*}$ for the first cyclization are respectively: $\alpha\alpha'$ (17–34) 1.0789; $\alpha\beta'$ (17–22) 1.0018; $\beta\alpha'$ (1–34) 0.9944; and $\beta\beta'$ (1–22) 0.9173.

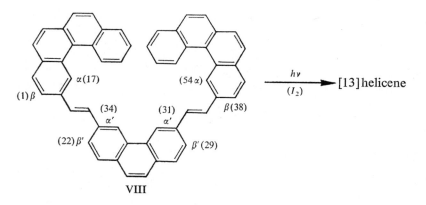

VIII

For the second cyclization, following the $\alpha\alpha'$ (17–34) cyclization, the $F_{r,s}^{*}$ are: $\alpha\alpha'$ (54–31) 1.1626; $\alpha\beta'$ (54–29) 1.0469; $\beta\alpha'$ (38–31) 1.0419; and $\beta\beta'$ (38–29) 0.9262 [Martin and coll., unpublished work]. From the large number of examples examined by Laarhoven and from our own results, it follows that cyclizations leading to non-overcrowded hydrocarbons require $F_{r,s}^{*} \geq 1$, and those leading to benzopenta- and higher benzohelicenes require $F_{r,s}^{*} \geq 1.072$.

These approximations, with their severe limitations, are the only ones on which the organic chemist can base his predictions. It is our hope that theoretical chemists will attempt, in collaboration with photochemists, to get a deeper understanding of this crucial problem.

STRUCTURAL DETERMINATIONS

1. *X-Ray Crystallography*

The structures of Wynberg's benzo[d]naphto[1,2-d']benzo[1,2-b:4,3-b']dithiophen (IX) [19] and of hexahelicene [20] have been confirmed recently by X-ray diffraction.

IX

2. *Independent Syntheses*

The synthesis of [7]helicene has been achieved by the photocyclization of the two isomeric 1,2-diarylethylenes X and XI.

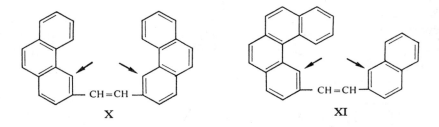

If rearrangements of the type mentioned earlier are excluded, only one common hydrocarbon, namely [7]helicene, can result from these experiments, thus proving the structure of the final product.

These structural determinations, based on independent syntheses, suffer from serious drawbacks:

They are time consuming.

They are not entirely unambiguous, for they are based on the assumption that no rearrangement takes place.

They are difficult to apply in many cases (e.g., the synthesis of highly condensed helicenes, substituted helicenes, azahelicenes).

3. *NMR Spectroscopy*

With the structure of [6]helicene firmly established by Newman's synthesis, NMR spectroscopy was used to determine the structure of the hydrocarbons isolated from the photolyses, which might lead to the higher helicenes.

A comparative study of the NMR spectra of the helicenes from phenanthrene to [9]helicene [21], including calculations of the four spin systems and the study of substituted derivatives, *epi* cross-ring couplings and specific solvent effects, fully confirmed the structures attributed to the hydrocarbons described as hepta-, octa- and nonahelicene in preliminary communications. In the synthesis of tridecahelicene [6], this technique allowed us to determine the structure of the monocyclized product, which proved to be an octahelicene substituted in position 2. NMR spectroscopy was also used to make a choice between the [13]helicene structure and the isomeric structure — first cyclization $\alpha\alpha'$ (17–34), second cyclization $\alpha\beta'$ (54–31) in VIII — which was not excluded by deuterium labelling experiments (see below).

Although very useful in relatively simple cases, NMR spectroscopy cannot be used alone for structural determinations when a large number of highly overcrowded isomeric hydrocarbons can result from a cyclization. On the other hand, there is no doubt that NMR spectroscopy is the best method to distinguish the isomeric azahelicenes, in which the nitrogen atom is in a terminal ring.

4. *Deuterium Labelling*

The use of deuterium labelling is justified by the following considerations: The photosyntheses of unsubstituted helicenes are limited practically to the 1,2-diarylethylenes (ArCH=CHAr'), in which the aromatic radicals are

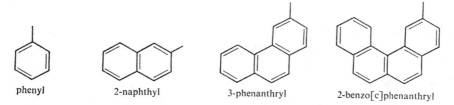

phenyl   2-naphthyl   3-phenanthryl   2-benzo[c]phenanthryl

The use of 1,2-diarylethylenes containing 2-[n]helicenyl radicals ($n \geq 5$) is unfavourable for the following reasons:

[5]Helicene (XII) gives benzo[g, h, i]perylene (XIII) on photolysis.

1,2-Diarylethylenes, with 2-[m]helicenyl radicals ($m \geq 6$), are still difficult to prepare.

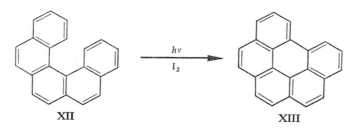

XII   XIII

It follows that [9]helicene is the highest helicene to be reached without difficulty by a monocyclization process. A few examples of double photocyclizations have been described recently by Dietz and Scholz [22]. As already mentioned, [13]helicene, the first member of the multilayered helicenes, was synthesized by the double photocyclization of VIII.

In this particular case, the 'normal' double photocyclization could give rise to ten isomers. It was therefore decided to use deuterium labelling to determine the structure of the final product. The method, first tested on hexahelicene [23], proved very successful. The presence or absence of deuterium in the final hydrocarbon is determined, on a mg scale, by mass spectrometry.

In the hydrocarbon series, no rearrangement or exchange of deuterium have been observed so far in the photolytic process. On the other hand, Wynberg et al. [24 and earlier literature cited therein] have shown that photorearrangements are quite common in the thiophen series. Deuterium labelling in the hydrocarbon series is, however, not entirely unambiguous if rearrangements leading to anthracenic structures are considered. This can be illustrated in the case of the synthesis of [7]helicene.

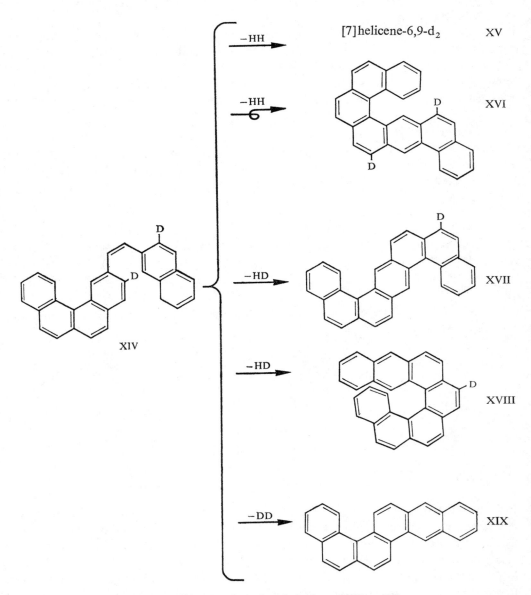

+ rearranged isomers formed with the loss of HD or DD

Two bis-deuteriated hydrocarbons — [7]helicene-6,9-d_2 and the rearranged product XVI — can be formed by the loss of 2H in the cyclization process. In order to remove this ambiguity, it is necessary to carry out another synthesis of [7]helicene, which would give a different rearranged bis-deuteriated hydrocarbon (XXII).

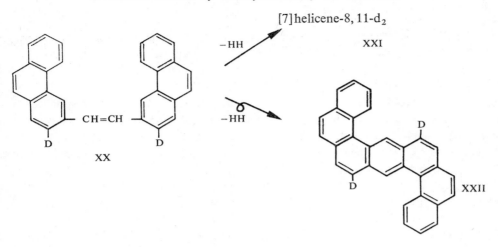

[7]helicene-8,11-d$_2$

XXI

XX

XXII

The synthesis of [7]helicene-6,9-d$_2$ (XV) has been achieved recently [Martin and coll., unpublished results]. It gave, as expected, a bis-deuteriated hydrocarbon ([7]helicene-d$_2$) and a mono-deuteriated isomer. The structure (XVII) of the last hydrocarbon had been deduced earlier from its NMR spectrum [3]. This conclusion is now fully confirmed by a new labelling experiment (XXIII → XXIV), designed to show which deuterium had been lost in the first experiment (XIV → XVII).

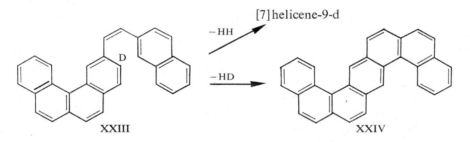

[7]helicene-9-d

XXIII   XXIV

Work on the synthesis of [7]helicene-8,11-d$_2$ (XXI) is progressing satisfactorily.

It is important to point out at this stage that the rearrangement observed by Cohen, Mijovic and Newman (V → VII) [13] is probably rather exceptional. This conclusion is based on the following observations:

The photolysis of the disubstituted stilbene derivative (4,4′-dimethoxycarbonyl) was carried out in butanol solution, in the absence of iodine.

The rearrangement was not observed when the photolysis of the same derivative was performed in benzene solution containing a trace of iodine [Martin and coll., unpublished results].

Up to now, no rearrangement of this type has been observed with unsubstituted hydrocarbons.

The possibility of a rearrangement occurring in a photo-induced cyclodehydrogenation should certainly not be overlooked, but the chances that it will occur in the hydrocarbon series seem rather slight.

5. Optical Rotatory Dispersion

The ORD curves of the known helicenes have the same general shape; they are progressively shifted to higher wavelengths by the successive annelation of benzene rings.

6. Mass Spectrometry

An intense (up to 90% of the M ion) fragmentation peak m/e 300, corresponding to the mass of coronene, appears in the mass spectrum of [6]helicene. The genesis of this peak has been discussed at great length by Dougherty [25].

In the mass spectra of hepta-, octa- and nonahelicene, the m/e 300 peak is still present, but is much weaker (8% to 16% of the M ion).

The MS of 4-aza[6]helicene shows a fragmentation ion m/e 301, corresponding to the mass of azacoronene. In the case of 2-aza[6]helicene, the two fragments, m/e 300 and m/e 301, are observed [Martin and coll., unpublished work].

It is thus clear that ORD and MS can yield useful structural informations in the helicene series.

New Syntheses of [6]Helicene

The successful synthesis of [13]helicene by a double photocyclization prompted us to investigate new and simpler reaction schemes for the preparation of [6]helicene. These unpublished results will be described briefly.

The remarkable 'Siegrist reaction' [10–12]

$$ArCH_3 + Ar'CH=N\phi \xrightarrow[DMF]{t\text{-BuOK}} ArCH=CHAr'$$

has been used to prepare XXVI and XXVIII by the following schemes:

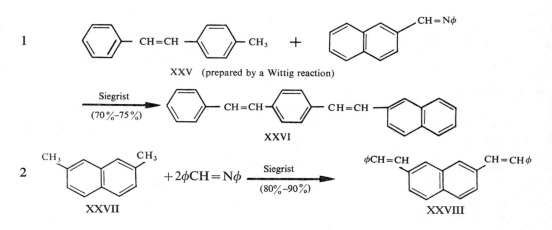

The double photocyclization of XXVI and XXVIII (benzene solutions containing a trace of iodine) gave [6]helicene in 55% to 60% yields (the reaction conditions have not been optimized). Using the first synthesis described above, [6]helicene can be prepared from p-tolualdehyde, benzyl chloride, 2-naphthaldehyde and aniline in a very short time. The second synthesis is actually shorter, but 2,7-dimethylnaphthalene is unfortunately difficult to get commercially nowadays.

122

SUMMARY OF THE BEST SYNTHESES OF THE HELICENES

[6] Helicene : Double photocyclization of XXVI (55%) or XXVIII (60%).

[7] Helicene : Photocyclization of 1,2-di(3-phenanthryl)ethylene (XX; D = H)
   (> 50%).

[8] Helicene : Photocyclization of 1-(3-phenanthryl)-2-(2-benzo[c]phenanthryl)ethyl-
   ene (62%).

[9] Helicene : Photocyclization of 1,2-di(2-benzo[c]phenanthryl)ethylene (70%).

[13] Helicene : Double photocyclization of VIII (52%).

1-Aza[6]helicene : Photocyclization of 1-(2-benzo[c]phenanthryl)-2-(3-pyridyl)ethyl-
   ene (15%).

2-Aza[6]helicene : Photocyclization of 1-(2-benzo[c]phenanthryl)-2-(4-pyridyl)ethyl-
   ene (12%).

4-Aza[6]helicene : Photocyclization of 1-(2-benzo[c]phenanthryl)-2-(2-pyridyl)ethyl-
   ene (18%).

Benzo[d]naphtho[1,2-d']benzo[1,2-b:4,3-b']dithiophene : Photocyclization of
   XXIX (70%) [8].

Naphtho[1,2-d]benzo[b'']thieno[4,5-d']benzo[1,2-b:4,3-b']dithiophene :
   Photocyclization of XXX (40%) [8].

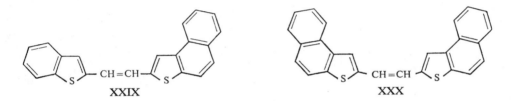

XXIX   XXX

CONCLUSIONS

The photochemical synthesis of helicenes, substituted helicenes, aza- and heterohelicenes has now reached a stage which makes these interesting tridimensional chiral aromatic molecules available for the systematic study of their chemical, optical, spectroscopic and physical properties.

REFERENCES

1 M. S. Newman & D. Lednicer (1956) *J. Am.*
   Chem. Soc., 78 : 4765.

2 D. Bogaert-Verhoogen & R. H. Martin (1967)
   Tetrahedron Lett., p. 3045.

3 R. H. Martin, M. Flammang-Barbieux,
   J. P. Cosyn & M. Gelbcke (1968) *ibid.*, p. 3507.

4 C. S. Wood & F. B. Mallory (1964) *J. Org.*
   Chem., 29 : 3373.

5 M. Flammang-Barbieux, J. Nasielski &
   R. H. Martin (1967) *Tetrahedron Lett.*, p. 743.

6 R. H. Martin, G. Morren & J. J. Schurter
   (1969) *ibid.*, p. 3683.

7 R. H. Martin & M. Deblecker (1969) *ibid.*, p. 3597.

8 H. Wynberg & M. B. Groen (1968) *J. Am.*
   Chem. Soc., 90 : 5339.

9 Idem (1969) *Chem. Comm.*, p. 964.

10 A. E. Siegrist (1967) *Helv. Chim. Acta,*
   50 : 906.

11 A. E. Siegrist & H. R. Meyer (1969) *ibid.*,
   52 : 1282.

12 A. E. Siegrist, P. Liechti, H. R. Meyer &
   K. Weber (1969) *ibid.*, 52 : 2521.

13 S. D. Cohen, M. V. Mijovic & G. H. Newman
   (1968) *Chem. Comm.*, p. 722.

14 H. Güsten & L. Klasinc (1968) *Tetrahedron,*
   24 : 5499.

15 F. B. Mallory, S. Wood & J. T. Gordon (1964)
   J. Am. Chem. Soc., 86 : 3094.

16 M. Scholz, M. Mühlstädt & F. Dietz (1967)
   Tetrahedron Lett., p. 665.

17 M. Scholz, F. Dietz & M. Mühlstädt (1967) *Z. Chem.*, 7 : 329.

18 W. H. Laarhoven, T. J. H. M. Cuppen & R. J. F. Nivard (1968) *Rec. Trav. Chim.*, 87 : 687.

19 G. Stulen & G. J. Vissert (1969) *Chem. Comm.*, p. 965.

20 I. R. Mackay, J. M. Robertson & J. G. Sime, *ibid.*, p. 1470.

21 R. H. Martin, N. Defay, H. P. Figeys, M. Flammang-Barbieux, J. P. Cosyn, M. Gelbcke & J. J. Schurter (1969) *Tetrahedron*, 25 : 4985.

22 F. Dietz & M. Scholz (1968) *ibid.*, 24 : 6845.

23 R. H. Martin & J .J. Schurter (1969) *Tetrahedron Lett.*, p. 3679.

24 R. M. Kellog & H. Wynberg (1967) *J. Am. Chem. Soc.*, 89 : 3495.

25 R. C. Dougherty (1968) *ibid.*, 90 : 5788.

Discussion

F. Bergmann (Hebrew University, Jerusalem):

Did anybody check whether the helicenes undergo a change with deuterium?

R. H. Martin:

We have never observed deuterium exchanges in our experiments.

F. Bergmann:

Did you ever expose these compounds to deuterium under irradiation?

R. H. Martin:

No, not yet.

F. Bergmann:

This might give an interesting pointer to the labelled positions selected.

R. H. Martin:

For the introduction of a deuterium atom in a specific position, we have used an excellent method published by B. Chenon, L. C. Leitch, R. N. Renaud and L. Pichat (1964) *Bull. Soc. Chim. Fr.*, 38. A mixture of $ArBr + D_2O + CaO + Zn$ dust is refluxed for two or three days; the deuteriated derivative (ArD) is isolated in quasi-quantitative yield. The reaction is very specific (we have never detected any isomer) and the isotopic purity is $> 97.5\%$. There is no visible exchange of deuterium during the photocyclization of the deuteriated 1, 2-diarylethylenes.

R. H. Schlessinger:

I think you said heptahelicene photo-decomposed. Is this correct? Half the helicene undergoes photo-decomposition?

R. H. Martin:

Yes. We have irradiated heptahelicene under the conditions used for the cyclization: benzene solution, 450 W medium pressure Hanovia mercury lamp and a trace of iodine. After three hours' irradiation, 85% of heptahelicene is 'destroyed'.

R. H. Schlessinger:

I realize this is an appropriate control experiment. My question was directed to a slightly different point. Was this system rigorously degassed—that is, was oxygen rigorously excluded?

R. H. Martin:

We have not tried the experiment in the absence of oxygen.

R. H. Schlessinger:

The reason for my question is this: Is this conceivably a singlet oxygen decomposition product, or are you getting some real interaction between the two overlapping benzene rings?

Discussion

R. H. Martin:

As we have not yet carried out the experiment, I cannot answer your question. In many cases, we have observed the oxidation of 1,2-diarylethylenes (ArCH=CHar) to the corresponding aldehydes (ArCHO) as a side reaction to the photo-induced cyclodehydrogenations (without degassing).

S. Sarel:

I am concerned with the question of the optical resolution of the compounds. I understand that the synthesis leads to racemic forms, and the drawback, compared to Newman's method, is that you cannot resolve the compounds into the antipodes.

R. H. Martin:

The resolution of the unsubstituted helicenes is not an easy problem. Newman's reagent, an optically active π-complexing agent, can be used for this purpose. On the other hand, we have observed the 'spontaneous resolution' by crystallization, from benzene solutions, in the case of hexa-, hepta-, octa- and nonahelicene. The 'spontaneous resolution' of tridecahelicene has not yet been observed. However, 'spontaneous resolution' is not a satisfactory preparative scale method for the resolution of the helicenes.

S. Sarel:

Is it possible to induce asymmetry by using an asymmetric environment, either a solvent or polarized light?

R. H. Martin:

Yes, certainly. Right- or left-handed circularly polarized light could be used for the photocyclization of appropriate 1,2-diarylethylenes. We are presently working on this problem in collaboration with Professor Kagan, Paris.

As an example, the product (dl mixture), resulting from the monocyclization in the synthesis of tridecahelicene, should be an excellent starting material for this kind of investigation.

126

On the Optical Activity of some Aromatic Systems: Hexahelicene, Heptahelicene and [n,n]-Vespirene

by G. WAGNIÈRE

Institute of Physical Chemistry of the University of Zurich

I. Introduction

THE APPROXIMATION consisting in treating the π electrons separately from the underlying σ core in the study of the long-wavelength spectra of conjugated systems has been convincingly justified. Simple Hückel theory has shown the striking correlation between dominant geometric features and the centre of gravity of the lower absorption bands. Further refinement, such as the method developed by Pariser, Parr [1] and Pople [2] (which, for simplicity, we here shall call the PPP method), renders in many cases a detailed assignment of transitions possible, taking satisfactorily into account the effect of electron interaction. The aim of our study consists in an investigation of the usefulness of the PPP method in calculating the optical activity, in particular the CD spectrum, of large asymmetric aromatic systems. Although the absence of a plane of symmetry does not here allow a strict separation between σ and π electrons anymore, the spectral features nevertheless seem still to justify such a distinction.

The contribution of a given transition $a \rightarrow b$ in a molecule to the optical activity is proportional to the rotational strength $R_{a \rightarrow b}$ [3]. For a molecule in solution, this is the imaginary part of the scalar product of the electric and magnetic transition moment,

$$R_{a \rightarrow b} = Im \langle a \,|\, \vec{R} \,|\, b \rangle \, \langle b \,|\, \vec{M} \,|\, a \rangle. \tag{1}$$

Within the general frame of the PPP method, the ground state a may be written as a single Slater determinant built from SCF MO's, which in turn are expressed as linear combinations of $2p_\pi$ AO's of the carbon atoms χ_p. As a result of configuration interaction, the excited state b will be a linear combination of functions describing single excitations $i \rightarrow j$:

$$a = \phi_G = |\varphi_1 \bar{\varphi}_1 \cdots \varphi_i \bar{\varphi}_i \cdots \varphi_{N/2} \bar{\varphi}_{N/2}|; \quad \varphi_i = \sum_p c_{ip} \chi_p,$$

$$\phi_i^j = \frac{1}{\sqrt{2}} \left\{ |\varphi_1 \bar{\varphi}_1 \cdots \varphi_i \bar{\varphi}_j \cdots \varphi_{N/2} \bar{\varphi}_{N/2}| + |\varphi_1 \bar{\varphi}_1 \cdots \varphi_j \bar{\varphi}_i \cdots \varphi_{N/2} \bar{\varphi}_{N/2}| \right\},$$

$$b = \sum_{ij} B_{ij} \phi_i^j \qquad \begin{cases} i_{min} \leq i \leq N/2 \\ (N/2 + 1) \leq j \leq i_{max} \end{cases}$$

127

Thus,

$$R_{a \to b} = 2 \, Im \left\{ \sum_{ij} \sum_{i'j'} B_{ij} B_{i'j'} \langle \varphi_i | \vec{r} | \varphi_j \rangle \langle \varphi_{j'} | \vec{m} | \varphi_{i'} \rangle \right\}, \tag{2}$$

where $\vec{r}$ and $\vec{m}$ stand for the one-electron electric and magnetic dipole operators, respectively. One notices that not only do the scalar products of $\vec{r}$ and $\vec{m}$ for every one-electron transition appear, but also all the possible crossterms, weighted by the products of the coefficients B_{ij} and $B_{i'j'}$. These crossterms may, in those instances where configuration interaction is important, make the determining contributions. As we shall see, this is possibly the case in the examples to be treated. The calculation of the matrix elements $\langle \varphi_i | \vec{r} | \varphi_j \rangle$ and $\langle \varphi_j | \vec{m} | \varphi_i \rangle$, including all interatomic terms, has been described in more detail previously [4].

II. Hexahelicene and Heptahelicene

Several theoretical investigations have already dealt with the interpretation of the long-wavelength optical activity of hexahelicene. Fitts and Kirkwood [5], based on Kirkwood's polarizability theory [6], predicted that a right-handed helix should exhibit positive rotation, $[\alpha]_D = +3010°$. Moscowitz [7–8] carried out a Hückel molecular orbital calculation of the π system to assess the optical activity. Tinoco and Woody have discussed the optical activity of a free electron on a right-handed helix [9], applying their model to hexahelicene. In spite of these attempts, it does not appear, however, that certainty has been achieved as to the absolute configuration [10–11]. Thus, a further refinement of the method of calculation is indicated, and a PPP-type calculation, invoking the interaction of a large number of singly-excited configurations, seems an appropriate next step. Possibly, it may bridge the gap between the exciton approach [5] and the one-electron approach [7, 9].

In our application of the PPP method to hexahelicene, the simplest possible parametrization is chosen, calibrated on the spectrum of benzene [12]. In particular, the effective ionization potential of carbon I_C is set equal to 9.00 eV and the nearest-neighbour resonance integrals are generally taken as -2.46 eV. The electron repulsion integrals are computed according to the charged-sphere approximation, with $\gamma_{pp} = 10.53$ eV. As a test, Fig. 1 shows the calculated and measured spectra of phenanthrene. In the adopted geometry for hexahelicene, the pitch of the helix is assumed to be 3Å (previous authors [5, 9] have taken the value 3.8Å; it seems to us very large), and the bond lengths in the slightly distorted benzene rings are all of the order of 1.40 ± 0.02 Å. In the calculation of the excited states, the interaction of the 64 lowest singly-excited configurations is taken into account. From measurements of the emission polarization, Weigang and collaborators [10] have proposed an assignment of the near-UV bands of hexahelicene. The predicted sequence of transitions (Fig. 2) is in good agreement with their conclusions. In general, the calculated frequencies are somewhat too high, in particular for the three lowest transitions (presumably corresponding to L_b, L_a and B_b, in order of increasing energy). The general features of the spectrum are, however, satisfactorily reproduced. Thus, the question appears legitimate if the CD spectrum may be predicted similarly.

128

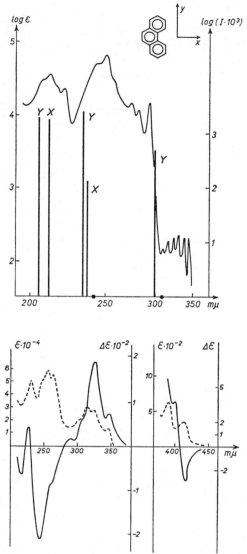

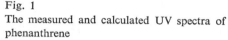

Fig. 1

The measured and calculated UV spectra of phenanthrene

The calculated intensities (*I*) are set equal to the square of the transition moments in Å², computed within the frame of the PPP method; the absolute scale of log (*I* · 10³) is arbitrary

Fig. 2

The UV and CD spectra of hexahelicene
Redrawn from Newman et al. [11]

Table 1 shows computed values for the product of the rotational strengths (in cgs units) times the transition energies (in eV) for the lowest transitions between one-electron states of right-handed [9] hexahelicene. The one-electron states are represented by the corresponding SCF MO's built from $2p_z$ AO's, the axes of which lie parallel to the helical axis. The computation of the electric moments is here carried out in the dipole velocity form [4]. Both in the electric and in the magnetic moments all interatomic contributions are explicitly included. We notice that the rotational strengths of all transitions of symmetry *A* show a negative sign, while those of symmetry *B* are either positive or weakly negative. Among the transitions listed, the ones with negative rotational strength thus appear to be more numerous and to outweigh the positive ones. It would, however, be erroneous to conclude from this that the long-wavelength optical activity of right-handed hexahelicene

129

Table 1

Calculated Values

for the Product of the Rotational Strengths (in cgs units)

Times the Transition Energies (in eV) for Transitions

between One-Electron States (SCF-MO's)

Electric transition moments are computed in the dipole velocity form, including all interatomic terms [4]

	Transition		Symmetry C_2			$R_{ij} \cdot \Delta\varepsilon_{ij} \cdot 10^{38}$
	i	j				
Hexahelicene	11	14	a	a	(A)	-3.087
	11	15	a	b	(B)	2.333
	11	16	a	a	(A)	-3.074
	12	14	b	a	(B)	0.323
	12	15	b	b	(A)	-7.264
	12	16	b	a	(B)	2.643
	13	14	a	a	(A)	-3.247
	13	15	a	b	(B)	-0.236
	13	16	a	a	(A)	-2.275
Heptahelicene	13	16	a	a	(A)	-0.706
	13	17	a	b	(B)	7.978
	13	18	a	b	(B)	-1.882
	14	16	a	a	(A)	-6.800
	14	17	a	b	(B)	-3.737
	14	18	a	b	(B)	8.886
	15	16	b	a	(B)	1.253
	15	17	b	b	(A)	-6.591
	15	18	b	b	(A)	-0.429

should be negative. Configuration interaction causes the opposite to be the case (see Table 2 and Fig. 3).

If one uses exact wave functions, the values obtained for the electric transition moments computed in the dipole velocity form should be identical to those obtained in the dipole vector form. Our wave functions are approximate and are normalized neglecting overlap. As a test, we consequently also have calculated rotational strengths between ground state a and excited states b, R'_{ab}, by using electric moments computed in the vector form, within the usual approximations of the PPP method (neglect of interatomic terms). The results coincide in sign with those derived from electric moments in the dipole velocity form, R_{ab}, and the absolute values agree sufficiently well to allow us to draw some general conclusions. Both in the calculation of R_{ab} and R'_{ab} the contribution of 64 singly-excited configurations was taken into account.

The calculations seem to agree with the measured spectra in the following points: The transition of lowest energy, L_b, is accidentally electric dipole forbidden. Thus, both its UV and CD intensity must be vibronically induced [13]. The next transition, L_a, does have some electric dipole intensity in the y direction, but in our calculation there is still 'destructive interference' in the parallel component of the magnetic moment. The influence of both L_a and L_b on the very long-wavelength optical activity is apparently minimal. Of importance are the following x, z-polarized transitions of symmetry B and positive rotational strength. We would assume these transitions to make up the B_b band system [10]. However,

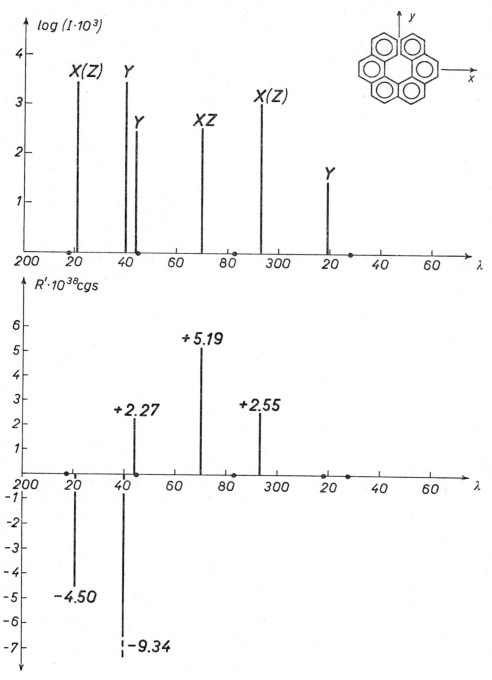

Fig. 3
Calculated UV and CD spectra of hexahelicene
The values of the UV intensities are computed within the frame of the PPP method; for the rotational strengths, the values R'_{ab} (see Table 2) are represented; vanishingly small values for intensities and (or) rotational strengths are indicated by dots

Table 2

Rotational Strengths between Ground State a and Excited State b

Interaction of 64 singly-excited configurations is fully taken into account for every excited state; for values R_{ab} electric moments are computed in dipole-velocity form; for values R'_{ab} they are calculated in dipole-vector form within approximations of the PPP-method; for both R_{ab} and R'_{ab} magnetic moments are determined in the same way, including all interatomic terms [4]

	Excited State b λ (mμ)	Symmetry C_2	$R_{ab} \cdot 10^{38}$	$R'_{ab} \cdot 10^{38}$
Hexahelicene	328	B	0.003	—
	319	A	−0.015	−0.047
	293	B	9.597	2.553
	283	A	−0.011	—
	270	B	8.582	5.189
	245	B	0.006	—
	244	A	0.009	2.274
	240	A	−3.524	−9.340
	221	B	−0.409	−4.503
Heptahelicene	333	A	−0.002	
	327	B	1.967	
	292	B	0.005	
	287	B	14.886	
	285	A	−0.021	
	254	A	−0.010	
	253	B	1.843	
	246	A	−5.386	
	237	A	−0.003	
	227	B	0.360	

this designation does not seem to coincide with the conclusions of Weigang and Trouard Dodson [13]. These bands obviously determine the sign of the optical activity at, say, the Na D-line. After these bands there is a change of sign in the CD spectrum, for which at least one y-polarized transition is responsible, with strongly negative rotational strength. Concerning the absolute configuration of hexahelicene, our results seem to indicate that (+)-hexahelicene is right-handed.

The calculated sequence of transitions for heptahelicene is similar to that of hexahelicene. The predicted transition energies are also rather too high, though. The CD spectrum should be dominated in the longer-wavelength part by strong positive Cotton effects from x, z-polarized transitions, followed by a change of sign, at very roughly 250 mμ, caused by the negative contribution of at least one y-polarized transition (Fig. 4). From these results we infer that (+)-heptahelicene should also be right-handed.

In our PPP calculations on heptahelicene, we have assumed the interaction between the terminal benzene rings to be basically electrostatic (by including all the γ integrals) and have set resonance integrals β between superposed carbon atoms equal to zero. Although the distance between these atoms is probably of the order of 3Å (our assumption) or even greater, some kind of resonance interaction, as in charge transfer complexes [14], cannot be excluded. This point should be further investigated, particularly also in the higher helicenes [15–16].

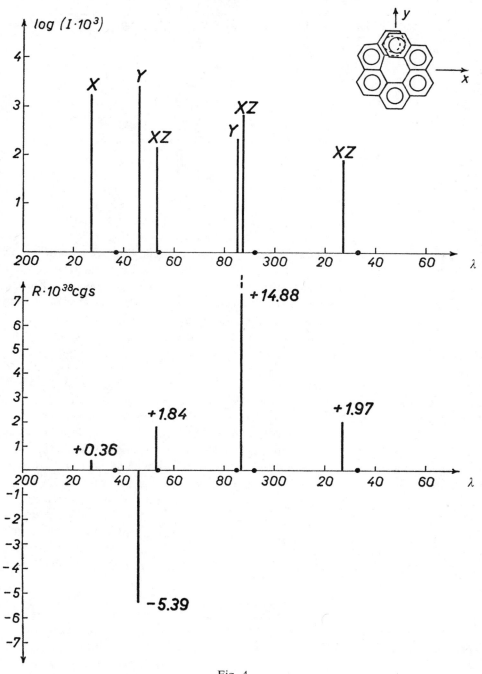

Fig. 4
Calculated UV and CD spectra of heptahelicene (see also Table 2)

133

III. [n, n]-Vespirene

Prelog and Haas [17] have recently synthesized some strongly optically active derivatives of 9, 9′-spirobifluorene, the [n,n]-vespirenes.

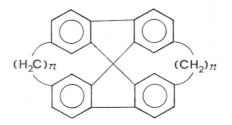

The high value of the measured optical activity may hardly be ascribed to the presence of the alkyl bridges *per se*, but rather to a concomitant distortion of the spirobifluorene system. While in the unsubstituted molecule the two fluorene fragments lie perpendicular to each other, in the vespirenes they are possibly tilted, this effect being more pronounced for smaller n. On this assumption a discussion of the CD spectrum has been given, based on the exciton model, as well as a prediction of the absolute configuration [18]. Our aim is to investigate the usefulness of the PPP method to gain a more detailed description of the electronic structure. Parametrization and methods of calculation are the same as for the helicenes. The molecule has been treated as a 24-electron system. All γ integrals between every carbon atom (except the central one) were included; spiroconjugation [18] was, however, neglected. Fig. 6 shows measured [19] spectra and the calculated transition frequencies and intensities. The location of a y, z-polarized band at about 270 mμ is apparently correctly predicted, but the long-wavelength part of the spectrum is not accounted for by the calculation. The calculated, very strong, and also y, z-polarized transitions around 200 mμ probably correspond to the bands between 200 and 240 mμ in the spectrum. It seems that the splitting of these bands due to exciton-type interaction is stronger in reality than in the calculation. The angle between the planes of the two fluorenyl fragments is set equal to 70°, as shown in Fig. 5, and the $2p_\pi$ orbitals are assumed to remain perpendicular to these planes. The rotational strengths are then predicted to be of the same order of magnitude as in the helicenes, in agreement with experimental observation. Through configuration interaction the electric moments of one fragment couple to the magnetic moments of the other one, leading to these relatively large values. The absolute configuration may be inferred from Fig. 5.

IV. Summary

The PPP method, invoking extensive interaction of singly-excited configurations, seems to predict satisfactorily the dominant features of the UV and CD spectra of hexahelicene and to indicate that (+)-hexahelicene should be right-handed. Similar results are obtained for heptahelicene.

The agreement of the calculations on [n,n]-vespirene — or rather, tilted 9,9′-spirobifluorene — with spectral data is only fair. The longest-wavelength transition is not accounted for, and exciton-type splittings are predicted to be too small. This may in part be due to the neglect of spiroconjugation. For an angle of 70° between the planes of the

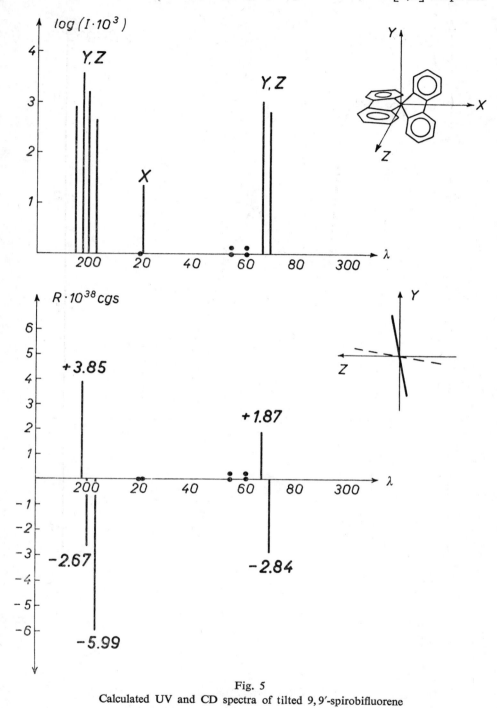

Fig. 5
Calculated UV and CD spectra of tilted 9, 9′-spirobifluorene

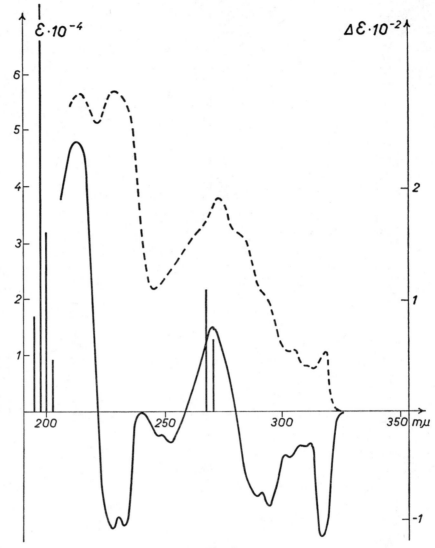

Fig. 6

Measured UV and CD spectra of [6, 6]-vespirene [19]. The relative values of the calculated intensities of the
y, z-polarized bands are also drawn, in an arbitrary absolute scale

fluorenyl fragments, the rotational strengths are computed to be of the same order of
magnitude as in the helicenes, in agreement with experiment.

ACKNOWLEDGEMENTS

The author thanks Prof. V. Prelog for some stimulating discussions, for letting him know
his paper [19] prior to publication and for permission to reproduce the UV and CD spectra
of [6,6]-vespirene. He is grateful to Mr W. Hug and Mr J. Kuhn for their assistance in some
parts of the calculations. The computations were performed at the Institute for Operations
Research and Electronic Data Processing of the University of Zurich.

REFERENCES

1 R. Pariser & R. G. Parr (1953) *J. Chem. Phys.*, 21 : 466, 767.

2 J. A. Pople (1953) *Trans. Faraday Soc.*, 49 : 1375.

3 H. Eyring, J. Walter & G. E. Kimball (1954) *Quantum Chemistry*, John Wiley & Sons, Inc., New York, Chap. XVII.

4 W. Hug & G. Wagnière (1970) *Theor. Chim. Acta*, 18 : 57.

5 D. D. Fitts & J. G. Kirkwood (1955) *J. Am. Chem. Soc.*, 77 : 4940.

6 J. G. Kirkwood (1937) *J. Chem. Phys.*, 5 : 479.

7 A. Moscowitz (1961) *Tetrahedron*, 13 : 48.

8 Idem (1962) *Adv. Chem. Phys.*, 4 : 67.

9 I. Tinoco & R. W. Woody (1964) *J. Chem. Phys.*, 40 : 160.

10 O. E. Weigang Jr, J. A. Turner & P. A. Trouard (1966) *ibid.*, 45 : 1126.

11 M. S. Newman, R. S. Darlak & L. Tsai (1967) *J. Am. Chem. Soc.*, 89 : 6191.

12 H. Labhart & G. Wagnière (1963) *Helv. Chim. Acta*, 46 : 1315.

13 O. E. Weigang Jr & P. A. Trouard Dodson (1968) *J. Chem. Phys.*, 49 : 4248.

14 R. Geiger & G. Wagnière (1970) *Chimia*, 24 : 37.

15 R. H. Martin, M. Flammang-Barbieux, J. P. Cosyn & M. Gelbcke (1968) *Tetrahedron Lett.*, 31 : 3507.

16 E. Vander Donckt, J. Nasielski, J. R. Greenleaf & J. B. Birks (1968) *Chem. Phys. Lett.*, 2 : 409.

17 G. Haas & V. Prelog (1969) *Helv. Chim. Acta*, 52 : 1202.

18 H. E. Simmons & T. Fukunaga (1967) *J. Am. Chem. Soc.*, 89 : 5208.

19 G. Haas, V. Prelog, G. Snatzke, P. B. Hulbert & W. Klyne, *Helv. Chim. Acta* [in press].

Discussion

S. Sarel:

I would like to ask if the contribution to the Cotton effect comes from $\pi \to \pi^*$ transitions, or if you have it also from $\sigma \to \pi$ transitions.

G. Wagnière:

Well, if we believe our results, it appears that the dominant contribution in the near UV is from $\pi \to \pi^*$ transitions and that there the transitions in which σ electrons participate are of negligible importance. Now, of course, the next step in an investigation of this problem would be to do a CNDO type of calculation, but somehow I don't have a very good feeling about doing something like that. Such a system would have about 130 orbitals. And then, if one would do the configuration interaction part, I think it would take a tremendous amount of computer time.

S. Sarel:

One of the possibilities for that is: What is the amplitude of your Cotton effect? If it is $\pi \to \pi^*$, it should be at least 10^5. From what I saw in the diagram, it seems that not all the bands are of the same amplitude.

G. Wagnière:

The $\Delta\varepsilon_{max}$ are of the order of 10^2, and the rotational strength which we get fits in with this order of magnitude. There is a formula which connects them, and it is roughly in agreement.

J. F. Labarre:

I have noticed that the first $\pi \to \pi^*$ transitions of hexa- and heptahelicene are at the same place in the UV spectrum. I think that this result allows us to neglect the space electronic effect which would perhaps occur between the two superimposable rings in heptahelicene. Such a three-dimensional conjugation would indeed give a strong bathochromic effect (as in rubrene, for example), if it did exist.

G. Wagnière:

Yes, they are computed to appear at 328 and 333 mμ, respectively. But it is found experimentally that they come out at pretty much longer wavelengths, namely, in both cases at about 400 mμ. It seems, however, that there is a certain dependence on the solvent one uses.

Mrs A. Pullman:

What is the distance between those two rings?

G. Wagnière:

Kirkwood assumed, in the picture of the helix, that this distance was 3.8 Å, and somehow I have a feeling that this is too much. We have assumed 3.0 Å.

Discussion

R. H. Martin:

A recent X-ray crystal analysis, carried out by I. R. McKay, J. M. Robertson and J. G. Sime (1969) *Chem. Comm.*, 1470, shows that the distance between C_1 and C_{16} in hexahelicene is 3.05 Å.

G. Wagnière:

May be we made a happy guess.

R. H. Martin:

The fact that your treatment and Moscowitz's treatment give different results shows how important it would be to determine the absolute configuration of a helicene. We are presently working on this problem.

Calculations of Ultraviolet Absorption Spectra and Ionization Potentials by a π-Electron Method

by O. W. ADAMS

National Science Foundation, Washington, D.C.

SOME TIME AGO Dr R. L. Miller* and I began a systematic investigation of the parameters used in semi-empirical molecular orbital calculations. We took as a starting point the Pariser-Parr-Pople π-electron theory, with the ultimate objective of including all-electron theories such as CNDO/2 and INDO.

The objective of the π-electron theory work was to achieve an internally consistent set of parameters that could be used with molecules containing not only carbon atoms, but also nitrogen and oxygen atoms (as well as hydrogen atoms, of course). The degree of success of these parameters was to be judged by the ability to predict several properties of the molecules. At present, these properties include the singlet–singlet transition energies and molecular ionization potentials (IP). It is well to point out that although the π theory has been replaced by all-electron methods for the calculation of many molecular properties, it still claims considerable utility in studying electronic spectra and IP's.

The parameters which were determined and the calculations by which they were tested were published earlier [1]. The purpose of this presentation is to extend the earlier work.

The unique feature of the method lies in handling the core integrals. Let us consider first how the basic core parameter equation was obtained from a calibration using the benzene molecule.

For benzene, explicit equations can be written for the singlet electronic transitions and the molecular IP (i.e., within Koopmans' approximation) in terms of electron repulsion integrals and core integrals. If we assume that we are dealing with an orthogonalized set of atomic orbitals (a.o.'s) and assume that non-nearest neighbour core integrals can be neglected, then these relations become quite simple [2], as shown in Table 1. We now adopt

Table 1
Calibration Equation for Benzene

$$^1\Delta E_1 = -H_{12}^\lambda + (\gamma_{12}^\lambda - 3\gamma_{13}^\lambda + 2\gamma_{14}^\lambda)/6$$
$$\mathrm{IP} = -H_{11}^\lambda - H_{12}^\lambda - \gamma_{11}^\lambda/2 - 5\gamma_{12}^\lambda/3 - 2\gamma_{13}^\lambda - 5\gamma_{14}^\lambda/6$$
$$\gamma_{ij} = \langle \lambda_i \lambda_i | \lambda_j \lambda_j \rangle$$

* Dean of Arts and Sciences, University of North Carolina, Greensboro, North Carolina.

the formalism of Nishimoto and Mataga [3] to evaluate the electron repulsion integrals. It is possible, therefore, by selecting the equations for the lowest singlet transition and the molecular IP to obtain two equations in the two unknowns, H_{11}^λ and H_{12}^λ. Here, λ refers to the fact that these are integrals over orthogonalized a.o.'s,

$$H_{ij}^\lambda = \langle \lambda_i | H_{\text{core}}(i) | \lambda_j \rangle. \tag{1}$$

We can therefore calculate H_{11}^λ and H_{12}^λ. However, these integrals are not desirable core parameters, since they are unique to benzene, and therefore not extendable to other molecules. This is true because $H_{\text{core}}(i)$ is unique to benzene, and, if we express the λ_i in terms of localized Slater orbitals,

$$\lambda = \chi s^{-\frac{1}{2}}, \tag{2}$$

it is clear that they are similarly characteristic of benzene, with its unique overlap matrix.

In order to obtain core parameters that are characteristic of atoms and bonds rather than molecules, we extend the benzene calibration process through two additional steps:

1. Utilizing the matrix relation $H = S^{\frac{1}{2}} H^\lambda S^{\frac{1}{2}}$, one converts the core integrals over orthogonalized orbitals to integrals over localized Slater orbitals, using $S^{\frac{1}{2}}$ over Slater orbitals (S is the overlap matrix). Therefore, knowing the matrices on the right, the H matrix is uniquely determined. For benzene, one obtains values for the integrals $H_{11}, H_{12}, H_{13}, H_{14}$, where

$$H_{ij} = \langle \chi_i | H_{\text{core}}(i) | \chi_j \rangle. \tag{3}$$

2. The Goeppert-Mayer and Sklar expansion of $H_{\text{core}}(i)$ is used [4], neglecting neutral atom penetration integrals. The core integrals over Slater orbitals can now be expressed as follows:

$$H_{pp} = \langle \chi_p | T(i) + U_p^{n_p}(i) | \chi_p \rangle - \sum_{r \neq p} n_r \gamma_{rp}^\lambda, \tag{4}$$

$$H_{pq} = \langle \chi_p | T(i) + U_p^{n_p}(i) + U_q^{n_q}(i) | \chi_q \rangle - \sum_{r \neq p, q} n_r (rr | pq)^\lambda. \tag{5}$$

The latter equation, with the Mulliken approximation for the electron repulsion integrals, gives

$$H_{pq} = \langle \chi_p | T(i) + U_p^{n_p}(i) + U_q^{n_q}(i) | \chi_q \rangle - (S_{pq}/2) \sum_r n_r [\gamma_{pr}^\lambda + \gamma_{qr}^\lambda], \tag{6}$$

where

$$\gamma_{pq}^\lambda = \langle \lambda_p \lambda_p | \lambda_p \lambda_q \rangle, \tag{6a}$$

$T(i)$ are kinetic energy operators, and $U_p^{n_p}(i)$ are potential energy operators for centres of charge n_p.

The desired core parameters are now taken as the first terms in Equations (5) and (6) and are designated H_{pp}^0 and H_{pq}^0.

The results of the benzene calibration are shown in Table 2.

Table 2
Calibration of Benzene Core Parameters

Input: $^1L_b = 4.72$ eV*; IP $= 9.25$ eV**; $\gamma_{11}^\lambda = 10.80$ eV

Results (in eV)		
	$H_{11}^\lambda = -31.72023$	$H_{11}^0 = -11.14128$***
	$H_{12}^\lambda = -2.29815$	$H_{12}^0 = -6.13067$
		$H_{13}^0 = -1.07433$
	$H_{11} = -32.84891$	$H_{14}^0 = -0.36345$
	$H_{12} = -10.16628$	
	$H_{13} = -1.68724$	$^1L_a = 5.94$ (exp. $= 5.90$*)
	$H_{14} = -0.64121$	$^1B = 6.76$ (exp. $= 6.74$*)

   * Petruska [5]
  ** Watanabe [6]
*** In subsequent calculations this parameter was assigned the value -11.16 eV [7]

It is clear that the one-centre parameter H_{11}^0 is almost identical to the valence state ionization potential (VSIP) of carbon [7]. In all subsequent work this integral was therefore set equal to the appropriate VSIP.

The parameters H_{12}^0, H_{13}^0 and H_{14}^0 were found to fit the following relation:

$$H_{pq}^0 = -11.1712 \, (S_{pq} - 0.0852 \, R_{pq} + 0.24561) - (S_{pq}/2)[\gamma_{pp}^\lambda + \gamma_{pq}^\lambda]. \tag{7}$$

Refitting to make the lending term equal $(VSIP)_C$ gives

$$H_{pq}^0 = H_{pp}^0 (S_{pq} - 0.0855 \, R_{pq} + 0.24639) - (S_{pq}/2)[\gamma_{pp}^\lambda + \gamma_{pq}^\lambda]. \tag{8}$$

The extension of this equation to the centres p and q is different, and with core charges n_p and n_q we get

$$H_{pq}^0 = \frac{(H_{pp}^0 + H_{qq}^0)}{2} [S_{pq} - 0.0855 \, R_{pq} + 0.24639] + \tag{9}$$

$$- n_p (S_{pq}/4)[\gamma_{pp}^\lambda + \gamma_{pq}^\lambda] - n_q (S_{pq}/4)[\gamma_{qq}^\lambda + \gamma_{pq}^\lambda].$$

Results presented in our earlier publication [1] provided a test of the utilization of these core parameters by calculating the lowest singlet transitions and ionization potentials of a large number of hydrocarbons, as well as the heterocyclic molecules pyridine, p-benzoquinone, pyrrole and furan. In calculating the spectra, up to 30 configurations corresponding to one-electron excitations were included in a configuration-interaction calculation performed after the SCF calculation had been completed.

All overlap integrals used in the calculations were evaluated by using a Slater-type orbital, with orbital exponents determined from Slater's rules (see Table 3).

A number of additional calculations have now been made, in particular on the more common azabenzenes, on aniline and on the pyrimidine bases uracil and cytosine. The purine bases, guanine and adenine, have also been examined. In these calculations, as in the earlier work, all of the valence-state data are those of Hinze and Jaffé [7], the values for which are shown in Table 3.

For the azabenzenes, two sets of calculations were performed, one using orbital exponents

142

Table 3
Hetero-Atomic Data

Atom*	VSIP**	VSEA***	γ_{ii}^{λ} $^{\triangledown}$	ζ $^{\triangledown\triangledown}$
$\overset{.}{N}$	14.12	1.78	12.34	1.95
$\overset{..}{N}$	28.775	12.305	16.47	2.125
$\overset{.}{O}$	17.70	2.47	15.24	2.275
$\overset{..}{O}$	34.07	15.22	18.85	2.45

* The dots refer to the number of electrons donated to the π system

** VSIP denotes valence-state ionization potential according to Hinze and Jaffé [7]

*** VSEA denotes valence-state electron affinity according to Hinze and Jaffé [7]

$^{\triangledown}$ γ_{ii}^{λ} = VSIP – VSEA.

$^{\triangledown\triangledown}$ Orbital exponent in Slater a.o.

for the Slater-type orbitals, determined by Slater's rules ($\zeta_N = 1.95$), and the other with the value $\zeta_N = 1.785$, which gives the best value for the lowest singlet transition of pyridine. These results are shown in Tables 4–5 for the singlet transitions, and in Table 6 for the ionization potentials.

Table 4
Results for Azabenzenes — Lowest Singlet Transitions * (eV)

Molecule		^1L_b		^1L_a		^1B_a		^1B_b	
		ΔE	f	ΔE	f	ΔE	f	ΔE	f
Pyridine	Calc.	4.67	0.067	5.97	0.030	6.86	0.944	6.91	1.105
	Exp.**	4.79		6.10		7.04		7.04	
Pyrazine	Calc.	4.45	0.185	5.97	0.149	7.32	1.031	7.36	0.847
	Exp.**	4.82		6.12		~7.44		~7.44	
Pyrimidine	Calc.	4.79	0.070	6.05	0.089	6.94	0.856	7.16	1.052
	Exp.**	5.01		6.46		~7.25		~7.25	
Pyridazine	Calc.	4.69	0.071	5.93	0.002	6.64		7.14	
	Exp.**	4.89		6.19		~7.10		~7.10	
s-Tetrazine	Calc.	4.68	0.204	6.58	0.029	7.28	~1.110	8.10	1.026
	Exp.**	4.92							
s-Triazine	Calc.	5.08	0.0	6.41	0.0	7.11	0.975	7.11	0.975
	Exp.**	5.41							

* Using $\zeta_N = 1.95$ ** Petruska [5]

Table 5

Results for Azabenzenes — Lowest Singlet Transitions* (eV)

Molecule		^1L_b		^1L_a		^1B_a		^1B_b	
		ΔE	f	ΔE	f	ΔE	f	ΔE	f
Pyridine	Calc.	4.80	0.053	6.02	0.076	6.95	1.020	7.02	1.053
	Exp.**	4.79		6.10		7.04		7.04	
Pyrazine	Calc.	4.77	0.168	6.02	0.225	7.47	0.925	7.63	0.958
	Exp.**	4.82		6.12		~7.44		~7.44	
Pyrimidine	Calc.	4.99	0.065	6.22	0.126	7.23	0.925	7.30	1.055
	Exp.**	5.01		6.46		~7.25		~7.25	
Pyridazine	Calc.	4.90	0.063	6.11	0.023	6.91	0.940	7.28	1.140
	Exp.**	4.89		6.19		~7.10		~7.10	
s-Tetrazine	Calc.	5.08	0.211	6.92	0.095	7.76	1.145	8.40	1.045
	Exp.**	4.92							
s-Triazine	Calc.	5.35	0.0	6.68	0.0	7.39	1.035	7.39	1.035
	Exp.**	5.41							

* Using $\zeta_\lambda = 1.785$ ** Petruska [5]

Table 6

Azabenzenes — Ionization Potentials (eV)

Molecule	Calc.		Exp. $^\triangledown$
	A^*	B^{**}	
Pyridine	9.18	9.19	9.28
Pyrazine	9.13	9.13	9.27
Pyrimidine	9.53	9.54	9.47
Pyridazine	9.58	9.59	8.91
s-Tetrazine	10.10	10.12	—
s-Triazine	10.52	10.50	—

* $\zeta_i = 1.785$ ** $\zeta_i = 1.950$ $^\triangledown$ Turner [8]

In general, the results of Tables 5–6 are quite good. The poorest result is that for the predicted ionization potential of pyridazine. It may be that in this case the electron being ionized is, indeed, not a π-electron, but perhaps one of the lone-pair electrons on the nitrogen. The next molecule to be examined, aniline, was chosen to illustrate the effect of a nitrogen atom not contained within the aromatic ring. For the initial calculations, the nitrogen atom of aniline was represented by the same valence-state data as used earlier for the pyrrole-type nitrogen atom.

The results obtained for aniline are shown in Table 7. It is clear that the results for the singlets of aniline are in fair agreement with experiment. The predicted ionization potential is excellent.

Table 7
Results for Aniline
Lower Singlet Transitions (eV)

Transition Energy		f	
Calc.	Obs.*	Calc.	Obs.*
4.20	4.40	0.111	0.028
5.37	5.39	0.248	0.140
6.54	6.40	1.042	0.510
6.58	6.88	0.714	0.570

Ionization Potential (eV)

Calc.	7.67
Obs.**	7.70

* Kimura et al. [9]   ** Watanabe et al. [10]

There are several good π-electron calculations for aniline, and it is not the purpose of this work to claim better results. The significant feature of the results is that it was achieved using the same parameter equations as were originally derived from the benzene calibration. Passing now to some more difficult tests of the parameters, we consider the purine and pyrimidine bases adenine, guanine, uracil and cytosine. The calculations are identical to those done earlier. For the pyridine-type nitrogen atom, again, the orbital exponent $\zeta_N = 1.785$ was used.

The lower singlets calculated for the pyrimidine bases are shown in Table 8. In this Table the experimental results are shown for both solution spectra [11] and vapour phase spectra [12] (in parentheses). The experimental oscillator strengths are from the solution results. The last column in the table shows those singlet configurational wave functions which give the major contribution to the particular calculated excited state wave function.

Table 8
Pyrimidine Bases — Lower Singlet Transitions (eV)

Molecule		Calc.		Exp.*		Major Singlet Component
		ΔE	f	ΔE	f	
Uracil	B_{2u}	4.82	0.34	4.8 (5.1)	0.2	V_{56}
	B_{1u}	5.50	0.03	5.4	w	V_{46}
		5.99	0.46	6.1 (6.0)	0.2	V_{57}
	E_{1u}	6.49	0.81	6.8 (6.6)		V_{47}
Cytosine	B_{2u}	4.05	0.12	4.5 (4.3)	0.2	V_{56}
	B_{1u}	5.13	0.01	5.2	0.2	V_{46}
		6.05	1.12	6.1	0.6	V_{57}
	E_{1u}	6.67	0.23	6.7		V_{47}

* Solution spectra: Clark & Tinoco [11]; vapour spectra: Clark et al. [12]

The results achieved for uracil are particularly encouraging. The cytosine calculation gives a lowest singlet transition that is somewhat too low, but the remaining transitions are quite close to experimental values. The lower singlet transitions for the purine bases are shown in Table 9. For adenine, five transitions are predicted over the measured wave-

Table 9
Purine Bases — Lower Singlet Transitions (eV)

Molecule	Calc.		Exp.		Major Singlet Components
	ΔE	f	ΔE	f	
Adenine	4.62	0.122	4.8	0.3	V_{67}, V_{68}
	4.88	0.398			V_{67}, V_{68}
	5.96	0.01			V_{69}
	6.27	1.049	6.0	0.4	V_{57}
	6.73	0.447	6.7		V_{58}
Guanine	3.58	0.338	4.5 (4.3)	0.1	V_{78}
	4.39	0.298			V_{79}
	5.32	0.158	4.9	0.3	$V_{7,10}$
	6.14	0.051	6.0		V_{68}
	6.57	0.065	6.6	1.1	$V_{7,11}$
	6.70	0.099			V_{58}, V_{69}
	7.43	0.973			V_{58}, V_{69}

length range. Aside from the very weak excitation predicted at 5.96 eV., these can be matched reasonably well with experiment, if it is assumed that the measured low-energy band actually contains contributions from two excitations. The results for guanine are not very good. In particular, the lowest singlet transition is much too low in energy. Also, the most intense band is predicted to lie at 7.43 eV. This should probably correspond to the intense observed transition at 6.6 eV.

Table 10 shows the ionization potentials calculated for the four bases. There appear to be no experimental results for comparison with these values; however, the order predicted — namely, G < A < C < U — appears to be correct [13].

Table 10
Purine and Pyrimidine Bases
Ionization Potentials (eV)

Guanine	6.23
Adenine	7.19
Cytosine	7.73
Uracil	8.71

In conclusion, additional results have been presented which indicate the value of the parameters we have adopted. It is encouraging that hetero-systems containing a large variety and large number of hetero-atoms can be successfully handled (uracil, cytosine and adenine). The reasons for the breakdown of the parameters in the case of guanine needs to be examined.

ACKNOWLEDGEMENT

Some of the results presented here were obtained by O. W. Adams at the Abbott Laboratories, North Chicago, Illinois. Permission to publish these results is greatly appreciated. Some of the more recent calculations were performed at the National Institutes of Health, Bethesda, Maryland.

REFERENCES

1 O. W. Adams & R. L. Miller (1968) *Theor. Chim. Acta* (Berlin), 12 : 151.

2 P. G. Lykos (1961) *J. Chem. Phys.*, 35: 1249.

3 K. Nishimoto & N. Mataga (1957) *Z. Phys. Chem.* (Frankfort), 12 : 335.

4 M. Goeppert-Mayer & A. L. Sklar (1948) *J. Chem. Phys.*, 6 : 645.

5 J. Petruska (1961) *J. Chem. Phys.*, 34 : 1120.

6 K. Watanabe (1957) *J. Chem. Phys.*, 26 : 542.

7 J. Hinze & H. H. Jaffé (1962) *J. Am. Chem. Soc.*, 84 : 540.

8 D. W. Turner (1966) *Adv. Phys. Org. Chem.*, 4 : 31.

9 K. Kimura, H. Tsubomura & K. Nagakura (1964) *Bull. Chem. Soc. Japan*, 37 : 1336.

10 K. Watanabe et al. (1959) *J. Quantum Spectrosc. Rad. Trans.*, 2 : 369.

11 L. B. Clark & I. Tinoco (1965) *J. Am. Chem. Soc.*, 87 : 11.

12 Idem (1965) *J. Phys. Chem.*, 69 : 3615.

13 H. Berthod, C. Geissner-Prettre & A. Pullman (1966) *Theor. Chim. Acta* (Berlin), 5 : 53.

Discussion

O. W. Adams (in replying to a question from Mr Heilbronner):

I have done only the hydrocarbons, but on one slide I have listed some typical non-alternant hydrocarbons; I did have heptafulvene and fulvene. As far as the calibration is concerned, the closest I came to that problem was trying to calibrate with hexamethylbenzene (HMB), to see if I could not deal with the methylbenzenes. This works — I mean, you can use HMB for the calibration, but the difficulty is that the experimental values for the lowest-singlet transition and ionization potential for hexamethylbenzene are only estimates, so I was never quite sure of the results and I never published them. If you do this calibration, you get quite good results for the remaining methyl-benzenes. However, looking at the parameters that result and comparing them to the benzene parameters, I am not at all sure that they make any sense, because, in principle, we are dealing with different types of carbon atoms. While these should be related logically, it did not seem to me at the time that they were very logically related.

E. D. Bergmann:

You have calculated the ionization potentials of fulvalene, heptafulvalene and heptafulvene. Do you know of any experimental measurements to corroborate your results?

O. W. Adams:

I do not know. Does anyone know?

E. Heilbronner:

I do not remember whether it has been published, but dimethylfulvene has been measured. We are trying now to measure fulvene and the other compounds.

Mrs A. Pullman:

There is no doubt that every theoretician who has used the Pariser-Parr-Pople method and has tried to rationalize the choice of the parameter values, has run into the same difficulty, namely, the impossibility of completely rationalizing it, even though the number of compounds studied is large enough to obtain all the necessary variables in principle. It seems to me that the reason for this is that the parameters in the PPP method cover, at the same time, a number of simplifications and approximations which are of quite different natures and which can vary in very different ways in going from one molecule to another. Clearly, this makes a complete rationalization impossible.

O. W. Adams:

I agree with this point of view. All you can do is to adopt parameters and try and see how far you can go with them; when they start breaking down, you have to discover a new way of getting parameters.

H. Kuroda:

We have carried out calculations by the PPP method on about 70 aromatic compounds, including heterocyclic systems, and have found that IP is always predicted low in comparison with the

Discussion

experimental values, if we assume Koopmans' theorem. This is improved when we take into account all β terms. Thus, I believe that we should take into account all β terms in order to make a parametrization of the semi-empirical parameters by using the experimental value of IP.

B. Pullman:

Your results, as far as the ionization potentials of cytosine are concerned, are quite bad, so that this disqualifies your procedure to some extent. The value of 6.3 is impossible for this type of molecule. Either your method is wrong or this particular calculation is wrong.

O. W. Adams:

It is really surprising, because the calculations for the other bases were reasonably good.

E. D. Bergmann:

Are you sure that your calculation gives you the ionization of a π electron? You might be getting something else, in this case, and then one might understand why you obtain a much lower value.

B. Pullman:

No, the lowest ionization is the π one. The other terms are higher.

Novel Molecules Bearing Substituents
within the Cavity of the π-Electron Cloud
A New Synthetic Approach

by V. BOEKELHEIDE *and* R.H. MITCHELL

Department of Chemistry, University of Oregon, Eugene

SINCE the first introduction of molecular orbital theory by Hückel [1] in 1937 as an explanation of aromaticity, a great effort has been made to test the predictions of the theory experimentally. In particular the brilliant work of Sondheimer has made available a large number of annulenes with varying numbers (both $4n$ and $4n + 2$) of π electrons [2]. In our own studies we have been concerned with the shape of the π-electron cloud, whether in fact it contains a vacant cavity and, if so, the steric and electronic interactions that would occur with substituents inserted into this cavity. Our first reports were concerned with *trans*-15,16-dimethyldihydropyrene (I), its chemistry and physical properties [3–7]. Later, the syntheses of *trans*-15,16-diethyldihydropyrene (II) and *trans*-15,16-di-*n*-propyl-dihydropyrene (III) were provided as well [8–9]. In addition, syntheses of hetero-substituted dihydropyrenes [10] and *cis*-dihydropyrenes [11–13] were accomplished. Of particular interest were the photoisomerizations of the dihydropyrenes [14] and the magnetic properties of their dianions [15].

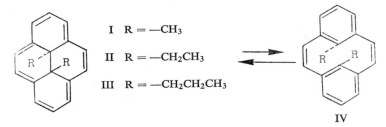

I	R = —CH$_3$
II	R = —CH$_2$CH$_3$
III	R = —CH$_2$CH$_2$CH$_3$

IV

The syntheses of the dihydropyrenes have usually required about thirteen steps, with overall yields of about 1%. Because of the high interest in these molecules and the need for larger quantities, we have been led to explore new approaches. Originally, we had been led to undertake the synthesis of the valence tautomers (IV) in the belief that they would spontaneously isomerize to I, a fact later established experimentally [14]. Unfortunately, however, all attempts to convert [2.2]metacyclophanes to their corresponding diolefins (IV) were completely unsuccessful; the bridging methylenes resisted all attempts directed toward the introduction of substituents [16]. The alternate of devising a metacyclophane synthesis that would lead to substituents in the bridges was partially successful [17], but failed for the case where the internal groups (R) were methyl.

Novel Molecules Bearing Substituents within the Cavity of the π-Electron Cloud

Recently, there has been considerable interest in the conformational inversions of 2,11-dithia[3.3]metacyclophanes [18–19]. Independently of Sato and Vögtle, we had also become interested in 2,11-dithia[3.3]metacyclophanes, not from an interest in their conformational inversions, but rather from the possibility of utilizing them as starting materials for syntheses of [2.2]metacyclophane-1,9-dienes. As shown below, to realize this goal requires the extrusion of a sulphur atom, with concomitant formation of a carbon–carbon double bond. The standard procedure for such a transformation is the Ramberg-Bäcklund reaction. However, preliminary experiments indicated to us that for our purposes there were difficulties in employing the Ramberg-Bäcklund reaction, and so our attention turned toward the possibility of developing a new procedure suitable for strained molecules.

V   IV

From consideration of molecular models, it was clear that any reaction involving extrusion of sulphur from V with concomitant carbon–carbon bond formation would require that this be an intramolecular process with frontside displacement. Interestingly, Brewster's study of the Stevens rearrangement of ammonium salts, using optically active molecules, has established that the Stevens rearrangement is an intramolecular process involving frontside attack with complete retention of configuration [20]. More recent studies confirm this and suggest a pathway involving a diradical intermediate [21–22]. It appeared that the Stevens rearrangement of sulphonium salts should be analogous, and the resulting products would be well suited for an elimination reaction to complete the transformation.

When 2,6-*bis*(bromomethyl)toluene (VI) was treated with sodium sulphide in ethanol, two isomeric products were obtained, which have been assigned structures VII and VIII.*
With internal groups as large as methyl, conformational inversion of the 2,11-dithia[3.3]metacyclophanes is no longer possible, and so the isomers corresponding to the *cis* and *trans* arrangement of the aromatic rings are stable and can be isolated and separated. The respective assignments of structure are based on NMR data. The internal methyl protons of VII feel the ring current of the opposite ring and are shifted to a higher field, appearing

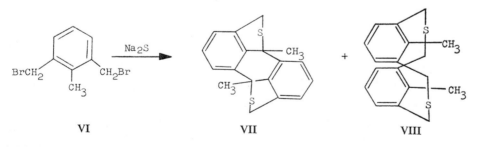

VI   VII   VIII

---

* Satisfactory elemental analyses, mass spectra and the other usual spectral data have been obtained for all new compounds. Experimental details of their preparation will be reported elsewhere.

at τ 8.70, whereas the internal methyl protons of VIII are normal and appear at τ 7.46. Similarly, the aromatic protons of VII show the normal metacyclophane pattern at τ 2.6–3.0, whereas the aromatic protons of VIII are shifted upfield to τ 3.34, a common consequence of superimposing aromatic rings [23–25].

The conversion of VII to its corresponding disulphonium salt IX was readily accomplished in quantitive yield, using the very convenient methylating reagent popularized by Borch [26], and gave the fluoroborate of IX as white crystals, m.p. 210° dec. When IX was treated with sodium hydride in tetrahydrofuran, the Stevens rearrangement occurred smoothly in 90% yield to give the corresponding disulphide X. Although X was a mixture of isomers, as would be expected, the NMR spectrum of the mixture (internal methlyl singlets between τ 9.0 and τ 9.5) clearly showed that all of these isomers were related to the *trans*-[2.2]meta-cyclophanes; thus no conformational interconversion occurred during the Stevens rearrangement.

This mixture of isomers was again methylated wi h dimethoxy-carbonium fluoroborate in methylene chloride at −30° to give the disulphonium salt XI. The final elimination of dimethyl sulphide from XI has been investigated, using a number of different bases, and, with potassium *t*-butoxide in boiling *t*-butanol, the conversion of XI to 15,16-dimethyl-dihydropyrene (XII) occurs in about 80% yield.

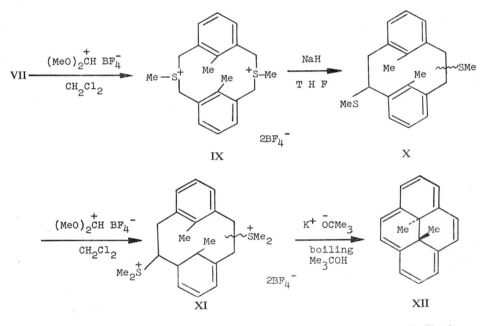

Thus, the overall conversion of 2,11-dithia[3.3]metacyclophanes to 15,16-dihydropyrenes *via* a Stevens rearrangement and elimination is a simple and high-yield procedure. This method has been applied to other systems and appears to be a general method for replacing a sulphide linkage by a carbon–carbon double bond.

The fact that the elimination of dimethyl sulphide from XI first yields the diene IV (R = —CH$_3$), which then undergoes valence tautomerization to 15,16-dimethyldihydro-pyrene (XII), has also been shown. When XI was treated with sodium hydride in tetra-hydrofuran at room temperature, it was converted in about 60% yield to a mixture of about equal parts of IV (R = —CH₃) and XII, as analysed from NMR data. Either heating or chromatography over silica gel converts the mixture entirely to XII.

It is of interest that the *cis* isomer VIII, after conversion to the corresponding sulphonium salt (*cis* geometry), also readily undergoes the Stevens rearrangement to provide a mixture of isomers corresponding to X, but with *cis* geometry. The evidence for *cis* geometry is based on the NMR spectrum of the mixture, which shows the internal methyl protons at τ 7.7–7.9 (the normal region for a toluene methyl) and the absence of ring current effects.

This is the first instance of the preparation of a [2.2]metacyclophane in which the benzene rings are *cis* to each other. Molecular models suggest that the proximity of the two aromatic rings must be comparable to that for the case of [2.2]paracyclophane. In fact, the ultraviolet spectra of our *cis* isomers show no long-wavelength absorption band comparable to that found for [2.2]paracyclophane [27], and this may be of some theoretical import [28].

In view of the fact that the conversion of 9,18-dimethyl-2,11-dithia[3.3]metacyclophane (VII) to *trans*-15,16-dimethyl-dihydropyrene (I) involves only two operations, both of which proceed in high yield, it was of interest to explore the generality of the method, and our attention fastened on the goal of synthesizing *trans*-15,16-dihydropyrene (XIX), the parent molecule for this whole series.

For this purpose we needed 2,11-dithia[3.3]metacyclophane (XIV), and, although its preparation in low yield has been described previously, we found that under suitable conditions the conversion of *m*-xylylene dibromide (XIII) to XIV could be accomplished in 50% yield. Since conformational inversion of XIV occurs at room temperature, only a single isomer is isolated.

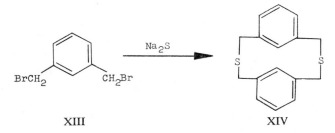

XIII   XIV

Treatment of 2,11-dithia[3.3]metacyclophane (XIV) with dimethoxycarbonium fluoroborate in methylene chloride gave the corresponding sulphonium salt (XV) in essentially quantitative yield. When the sulphonium salt (XV) was dissolved in dimethyl sulphoxide containing potassium *t*-butoxide, it was converted in a few minutes and in high yield to a mixture of isomers corresponding to XVI. Remethylation of XVI with dimethoxycarbo-

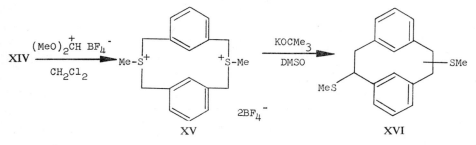

XV   XVI

153

nium fluoroborate gave again the sulphonium salt XVII as a mixture of isomers. Treatment of this sulphonium salt with various bases has been studied in some detail. In each case, a deep green solution results. However, on work-up, the hydrocarbon fraction is colourless to pale yellow and contains pyrene. Careful chromatography, though, over silica gel leads to the isolation of a white crystalline hydrocarbon in about 30% yield, to which we assign the structure of the *trans*-[2.2]metacyclophane-1,9-diene (XVIII).

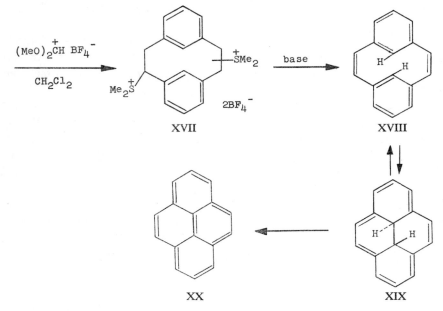

The assignment of 'a' structure to XVIII is based on evidence derived from mass spectral, NMR, ultraviolet and infra-red absorption data. In addition, it was shown to undergo catalytic hydrogenation to give *trans*-[2.2]metacyclophane. Its mass spectrum shows the parent molecular ion at 204 ($C_{16}H_{12}$). Its NMR spectrum shows aromatic absorption and vinyl hydrogen absorption in the ratio of 2:1. In the ultraviolet there is a strong absorption at 280 nm, with nothing at longer wavelengths. In the infra-red, the absorption due to a *cis*-olefinic bond is apparent. The only puzzling part of the assignment is the stability of the compound and the ease with which it can be handled. In the case of all the other dihydropyrenes that we have studied, valence tautomerization from the [2.2]meta-cyclophane-1,9-diene to the corresponding dihydropyrene is a spontaneous and relatively rapid process. The stability of XVIII demonstrates a large, and unexpected, energy barrier to valence tautomerization.

However, when XVIII was taken up in degassed cyclohexane using high vacuum techniques and irradiated using a low-pressure Hanovia lamp, it was converted to the desired *trans*-15,16-dihydropyrene (XIX). This was confirmed by its NMR spectrum, which shows the internal protons as a sharp singlet at $\tau 15.49$. The external protons show the normal dihydropyrene pattern. Also, the ultraviolet and visible spectrum of XIX is similar to that of the other dihydropyrenes, except that it has more fine structure.

When it is considered that the internal protons of XIX are allylic, in contrast to the other known dialkyldihydropyrenes (I to III), having internal protons that are saturated, the observed chemical shift ($\tau 15.49$) is remarkably higher than that of the other examples

(τ 14.25). Undoubtedly, the internal protons of XIX must be very close to the region of highest ring current effect.

In degassed cyclohexane solutions *trans*-15,16-dihydropyrene is stable indefinitely. When such solutions are opened and a stream of oxygen is bubbled through, the *trans*-15,16-dihydropyrene is converted rapidly (five minutes), but not instantaneously, to pyrene (XX). Likewise, prolonged irradiation of *trans*-15,16-dihydropyrene leads to a clean conversion to pyrene.

Although *trans*-15,16-dihydropyrene might be expected to undergo a rapid sigmatropic rearrangement of the internal hydrogens to give a naphthalene derivative, the molecule is surprisingly stable at room temperature. At temperatures above 60° a thermal rearrangement does occur, and it may well be a sigmatropic rearrangement.

The ease with which *trans*-15,16-dimethyldihydropyrene and *trans*-15,16-dihydropyrene itself have been synthesized demonstrates the power of this new synthetic approach. We have explored to some extent the generality of this method and have found it an extremely facile route for preparing a number of highly strained molecules. These results will be reported elsewhere. We are continuing our studies on the applications of this method both for the synthesis of dihydropyrene derivatives and for the preparation of other unusual molecules of high theoretical interest.

ACKNOWLEDGEMENTS

We thank the National Science Foundation for their support of this study. We also thank the Israel Academy of Sciences and Humanities for the privilege of presenting this work.

REFERENCES

1 E. Hückel (1937) *Z. Electrochem.*, 43 : 752.
2 F. Sondheimer, I. C. Calder, J. A. Elix, Y. Gaoni, P. J. Garrat, K. Grohmann, G. Di Maio, J. Mayer, M. V. Sargent & Wolovsky (1967) in: *Aromaticity (Special Publication, No. 21)*, The Chemical Society, London, p. 75.
3 V. Boekelheide & J. B. Phillips (1963) *J. Am. Chem. Soc.*, 85 : 1545.
4 Idem (1964) *Proc. Natn. Acad. Sci. USA*, 51 : 550.
5 Idem (1967) *J. Am. Chem. Soc.*, 89 : 1695.
6 J. B. Phillips, R. J. Molyneux, E. Sturm & V. Boekelheide, *ibid.*, p. 1704.
7 H. Blaschke, C. E. Ramey, I. C. Calder & V. Boekelheide, *ibid.* [in press].
8 V. Boekelheide & T. Miyasaka (1967) *ibid.*, 89 : 1709.
9 V. Boekelheide & T. A. Hylton (1970) *ibid.*, 92 : 3669.
10 V. Boekelheide & W. Pepperdine (1970) *ibid.*, 92 : 3684.
11 B. A. Hess Jr, A. S. Bailey & V. Boekelheide (1967) *ibid.*, 89 : 2746.
12 B. A. Hess Jr, A. S. Bailey, B. Bartusek & V. Boekelheide (1969) *ibid.*, 91 : 1665.
13 B. A. Hess Jr & V. Boekelheide, *ibid.*, p. 1672.
14 H. R. Blattmann, D. Meuche, E. Heilbronner,

R. J. Molyneux & V. Boekelheide (1965) *ibid.*, 87 : 130.
15 R. H. Mitchell, C. E. Klopfenstein & V. Boekelheide (1969) *ibid.*, 91 : 4931.
16 W. S. Lindsay, P. Stokes, L. G. Humber & V. Boekelheide (1961) *ibid.*, 83 : 943.
17 T. Hylton & V. Boekelheide (1968) *ibid.*, 90 : 6887.
18 F. Vögtle & L. Schunder (1969) *Chem. Ber.*, 102 : 2677.
19 T. Sato, M. Wakabayashi, M. Kainosho & K. Hata (1968) *Tetrahedron Lett.*, p. 4185.
20 J. H. Brewster & M. W. Kline (1952) *J. Am. Chem. Soc.*, 74 : 5179.
21 A. R. Lepley (1969) *Chem. Comm.*, p. 1460.
22 U. Schöllkopf, U. Ludwig, G. Ostermann & M. Patsch (1969) *Tetrahedron Lett.*, p. 3415.
23 R. H. Mitchell & F. Sondheimer (1968) *J. Am. Chem. Soc.*, 90 : 530.
24 D. J. Cram, C. K. Dalton & G. R. Knox (1963) *ibid.*, 85 : 1088.
25 R. H. Martin, G. Morren & J. J. Schurter (1969) *Tetrahedron Lett.*, p. 3683.
26 R. F. Borch (1969) *J. Org. Chem.*, 34 : 627.
27 D. J. Cram, N. L. Allinger & H. Steinberg (1954) *J. Am. Chem. Soc.*, 76 : 6132.
28 R. Gleiter (1969) *Tetrahedron Lett.*, p. 4453.

Discussion

R. H. Martin:

First of all, I would like to congratulate you on this magnificent work.

In your photocyclizations you isolated the *trans*-dihydropyrene derivative, on the one hand, and the *cis*-difluoro derivative, on the other hand. Now, assuming a concerted reaction, the Woodward-Hoffmann rules predict a conrotatory cyclization (*trans* configuration) *via* a first excited state and a disrotatory cyclization (*cis* configuration) *via* higher vibrational levels of the ground state. This is a very important point, since there are only speculations concerning the stereochemistry of the dihydro intermediates in the photo-induced cyclodehydrogenations of 1,2-diarylethylenes.

May I therefore ask you if you have compelling evidence for the *trans* configuration of your dihydro derivative?

V. Boekelheide:

Yes, we do. We have a complete X-ray analysis of the precursor molecule, [2,2] metacyclophane-1,9-diene, and this clearly has the *anti* conformation of the benzene rings. Furthermore, it is evident from models and from other studies that the energy barrier is too great for conformational flipping between the *anti* and *syn* conformers. Thus, on irradiation the only isomer that can possibly form is the *trans*-15,16-dihydropyrene.

Another thing is that you have very close correspondence in spectral properties with the *trans*-dialkyl compounds.

Now, the evidence is still incomplete, but our *cis*-dimethyl compound has a somewhat different UV spectrum, and I would expect to see this difference also in the *cis*-dihydrogen compound. There are some things that I have not told you about, and one is that you can, of course, start with both *cis* and *trans* arrangements, and so we should be able to make the corresponding *cis* compound. But as [2,2]paracyclophane has about 36 Kcal/mole of strain energy, so do these *cis*-arranged metacyclophanes, and there is a tremendous tendency for them to get over into the other arrangement. We find that this flipping, even if it is not possible in the free hydrocarbons themselves, does occur with a number of reagents. You can rationalize this as being an attack on the ring by a nucleophile and then a loss of the latter after a flip-through and relief of the strain. Or you can say that in highly strained systems the central proton might become very acidic; so the anion is formed and this can flip through.

But we are troubled by the fact that when we start with what we know to be a *cis* compound, in the *cis*-dihydrogen case, we come out with what we know is the *trans*-dihydropyrene. We don't have any *cis*-dihydropyrene as yet.

R. West:

I wonder if you have taken either the metacyclophane-diene or the extremely interesting new triene compounds, and tried to react them, for instance, with sodium methoxide, because the anion radicals of the compounds would, of course, be expected to be of unusual interest.

156

Discussion

V. Boekelheide:

We obtain a very fine spectrum of the radical anion, but it appears to be a little more difficult to interpret — in the same way as with naphthalene; if you are not careful, you can wind up with a radical anion of *two* molecules. Much of what I said covers things that happened in the last three to five months.

R. H. Martin:

You probably have not had time to study the charge transfer complexes?

V. Boekelheide:

No.

R. H. Martin:

Just a final question. You have not tried to make the *cis* compound by making a bridge compound, and then opening that one bridge?

V. Boekelheide:

We have thought about it.

B. Binsch:

On the first slide you showed us the irreversible thermal rearrangement. Now, I think you ought to look at this as a 1,5- or may be as a 1,9- shift of the methyl group.

V. Boekelheide:

I think that I termed it a 1,5-sigmatropic rearrangement.

B. Binsch:

In either case, you would expect that the group migrates under retention of configuration. Is that correct?

V. Boekelheide:

Yes, and I would like to know how to make such a compound to test this hypothesis.

B. Binsch:

This is the question I wanted to ask you.

Non-Classical Sulphur Heterocycles
The Chemistry of Simple Thiepin Systems

by R.H. SCHLESSINGER

Department of Chemistry, University of Rochester, New York

I. INTRODUCTION

THE REQUIREMENTS for aromaticity in heterocyclic compounds which do not satisfy the Hückel rule are of considerable interest. In this regard numerous studies have been reported for azepins and oxepins [1]; however, little is known about the possible aromatic character of simple thiepins [2].

When we began our efforts in this field, only annulated and highly substituted derivatives of thiepin had been isolated [3–4]. All of these heterocycles showed a tendency to extrude sulphur, and progress towards the preparation of simple thiepins had been seriously hampered because thermal instability was found to increase sharply with decreasing substitution [5]. For example, the tri-substituted benzo[*b*]thiepin (I) extrudes sulphur only when heated [6–7], while an attempt to synthesize the parent benzo[*b*]thiepin (II) under fairly mild conditions resulted only in the formation of naphthalene [8].

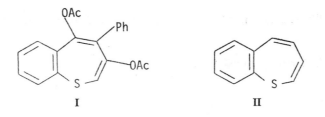

It is generally accepted that sulphur extrusion occurs by valence isomerization of the thiepin to the corresponding benzene sulphide, followed by irreversible loss of sulphur [9]. Evidence cited in favour of this mechanism was the observation that the benzo[*d*]thiepin-diacid (III) [10] undergoes loss of sulphur much more readily than its naphthalene homologue, IV [11]. Sulphur extrusion from III requires conversion of a benzene ring into the quinonoid intermediate V. It was argued that the increased stability of IV may be due to the additional energy necessary to convert two aromatic rings into the quinonoid intermediate VI [9].

This explanation suggested that it might be feasible to prepare a simple thiepin if the energy difference between the ground state of the heterocycle and the benzene sulphide intermediate were sufficiently great. On this basis, we decided to attempt the preparation

158

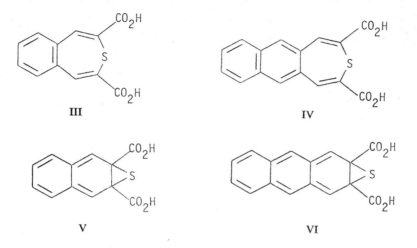

of thieno [3,4-*d*] thiepin (VII).* It was felt that thiepin VII might be stable enough to isolate, since simple molecular orbital calculations showed that considerable ground-state stabilization could be gained from charge-separated species, such as VIIa or VIIb.** Furthermore, some independent chemical evidence was available that indicated that sulphur extrusion from VII might be a high-energy process because of the tetravalent sulphur atom present in the quinonoid portion of the benzene sulphide-like intermediate VIII.***

II. Discussion

The preparation of VII (Scheme 1) has been realized, and the heterocycle was found to be remarkably stable [14]. In addition, oxidation of VII has afforded the thiepin sulphoxide IX, as well as the thiepin sulphone X [15].

* The dihydro analogue of VII, first prepared by Eglinton, Lardy, Raphael and Sims [12], was used as the starting material in this synthesis. It is interesting to note that these authors unsuccessfully treated the dihydro compound with N-bromosuccinimide for 72 hrs, in the hope of isolating benzo [*c*] thiophene (XI). Presumably, XI would have been formed in this reaction by extrusion of sulphur from the thiepin VII.

** Originally, only the 'azulene-like' charge-separated form VIIa was considered. However, simple Hückel calculations, done by Prof. L. Friedrich of this Department, indicate that both charge-separated species could contribute to stabilization.

*** For examples of high-energy tetravalent sulphur species, see Ponticello and Schlessinger [13] and references cited therein.

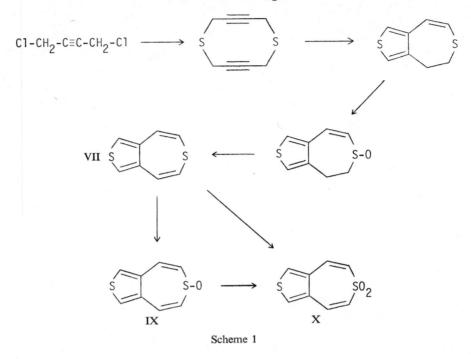

$Cl-CH_2-C\equiv C-CH_2-Cl \longrightarrow$

VII

IX   X

Scheme 1

Several interesting physical and chemical properties have been found for these thiepins [15]. The mass spectra (20 and 75 eV) of compounds VII, IX and X all gave rise to a peak at m/e 134, the molecular ion of benzo[c]thiophene (XI), in addition to parent ions at m/e 166, 182 and 198, respectively. Metastable (M^*) peaks, representing the loss of S, SO and SO_2, were observed in these spectra. In all three cases, an increase in the 134 peak, accompanied by the disappearance of the parent peak, occurred upon raising the source temperature.

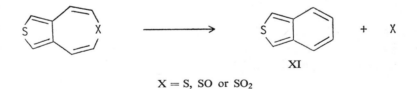

XI

$X = S$, SO or SO_2

The bright yellow thiepin VII showed maxima in the ultraviolet at $\lambda_{max}^{CH_3OH}$ 251 nm (ε 25,700) and 260 nm (ε 25,700), accompanied by well-defined absorption as far out as 390 nm. The ultraviolet spectrum of sulphone X gave an absorption at $\lambda_{max}^{CH_3OH}$ 246 nm (ε 36,800), with tailing to 320 nm, and a similar spectrum was also observed for sulphoxide IX, with a $\lambda_{max}^{CH_3OH}$ value of 244 nm (ε 22,400) and tailing to 330 nm.
The NMR values found for compounds VII, IX and X showed an interesting trend when compared to their dihydro analogues. Sulphone X showed proton resonance at field values between 0.25 and 0.55 ppm lower than sulphone XII. Sulphoxides IX and XIII gave resonance at very similar values. In marked contrast, thiepin VII gave proton resonance at field values between 0.25 and 0.70 ppm higher than its dihydro analogue, sulphide XIV. It is possible that this up-field shift may be due to the presence of a paramagnetic ring

160

current (12 π electrons) in VII. This phenomenon has been observed for a number of hydrocarbons containing $4n$ π electrons, but not for heterocyclic systems of this type.*

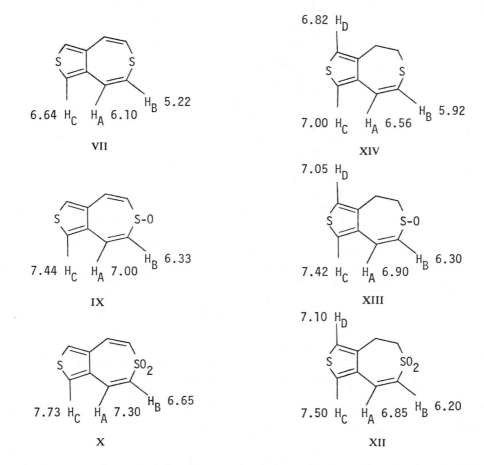

In order to ascertain more fully the nature of the π-electron delocalization in VII, an X-ray determination of the molecular geometry of the thiepin, along with the thiepin sulphone X, was undertaken [21]. Fig. 1 is a projection of the thiepin sulphone down the b axis, showing the bond distances and angles of X as determined from this analysis. Estimated standard deviations are: C–C = 0.006Å, C–S = 0.005Å, S–O = 0.003Å, and 0.4° for angles. These results were obtained from co-ordinate standard deviations, estimated from inverse elements of the inverse matrix after block diagonal anisotropic least-squares refinement. The seven-membered ring of X exists in the boat conformation in much the same fashion as in thiepin 1,1-dioxide [22]. The dihedral angles describing the ring puckering of sulphone X are 45.2° and 19.8°.

---

* A theoretical basis for the presence of a paramagnetic ring current in $4n$ π-electron monocyclic systems has been given by Pople and Untch [16]. For some examples of this phenomenon in hydrocarbons, see [17–20].

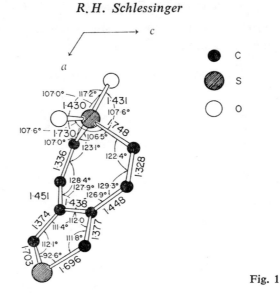

Fig. 1

The structure of thiepin VII was found to be disordered in the crystal [21]. Fig. 2 shows a portion of the structure as viewed down the *c* axis. The structure consists of thiepin molecules, which randomly occupy either site in the crystal lattice. This results in the centre of symmetry shown by the asterisk. Very similar disorder exists in the azulene crystal structure [23], whose unit cell constants and fractional atomic co-ordinates closely resemble this thiepin structure. Least-squares planes have been calculated for the thiepin ring, and these show the molecule to be almost planar. The two sulphur atoms of VII deviate significantly from the molecular least-squares plane (0.06Å), and this gives the molecule a slight boat shape. Similar, but smaller, deviations from planarity have been found in azulene.* Although the crystal disorder limits the accuracy of this analysis, the transannular carbon–carbon bond distance has been determined with reasonable accuracy and appears to be predominantly single-bonded in character (1.46 ± 0.02Å).**

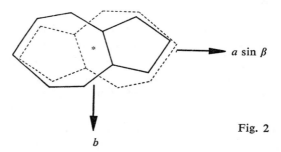

Fig. 2

Sulphide XIV, along with the sulphoxides IX and XIII, was found to undergo catalytic hydrogenation (palladium on carbon in methanol), with rapid and quantitative uptake of hydrogen, to give the tetrahydrothiepin XV in high yield. The tetrahydrosulphone XVI was obtained in the same manner from compounds X and XII. Thiepin VII was found to be completely inert to a variety of catalytic hydrogenation conditions.

* Deviations of carbon atoms in analogous positions in azulene are 0.0088Å [24].
** In azulene, this bond length is 1.483Å, after full-matrix least-squares refinement.

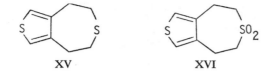

All three thiepins (VII, IX and X) undergo the Diels-Alder reaction with N-phenyl-maleimide to give either the *exo* adduct XVII, or mixtures of XVII and its corresponding *endo* isomer XVIII.* It has been shown that XVII and XVIII arise by the initial combination of the thiepins with a dienophile to give adducts XIX and XX (X = S, SO or SO₂), followed by extrusion of sulphur, sulphur monoxide or sulphur dioxide from these intermediate adducts, to give the final adducts XVII and XVIII. Evidence favouring this reaction pathway was obtained by studying, at a variety of temperatures, the Diels-Alder behaviour of thiepins VII, IX and X, along with the thiophene XI.

Using the dienophile as the solvent, at 90° and 150°, X gave in 85% yield a 1:1 mixture of XVII and XVIII. This ratio changed to 4:1, with the *exo* adduct predominating at 240°. At 90°, IX gave both adducts in equal amounts, while a 3:1 ratio in favour of the *exo* adduct was observed at 150°. Thiepin VII gave only the *exo* isomer at 150°. However, at 90° and 120°, VII gave rise to a mixture of the adducts, the *exo* form again predominating in a ratio of 3:1. Thiophene XI has been found to give approximately equal amounts of XVII and XVIII at temperatures ranging from 90° to 240°. Prolonged heating of 1:1 mixtures of the adducts at temperatures of 90°, 120°, 150° and 240° did not significantly change their composition. Additionally, VII, IX and X have been shown to be thermally stable at temperatures sufficient to effect their combination with dienophile. Interestingly, both IX and X undergo the Diels-Alder reaction with N-phenylmaleimide at a rate faster than that found for thiepin VII.

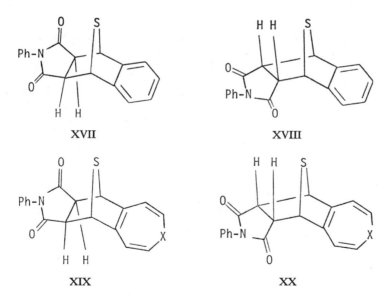

* These adducts have been prepared from benzo[c]thiophene and N-phenylmaleimide by Cava and Pollack [25].

III. Conclusions

It is evident from the foregoing physical and chemical data that thiepin VII is an extensively delocalized $4n$ π-electron system, in which charge-separated resonance forms contribute significantly to the ground-state electronic make-up of the heterocycle.

Acknowledgements

Grateful thanks to my collaborator Dr G. S. Ponticello, who carried out the experimental portion of this work, and to Prof. H. M. Sobell and Dr T. D. Sakore for the X-ray determinations. This work was partially supported by a Fredrick Gardner Cottrell grant from the Research Corporation and by the Petroleum Research Fund (Grant No. 756 – G and 4767 – AC 1).

References

1 E. Vogel & H. Gunther (1967) *Angew. Chem. (Int. Edn)*, 6 : 385.

2 R. Zahradnik (1965) in : *Advances in Heterocyclic Chemistry*, V (ed. A. R. Katritzky), Academic Press, New York, p. 21.

3 K. Dimroth & G. Lenke (1965) *Chem. Ber.*, 89 : 2608.

4 F. Dallacker, K. W. Glombitza & M. Lipp (1961) *Ann.*, 643 : 82.

5 J. D. Loudon (1961) in : *Organic Sulfur Compounds*, I (ed. N. Kharasch), Pergamon, New York, p. 299.

6 H. Hofmann & H. Westernacher (1967) *Angew. Chem. (Int. Edn)*, 6 : 255.

7 *Ibid.* (1965) 4 : 872.

8 V. J. Traynelis & J. R. Livingston (1964) *J. Org. Chem.*, 29 : 1092.

9 B. P. Stark & A. J. Duke (1967) *Extrusion Reactions*, Pergamon, New York, p. 91.

10 G. P. Scott (1953) *J. Am. Chem. Soc.*, 75 : 6332.

11 J. D. Loudon & A. D. B. Sloan (1962) *J. Chem. Soc.*, p. 3262.

12 G. Eglinton, I. A. Lardy, R. A. Raphael & G. A. Sims (1964) *ibid.*, p. 1154.

13 I. S. Ponticello & R. H. Schlessinger (1968) *J. Am. Chem. Soc.*, 90 : 4190.

14 R. H. Schlessinger & G. S. Ponticello (1967) *ibid.*, 89 : 7138.

15 Idem (1968) *Tetrahedron Lett.*, p. 3017.

16 J. A. Pople & K. G. Untch (1966) *J. Am. Chem. Soc.*, 88 : 4811.

17 I. C. Calder & F. Sondheimer (1966) *Chem. Comm.*, p. 904.

18 G. Schroeder & J. F. M. Oth (1966) *Tetrahedron Lett.*, p. 4083.

19 B. M. Trost & G. M. Bright (1967) *J. Am. Chem. Soc.*, 89 : 4244.

20 K. G. Untch & D. C. Wysocki (1967) *ibid.*, 89 : 6386.

21 T. D. Sakore, R. H. Schlessinger & H. M. Sobell (1969) *ibid.*, 91 : 3995.

22 H. L. Ammon, P. H. Watts, J. M. Stewart & W. L. Mock (1968) *ibid.*, 90 : 4501.

23 J. M. Robertson, H. M. M. Shearer, G. A. Sim & D. G. Watson (1962) *Acta Crystallogr.*, 15 : 1.

24 A. W. Hanson (1966) *ibid.*, 19 : 19.

25 M. P. Cava & N. M. Pollack (1966) *J. Am. Chem. Soc.*, 88 : 4112.

Discussion

E. D. Bergmann:

In connection with the seven-membered ring sulfone you described, it might be of interest to record that 2, 3, 6, 7-dibenzothiepin also gives a sulfone easily upon oxidation (E. D. Bergmann & M. Rabinovitz, 1960, *J. Org. Chem.*, 25:828).

Secondly, I would like to ask whether the type of 'dimer' which exists in the crystal of your compound persists also in solution.

Thirdly, are you not surprised at the similarity—in the solid state—between azulene and your thieno-thiepin? If one follows the principles of isosterism, this should not be so.

R. H. Schlessinger:

Quite honestly, we did not expect it. Incidentally, I should acknowledge two very important people. These X-ray structures were done only in part by us—they were done by colleagues of mine on the Faculty of the University of Rochester.

Obviously, our compound is not iso-electronic with azulene. Hückel calculations indicate that charge separation may in fact exist in the molecule to a limited extent.

Regarding the dimer, it is formed using one of the double bonds in the seven-membered ring. This compound will form complexes, but it does not form a complex with azulene, as we had hoped. Azulene is known to complex with a variety of compounds and still show disorder; thus, we do not know whether the solid is randomly disordered with respect to every other molecule, or exactly the type of disorder that exists.

H. Prinzbach:

Could you isolate the second adduct? Because, if this compound is sufficiently stable, it would be an elegant starting material for just enlarging one ring to a seven-membered ring (dithiaheptalene). We had a paper, together with Tochtermann from Heidelberg, showing exactly the same system, only with two additional benzene rings.

R. H. Schlessinger:

You are perfectly correct. We have had this compound in our hands and we have already started X-ray analysis.

J. F. Labarre:

Have you done some calculations of the S=O electronic structure in the sulphoxide and the sulphone you got? The actual structure of such a bond is still a tricky problem, and I suggest that you might use the new $\sigma + \pi$ treatment we have proposed, which will be published in the *Journal de Chimie Physique* by P. Castan, P. Dagnac-Amans and myself. It will be the first time, since the famous Moffitt paper, that such an approach will be attempted.

R. H. Schlessinger:

I am not sure, although I profess ignorance—we really have not looked.

B. Pullman:

I would like to add that S=O bonds have presented a serious question in quantum chemistry. There are no recent calculations on the subject, although many years ago Moffitt did some work on this, but that is old work—which does not mean it is wrong.

165

Mesoionic 5-Oxazolones

by G.V. BOYD

Department of Chemistry, Chelsea College of Science and Technology, London

MESOIONIC COMPOUNDS, because of their potential aromaticity and manifold mode of reaction, have aroused the interest of chemists for many years [1], and it is appropriate that they are mentioned at this Symposium. Our contribution concerns the oxazolium 5-oxides I, which, together with the sydnones II [2], isosydnones III [3–5] and oxatriazolones IV [6–8], form a series of mesoionic systems containing the O–C=O unit (see Fig. 1).

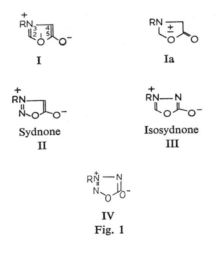

I

Ia

Sydnone
II

Isosydnone
III

IV
Fig. 1

The first oxazolium oxide appears to have been the acetyl derivative V prepared in the course of the Anglo-American co-operative research effort on penicillin during the 1939–1945 war [9]. In 1958 Lawson and Miles [10–11] described the condensed compound VI and recognized its mesoionic character (see Fig. 2).

Huisgen and his colleagues in Munich studied the 'münchnone' VII, the simplest compound of this type yet made, and demonstrated that it functioned as a dipole VIIa in cycloadditions, and that it reacted as the valence-tautomeric ketene VIIb in the formation of β-lactams when heated with Schiff's bases [12–16]. At a previous symposium on aromaticity [17], held at Sheffield four years ago, Professor Huisgen gave us a fascinating account of cycloaddition reactions of compound VII and other münchnones. In 1964 Singh and Singh [18] discovered that stable trifluoroacetyloxazolones (VIII) are formed when *N*-

166

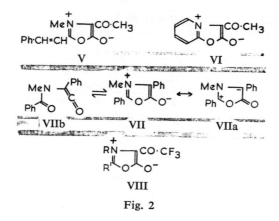

V   VI

VIIb   VII   VIIa

VIII

Fig. 2

substituted α-acylamino acids are treated with trifluoroacetic anhydride. We were able to generate a number of simple mesoionic oxazolones and, since cycloaddition reactions are being explored so thoroughly at Munich, we have concentrated our efforts on studying the stability of these compounds and their behaviour towards electrophilic and nucleophilic reagents [19–20].

Our interest in oxazolium oxides originated in the observation [21] that α-acylamino acids, on treatment with acetic anhydride and perchloric acid, are transformed into 5-oxazolonium perchlorates (e.g., IX), which absorb at the remarkably high frequency of 1,880–1,890 cm^{-1}. The salts are readily converted into azlactones.

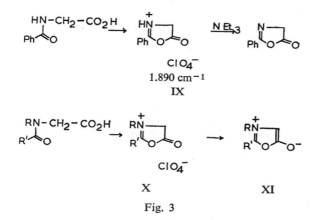

ClO_4^-
1.890 cm^{-1}
IX

X   XI

Fig. 3

When the nitrogen atom of the original amino acid carries an alkyl or aryl substituent, the resulting salts X on deprotonation yield mesoionic oxazolones XI. 2-Pyridone-1-acetic acid (XII) similarly gives the perchlorate XIII and thence the bicyclic base XIV.

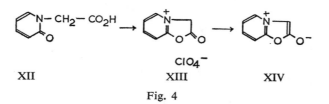

XII   XIII   XIV

ClO_4^-

Fig. 4

167

G. V. Boyd

The oxazolones, unsubstituted at C_4, generated in this way are much less stable than the corresponding sydnones and, indeed, cannot be isolated. They can, however, be studied in a freshly prepared solution. If the solutions are kept for longer than a few minutes, dimeric decomposition products are formed which are discussed below.

Electrophilic substitution reactions at C_4 are successful if the products contain strongly electron-withdrawing groups which stabilize the mesoionic system by delocalizing the negative charge. Examples are the uncatalysed acylations of the methylphenyloxazolium oxide XV by trifluoroacetic anhydride and p-nitrobenzoyl chloride (see Fig. 5).

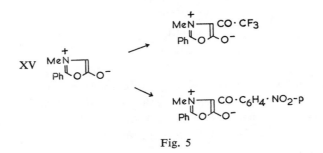

Fig. 5

Even very labile oxazolones can be trapped as the trifluoroacetyl derivatives when deprotonation is carried out in the presence of the acylating agent (see, e.g., Fig. 6).

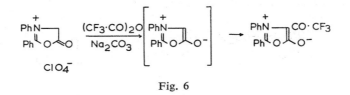

Fig. 6

The condensed oxide XIV readily yielded a number of alkyl and aryl ketones and could also be coupled with diazonium salts. However, stable azo-derivatives result only if the product contains a nitro group in the ortho or para position (see Fig. 7).

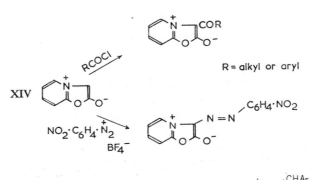

Fig. 7

168

The condensation of oxazolonium perchlorates with aromatic aldehydes to yield coloured arylidene derivatives XVI must involve the mesoionic oxazolones as reactive intermediates. We have not yet observed a substitution reaction at C_2 of the mesoionic system: attempts to introduce a trifluoroacetyl group into compound XVII failed. The proton, the simplest electrophile, invariably attaches itself at C_4, yielding the cations XVIII. The NMR spectra of the salts XIX, in which conjugation with the phenyl group would favour the 2(H)-form XIXa, show that they exist solely as the 4(H)-tautomers in solution. HMO calculations of electrophilic localization energies are in agreement with the observed electrophilic attack at C_4 (see Figs. 8–9).

XVII

Fig. 8

XVIII

XIX , NOT XIXa

Fig. 9

The mesoionic oxazolones are highly susceptible to attack by nucleophilic reagents which open the ring with the formation of acylamino acids or their derivatives. It is conceivable that these reactions proceed *via* the acylaminoketenes; but we have no evidence for their intermediacy (Fig. 10).

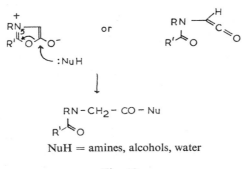

NuH = amines, alcohols, water

Fig. 10

Electrophilic substitution, followed by nucleophilic ring opening, occurs when the condensed oxazolone XIV is treated with salicylaldehyde and its derivatives: coumarins (e.g.,

XXI) are produced in high yield. The intermediate oxide XX can be isolated as its conjugate acid XXII when the reaction is performed in the presence of perchloric acid (see Fig. 11).

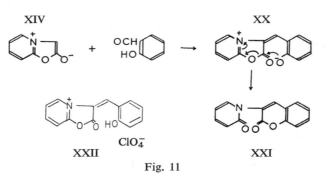

Fig. 11

Cycloaddition of dimethyl acetylenedicarboxylate to the dipolar system embodied in the mesoionic oxazolone structure proceeds at room temperature and gives pyrroles XXIII in almost quantitative yield. This reaction offers a ready means for trapping unstable compounds of this type; the sole failure was encountered with the pyridine oxazolone XIV, presumably because formation of the intermediate XXIV would involve destruction of the aromatic pyridinium structure. Instead, this compound prefers to dimerize (see Fig. 12).

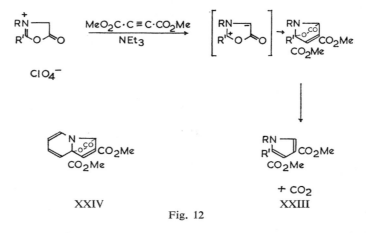

Fig. 12

When the oxazolium oxides decompose, they preserve their type and stabilized acyl derivatives XXV are formed, whose structures were established by infra-red, NMR and mass spectroscopy. These dimers arise by a substitution process, in which one molecule acts as a nucleophile and the other as the electrophile (see Fig. 13).

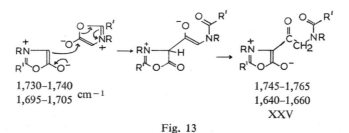

Fig. 13

170

The dimerizations were followed spectroscopically and were found to obey second-order kinetics. The relative rate constants show that fusion to a pyridine ring, electron donation to the nitrogen atom and the presence of an aryl group at C_2 slow down this process. The same order is observed in calculated electrophilic localization energies at C_4 and in nucleophilic localization energies at C_5, so that an oxazolone decomposes quickly because it is both a reactive nucleophile and electrophile (see Fig. 14).

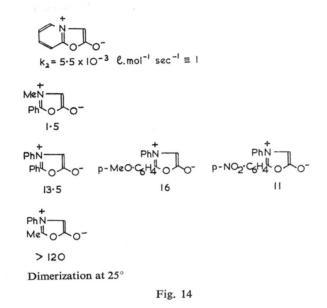

$k_2 = 5 \cdot 5 \times 10^{-3}$ $\ell.mol^{-1}$ $sec^{-1} \equiv 1$

1·5

13·5   16   11

> 120

Dimerization at 25°

Fig. 14

We have little to contribute to a discussion of the aromaticity of the oxazolium oxides. Their infra-red spectra, which exhibit oxazolone absorption at 1,730 to 1,740 cm^{-1}— a greatly reduced frequency from saturated azlactones (1,830 cm^{-1}), their salts (1,890 cm^{-1}) and unsaturated azlactones (1,790 cm^{-1})—indicate some negative charge on the exocyclic oxygen atom. In the stable azo and acyl compounds this band appears at higher frequencies (1,750 to 1,800 cm^{-1}), and the bulk of the negative charge seems to reside on the substituent group, since no carbonyl band above 1,600 cm^{-1} can be seen in the spectra of the keto derivatives.

ACKNOWLEDGEMENT

Much of this work was done by my collaborator, Peter Wright; it is a pleasure to acknowledge his contribution.

REFERENCES

1 W. Baker & W. D. Ollis (1957) *Q. Rev.*, 11 : 15.

2 F. H. C. Stewart (1964) *Chem. Rev.*, 64 : 129.

3 M. Hashimoto & M. Ohta (1961) *Bull. Chem. Soc. Japan*, 34 : 668.

4 C. Ainsworth (1965) *Can. J. Chem.*, 43 : 1607.

5 W. D. Ollis et al. (1969) *J. Chem. Soc.* (B), p. 1185.

6 G. Ponzio (1933) *Gazzetta*, 63 : 471.

7 J. H. Boyer & J. A. Hernandez (1956) *J. Am. Chem. Soc.*, 78 : 5124.

8 W. V. Farrar (1964) *J. Chem. Soc.*, p. 906.

9 J. W. Cornforth (1949) in : *The Chemistry of Penicillin* (ed. H. T. Clarke), Princeton University Press, Princeton, pp. 758, 830.

10 A. Lawson & D. H. Miles (1958) *Chemy Ind.*, p. 461.

11 Idem (1959) *J. Chem. Soc.*, p. 2865.

12 R. Huisgen, H. Gotthardt & H. O. Bayer (1964) *Angew. Chem.*, 76 : 185.

171

13 R. Huisgen, H. Gotthardt & H. O. Bayer
    (1964) *Tetrahedron Lett.*, p. 481.
14 H. Gotthardt, R. Huisgen & F. C. Schaefer,
    ibid., p. 487.
15 R. Huisgen & E. Funke (1967) *Angew. Chem.*,
    79 : 320.
16 *Ibid.*, p. 321.

17 R. Huisgen (1967) in : *Aromaticity* (*Chem. Soc.
    Special Publ.*, *No. 21*), p. 51.
18 G. Singh & S. Singh (1964) *Tetrahedron Lett.*,
    p. 3789.
19 G. V. Boyd & P. H. Wright (1969) *Chem. Comm.*,
    p. 182.
20 Idem (1970) *J. Chem. Soc.* (C), p. 1485.
21 G. V. Boyd (1968) *Chem. Comm.*, p. 1410.

Discussion

L. Friedman (Ohio):

You mentioned that the compounds can be handled in solution and that the bicyclic mesoionic compound has a reasonably long half-life. Have you studied the thermal, unimolecular decomposition?

G. V. Boyd:

I do not think the compounds would decompose in this manner.

E. D. Bergmann:

I would like to remark that your oxazolones exhibit a great similarity to the fulvenes; the latter also have some zwitterionic character and are given to substitution reactions.

All-Electron Calculations on Benzene Isomers

by G. BERTHIER,* A. Y. MEYER** *and* L. PRAUD*

*Laboratoire de Chimie Quantique de la Faculté des Sciences, Paris,**
*and Department of Organic Chemistry, the Hebrew University of Jerusalem***

I. INTRODUCTION

WITH THE present computing facilities, it is possible to perform non-empirical calculations including all the electrons for an organic compound of medium size, i.e., a molecule containing about six carbon atoms. In the first section of this report we will describe the results of calculations of this type for the planar conjugated C_6H_6 molecules: benzene (I), fulvene (II), dimethylenecyclobutene (III) and trimethylenecyclopropane (IV). In the second section we will present the results of similar calculations for the non-planar valence isomers of benzene: bicyclo[2.2.0]hexa-2,5-diene ('Dewar benzene', V), benzvalene (VI) and prismane (VII).

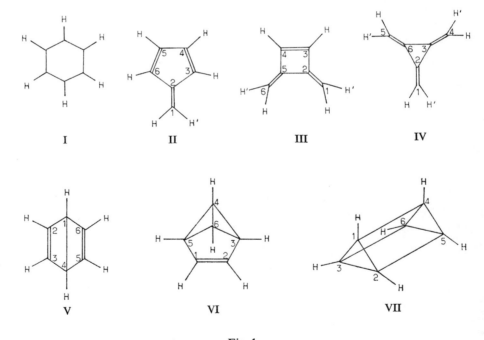

Fig. 1
The benzene isomers

174

As a first step we computed the molecular orbitals and energy levels of σ and π electrons of the benzene isomers, using the general program IBMOL [1]; the molecular orbitals were expanded in terms of Gaussian atomic orbitals, whose coefficients were determined by the LCAO-MO-SCF method [2]. As in previous work on heterocycles [3–4], the original atomic basis set contained seven s functions and 3×3 p functions for carbon, and three s functions for hydrogen. However, the number of independent functions involved in the variational process was reduced by the usual contraction technique to two s and three p functions for each carbon, and to one s function for each hydrogen, giving a basis set more or less equivalent to a Slater minimal basis set of $1s$ $2s$ $2p_x$ $2p_y$ $2p_z$. For the planar isomers, the interatomic distances chosen for the SCF calculation were determined experimentally [5–7]; for the non-planar isomers, they were deduced from the geometry of related compounds, supplemented by theoretical considerations (see Sect. III).

We next estimated, through a perturbation calculation, the effect of the configuration interaction on the relative stability of benzene isomers. According to the Epstein-Nesbet perturbation theory [8–9], the second-order correction, $E^{(2)}$, to the total SCF energy, $E_{SCF}^{(0)}$, is given by the expression

$$E^{(2)} = - \sum_k \frac{\left| \int \Psi_k^* H_0 \Psi_0 \right|^2}{E_k - E_0} d\tau,$$

where the summation is taken over the Hamiltonian matrix elements coming from all the diexcited wave functions Ψ_k with respect to the SCF ground state Ψ_0. For a C_6H_6 compound, a minimal basis set yields 27,720 terms, to be included in the sum over k. In order to simplify the analysis of the various contributions to $E^{(2)}$, it is convenient to replace the usual SCF molecular orbitals by quasi-localized equivalent orbitals. The latter were constructed from the occupied and virtual SCF molecular orbitals by means of unitary transformations satisfying the Boys criterion [10]. The total correction, $E^{(2)}$, gives a second-order approximation for the correlation energy contained in a set of 36 basis atomic orbitals.

II. PLANAR CONJUGATED ISOMERS OF BENZENE

The comparative study of the cross-conjugated benzene isomers is a good procedure for testing the validity of the various methods of quantum chemistry, because of the very particular properties of these compounds as compared with usual aromatic molecules. For example, the electronic structure of fulvene has often been recalculated in order to see whether the non-alternant character predicted by the Hückel method is retained in a more elaborate theory. The interest in all-electron calculations lies in the fact that they enable us to discuss the σ–π interaction, a problem that cannot be solved by pure π-electron methods.

1. Charge Distribution

The charge distribution resulting from a population analysis of the SCF molecular orbitals is shown in Fig. 2.

The main features of these diagrams are the usual charge transfer C^- H^+ of σ electrons from hydrogens to carbons and the shift of π electrons inside or outside the ring. Even if the σ net charges are over-estimated by any LCAO-SCF-MO calculation without an optimization of orbital exponents, they are not in obvious discord with experiment for two rea-

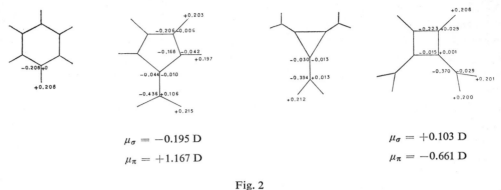

$$\mu_\sigma = -0.195 \text{ D}$$
$$\mu_\pi = +1.167 \text{ D}$$

$$\mu_\sigma = +0.103 \text{ D}$$
$$\mu_\pi = -0.661 \text{ D}$$

Fig. 2

σ Net charges and π net charges

sons: they are not directly connected with the polarity of the CH bonds, which are rather related to the dipole moment $C^+ H^-$ of the corresponding localized orbital [11–12]; they do not contribute appreciably to the total dipole moment, which is essentially determined in fulvene and dimethylene-cyclobutene by the π electrons. In fulvene, π charges are transferred from the methylene group to the ring, in conformity with the electron distribution predicted by the Hückel method for this type of non-alternant hydrocarbon. The resulting dipole moment (0.97 D) is not very different from that found by a non-empirical SCF calculation limited to π electrons (1.13 D) [13]. On the other hand, a π-electron transfer in the same direction is obtained for trimethylenecyclopropane, in disagreement with the predictions of the Hückel method for non-alternant hydrocarbons with triangular or heptagonal rings. This suggests that the dipole moments of methylenecyclopropene or heptafulvene might be reversed with respect to the results of the pure π-electron SCF calculation [14–15]. In dimethylenecyclobutene, π charges are transferred from the four-membered ring to the methylene groups, a result which cannot be obtained by the simple Hückel method.

It should also be stated that the total dipole moments calculated for dimethylenecyclobutene and fulvene by using the dipole length operator were in accordance with experiment until the analysis of their microwave spectrum was performed: Stark-effect measurements gave 0.62 D for the first compound [16] and 0.44 D for the second [17], at variance with earlier determinations. The fact that theory and experiment are in good agreement for dimethylenecyclobutene but not for fulvene is as yet unexplicable.

2. Energy Levels

The six lowest molecular orbitals, whose energies and main components are indicated in Table 1, are formed essentially from the K-shell orbitals of carbon atoms; their SCF orbital energies, e_i, can be identified with the energy needed to extract one $1s$ electron from carbon and can be correlated with X-ray excitation spectra.

Originally, the chemical shift of X-ray energy levels for a given atom in a series of molecules was attributed to the screening effect of the valence electrons on the inner shells, the lowest inner-shell energy being assigned to the atom with the most positive net charge [18]. Actually, in the series of benzene isomers, no correlation can be found between the $1s$ orbital energies and the total net charges given by the population analysis. More probably, the ionization

Table 1

Energies (in a.u.) and Localization of Inner-Shell Orbitals

	K_1	K_2	K_3	K_4	K_5	K_6
Benzene	-11.4225 ($C_1\ldots C_6$)	-11.4221 ($C_1\ldots C_6$)		-11.4210 ($C_1\ldots C_6$)		-11.4205 ($C_1\ldots C_6$)
Fulvene	-11.4390 (C_1)	-11.4330 (C_2)	-11.4059 ($C_4 C_5$)	-11.4052 ($C_4 C_5$)	-11.3978 ($C_3 C_6$)	-11.3978 ($C_3 C_6$)
Dimethylene-cyclobutene	-11.4395 ($C_2 C_5$)	-11.4389 ($C_2 C_5$)	-11.4276 ($C_3 C_4$)	-11.4262 ($C_3 C_4$)	-11.4021 ($C_1 C_6$)	-11.4021 ($C_1 C_6$)
Trimethylene-cyclopropane	-11.4626 ($C_2 C_3 C_6$)	-11.4614 ($C_2 C_3 C_6$)		-11.4256 ($C_1 C_4 C_5$)		-11.4255 ($C_1 C_4 C_5$)

energies of the inner-shell lone pairs are determined by the whole charge distribution [19], as in the case of valence-shell lone pairs [20].

The ordering of the highest occupied molecular orbitals is shown in Fig. 3. The π and σ degenerate energy levels of benzene are split in less symmetrical compounds, like fulvene and dimethylenecyclobutene, and the magnitude of the splitting seems to be due to the particular symmetry properties of each molecule rather than to their aromaticity.

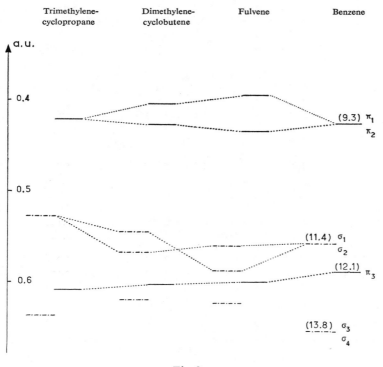

Fig. 3

Orbital energies (in eV) of the highest molecular orbitals

According to Koopmans' theorem, these energy levels should be related to successive ionization potentials, whose experimental values are indicated in brackets for benzene. The trend of the experimental results is reproduced, but the theoretical ionization energies are too high, as in the preceding LCAO-SCF-MO studies of aromatic molecules [4, 21–22]. For a more exact calculation, it is necessary to optimize the atomic basis set for the neutral molecule and the ion.

3. *Localization and Relative Stability*

If the electronic structure of a molecule is described by a set of SCF equivalent orbitals corresponding to the bonds and lone pairs of the chemical formula, each localized electron pair can be characterized by a total energy, which is obtained by adding the core and orbital energies of the molecular orbital under consideration and an appropriate part of the total nuclear repulsion energy [23]. The sum of the total bond and lone pair energies reproduces the SCF total energy exactly. The values obtained in this way are collected in Table 2.

Table 2
Localization of the Total Energy (in a.u.)

		Trimethylene-cyclopropane		Dimethylene-cyclobutene		Fulvene		Benzene	
K-shells of C	1		−29.314		−29.290		−29.325		−29.307
	2		−29.347		−29.325		−29.316		
	3		−29.347		−29.312		−29.283		
	4		−29.314		−29.312		−29.290		
	5		−29.314		−29.325		−29.290		
	6		−29.347		−29.290		−29.283		
σ Bonds C–H		1.1	−2.494	1.1	−2.483	1.1	−2.479	1.1	−2.230
		1.1′		1.1′	−2.436	1.1′	−2.479	2.2	
		4.4		3.3	−2.222	3.3	−2.001	3.3	
		4.4′		4.4	−2.222	4.4	−2.177	4.4	
		5.5′		6.6	−2.483	5.5	−2.177	5.5	
		5.5′		6.6′	−2.436	6.6	−2.001	6.6	
σ Bonds C–C		1.2	−5.038	1.2	−5.064	1.2	−5.076	1.2	−5.145
		2.3	−4.580	2.3	−4.855	2.3	−5.020	2.3	
		3.4	−5.038	3.4	−5.118	3.4	−5.182	3.4	
		3.6	−4.580	4.5	−4.855	4.5	−5.013	4.5	
		5.6	−5.038	5.6	−5.064	5.6	−5.182	5.6	
		6.2	−4.580	2.5	−4.869	6.2	−5.020	6.1	
π Bonds C=C	1.2		−3.238		−3.171		−3.278		−3.203
	3.4				−3.272		−3.179		
	5.6				−3.171		−3.179		
Total energy			−229.513		−229.576		−229.630		−229.701

The variation of the total energy in the series of the conjugated isomers of benzene is found to be due not to the π-electron system, but to the σ electrons, whose total energy decreases (in absolute value) from benzene to trimethylenecyclopropane. This overall-destabilization effect is due to a destabilization of the σ C–C bond system, partially balanced by the energy lowering due to the substitution of two C–H secondary bonds by one methylene group; the destabilization of the C–C bond can be ascribed to a strain effect.

Table 3

Relative Energies (in a.u. and Kcal/mole) of Conjugated Isomers of Benzene

	Total Energy		Energy with Respect to Benzene		
	$E^{(0)}_{SCF}$	$E^{(2)}$	$\Delta E^{(0)}_{SCF}$	$\Delta E^{(2)}$	Total
Benzene	−229.7007	−0.4370			
Fulvene	−229.6302	−0.4523	0.0705	−0.0153	0.0552
			(44.2)	(−9.6)	(34.6)
Dimethylene-cyclobutene	−229.5764	−0.4619	0.1243	−0.0249	0.0994
			(77.9)	(−15.6)	(62.3)
Trimethylene-cyclopropane	−229.5128	−0.4575	0.1879	−0.0205	0.1674
			(117.8)	(−12.8)	(105.0)

The preceding results are not appreciably altered if the second-order perturbation is taken into account, the correction term, $E^{(2)}$, being of the same order of magnitude for all the benzene isomers (Table 3). Actually, the largest contributions to $E^{(2)}$ come from diexcitations involving the same bond, namely, $\pi\pi \to \pi^*\pi^*$, $\pi\sigma \to \pi^*\sigma^*$ and CH CH $\to$ CH*CH* intrabond diexcitations; however, each of these is only a few per cent of the total second-order energy. It is gratifying to see that the theoretical predictions concerning the relative stability of fulvene with respect to benzene are improved after the second-order perturbation calculation: 34 Kcal/mole, as compared to 27 Kcal/mole for the experimental isomerization energy of dimethylfulvene [24].

III. NON-PLANAR VALENCE ISOMERS OF BENZENE

The newly aroused interest in non-planar benzene isomers [14–15, 26] concerns not only the preparation of Dewar benzene (V), benzvalene (VI), prismane (VII) and their derivatives, but also their reaction products, e.g., with hydrogen and complexing agents [27–28]. Although the problem of the electronic structure of these compounds has already attracted attention [29–31], no complete all-electron calculation has been reported as yet.

1. *Determination of the Geometry*

In such a case, the first difficulty encountered is the choice of the nuclear configuration for which the calculation should be performed. As an all-electron treatment for a 42-electron system is particularly time-consuming, it is not feasible to perform calculations for several nuclear configurations and look for the geometry that corresponds to the energy minimum. As a matter of fact, reasonable values can be estimated for the distances (C–C, C–H) between linked atoms and the angles (C′C, CC″) between the segments connecting adjacent atoms in a ring; the former were taken from related simpler compounds, the latter were chosen in conformity with the symmetry of the molecule (e.g., 60° in a cyclopropane-like ring), or were assumed. However, the angles between C–H bonds and other parts of the molecule were not determined by symmetry; they were made to depend on hybridization indices within the bent-bond model for small-ring compounds [32].

We began by an iterated all-valence electron treatment, calculating hybridization indices at each iteration and constructing a more plausible geometry for the next iteration. The program was an appropriately modified version of the CNDO/2 procedure, authored by Clark and Ragle and distributed by QCPE [33]. The molecular orbitals were calculated from Slater atomic orbitals (with $Z^*_{1s} = 1.2$ for hydrogen) and Pople-Segal parameters [34], then renormalized with overlap integrals and subjected to a Mulliken population analysis [35]. The C–C–H angles were obtained by iteration, starting with a set of assumed co-ordinates. From the expansion coefficients of the occupied molecular orbitals in terms of orthogonal basis function, Wiberg's bond indices, W_{kl}, i.e., the squared bond orders P^2_{kl} [36], were calculated. The amount of charge in the orbital $2s_C$ which is involved in a given C–H bond is identified with the bond index between the orbitals $2s_C$ and $1s_H$:

$$R_{CH} = W_{s_C-s_H} = p^2_{2s_C-1s_H}.$$

In a similar way, the amount of charge in the orbital $2s_C$, which is involved in a given C–C′ bond, is identified with the sum

$$R_{CC'} = W_{s_C-s_{C'}} + W_{s_C-x_{C'}} + W_{s_C-y_{C'}} + W_{s_C-z_{C'}},$$

where s_C, x_C, y_C and z_C are the orbitals $2s$ and $2p$ of carbon C′ [37–38]. The sum of the bond indices between any one orbital, k, and all the other orbitals in the molecule can be written as

$$\sum_{l \neq k} W_{kl} = 2p_{kk} - p^2_{kk}$$

(p_{kk} being simply the charge q_k of the orbital k), and is approximately equal to 1 for covalent bonds. Assuming that this sum is mainly formed by neighbouring-atom contributions, we have, approximately,

$$\sum_{k \in C} W_{k_C-s_H} = 1 \qquad (k = s, x, y, z);$$

$$\sum_{k \in C} \sum_{l \in C'} W_{k_C-l_{C'}} = 1 \qquad (k, l = s, x, y, z).$$

Consequently, the hybridization indices of carbons C in C–H and C–C′ bonds are

$$\lambda^2_{CH} = \frac{1 - R_{CH}}{R_{CC}};$$

$$\lambda^2_{CC} = \frac{1 - R_{CC}}{R_{CC}}.$$

These quantities enabled us to calculate the angle θ_{IJ} between hybrid orbitals emanating from carbon C and pointing towards its neighbours I and J:

$$\theta_{IJ} = \text{Arc cos}\left(-\frac{1}{\lambda_{CI}\lambda_{CJ}}\right).$$

The angles θ_{IJ} were temporarily considered as the angles between the segments joining the nuclei CI and CJ, and were used to construct an improved geometry, which allowed a new calculation to be performed. In practice, the process converges rapidly (e.g., three iterations were sufficient to fix the direction of the C–H bonds in prismane).

In prismane, the rings were assumed to be regular and the bond lengths to be equal to the standard values: $r_{C-C} = 1.537$ Å, $r_{C-H} = 1.086$ Å. As for the bonds between the two three-membered rings, it may be remarked that a bent bond, like C_1C_4, which is formed by the overlap of two hybrid orbitals centred on carbons C_1 and C_4, can be considered either in relation to C_1C_2 or to C_1C_3; in the former case, it is expected to make an angle, θ, with the segment C_1C_4 in the plane $C_1C_2C_5C_4$, while in the latter it makes the same angle in the $C_1C_3C_6C_4$ plane. Of course, symmetry requires that the bent bond C_1C_4 lies in still another plane, namely, the plane $C_1A_1A_2C_2$ passing through the middle (A_1A_2) of the segments C_2C_3 and C_5C_6. Its angle φ with the segment C_1C_4 is related to the interplanar angle α (60°) by

$$\cos^2 \varphi = (1 + tg^2\theta \cos^2 \alpha)^{-1}.$$

The exact value of φ is not merely a small problem of electronic structure, but is also of geometrical significance, because it relates the position in space of the CH segments with the angles θ_{IJ}, defined in terms of hybridization indices. All-valence electron calculations lead to the values $\theta \sim 46°$, $\varphi \sim 27°$, that is $\sim 131°$ for the angles (C_4C_1, C_1H_1) and (C_2C_1, C_1H_1). The hybridization of carbon at the apices of prismane is found to be $sp^{2.02}$ (C_1H_1), $sp^{2.72}$ (C_1C_4) and $sp^{4.44}$ $(C_1C_2$ or $C_1C_3)$, to be compared to sp^2 and sp^5 for the C–H and C–C bonds, respectively, of cyclopropane [39]. At first sight, it can be said that carbon C_4 takes the place of one of the hydrogens fixed to carbon C_1 in cyclopropane. However, prismane cannot be considered as 'two-bridged cyclopropanes', but rather as 'three-bridged ethanes', according to the population analysis results for junction bonds.

For benzvalene, it was assumed that the molecule has two perpendicular symmetry planes, one containing the ethylenic bond and its neighbouring groups, C_3–H_3 and C_5–H_5, the other passing through the out-of-plane groups, C_4–H_4 and C_6–H_6. The following bond lengths have been chosen: $r_{C=C} = 1.345$ Å, $r_{C-C} = 1.537$ Å, $r_{C-H} = 1.086$ Å, and the dihedral angle of the two cyclopropane rings (i.e., the angle C_3AC_5, where A is the middle of the segment C_4C_6) has been taken as equal to 120°. This fixes the angles around the double bond: a value of about 108° is calculated for the angles (C_1C_2, C_2C_3) or (C_5C_1, C_1C_2), which are supposed to be bisected by the bonds C_1–H_1 or C_2–H_2. Iterated all-valence electron calculations were carried out for other C–H bonds; it was found that (C_2C_3, C_3H_3) $\sim 126°$, $(C_6C_4, C_4H_4) \sim 99°$ — or, equivalently, $(C_6H_4, C_4H_4) \sim 100°$. The computed hybridization parameters suggest that the three-membered rings retain cyclopropane characteristics to a certain extent. Thus, hybridization at carbons C_3C_5 is $sp^{2.12}$ (C_3H_3), $sp^{2.61}$ (C_3C_2) and $sp^{4.39}$ $(C_3C_4$ or $C_3C_6)$, but the situation at carbons C_4C_6 is different: $sp^{2.92}$ (C_4H_4), $sp^{3.10}$ $(C_4H_3$ or $C_4C_5)$ and p (C_4C_6), probably because of the vertical positions of C_4 and C_6.

In the case of Dewar benzene, it was assumed that the dihedral angle of the two four-membered rings is equal to 120°. The following set of geometrical data has been used: $r_{C=C} = 1.345$ Å, $r_{C-C} = 1.48$ Å, $r_{C-H} = 1.086$ Å, $(C_1C_2, C_2C_3) \sim 95°$, and the hydrogens of the double bond were symmetrically arranged. The bonds C_1–H_1 and C_4–H_4 lie in the symmetry plane passing through C_1C_4, and are found, according to the hybridization of carbon atoms with this geometry, almost perpendicular ($\sim 92°$) to the segment C_1C_4.

The nuclear coordinates actually used as input data in our all-electron calculations are indicated in the appendix.

2. Charge Distribution

The charge distributions of the C_6H_6 isomers are compared in Fig. 4.

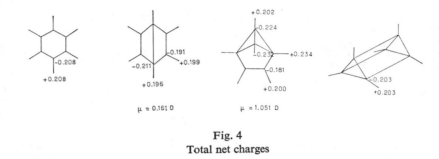

Fig. 4
Total net charges

The general picture of the molecule, as concerns the charge transfer C^-H^+, is nearly the same in the various isomers. Dewar benzene seems to be slightly less polar than benzene, benzvalene slightly more polar than benzene, and prismane very much like benzene. Actually, a very small dipole moment (0.16 D) was found for Dewar benzene; it is perpendicular to the plane of the two C=C bonds and to the edge C_1C_4 of the molecule and has its negative end towards the double bonds. A much larger charge transfer (1.05 D) was found for benzvalene; the dipole moment lies in the plane of the double bond along the symmetry axis, passing through the projection A of carbons C_4C_6, and has its negative end towards the double bond. On symmetry grounds, the mechanism giving rise to the dipole moment of benzvalene can be identified as a hyperconjugation effect of the out-of-plane methylene groups. The value of 1.05 D is not too different from the experimental dipole moment of related monocycles, such as cyclopentene (0.98 D) or cyclohexene (0.76 D) [40]. However, the present theoretical results are somewhat open to criticism, because the charge transfers in bonds $C_1–H_1$ (or $C_2–H_2$) and $C_4–H_4$ (or $C_6–H_6$) could more or less balance each other, according to the geometry assumed for benzvalene. The same reservations have to be made for Dewar benzene.

3. Energy Levels

Table 4 gives the energy and localization of the various K-shell orbitals, and Fig. 5 shows the ordering of the highest occupied energy levels.

Table 4
Energies (in a.u.) and Localization of Inner-Shell Orbitals

	K_1	K_2	K_3	K_4	K_5	K_6
Benzene	-11.4225 ($C_1\ldots C_6$)	-11.4221 ($C_1\ldots C_6$)		-11.4210 ($C_1\ldots C_6$)	-11.4205 ($C_1\ldots C_6$)	
Dewar benzene	-11.4165 ($C_2 C_3 C_5 C_6$)	-11.4165 ($C_2 C_3 C_5 C_6$)	-11.4151 ($C_2 C_3 C_5 C_6$)	-11.4151 ($C_2 C_3 C_5 C_6$)	-11.4035 ($C_1 C_4$)	-11.4028 ($C_1 C_4$)
Benzvalene	-11.4465 ($C_3 C_5$)	-11.4465 ($C_3 C_5$)	-11.4055 ($C_4 C_6$)	-11.4050 ($C_4 C_6$)	-11.3985 ($C_1 C_2$)	-11.3970 ($C_1 C_2$)
Prismane	-11.4079 ($C_1\ldots C_6$)	-11.4075 ($C_1\ldots C_6$)	-11.4072 ($C_1\ldots C_6$)		-11.4067 ($C_1\ldots C_6$)	

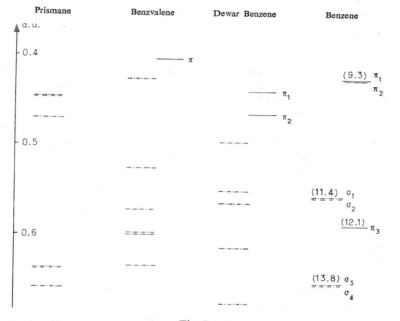

Fig. 5
Orbital energies (in eV) of the highest molecular orbitals

Although the σ–π separation cannot be completely achieved in any SCF calculation on non-planar molecules, it is possible to recognize quasi-π molecular orbitals among the occupied and virtual orbitals of the two unsaturated non-planar isomers of benzene: the highest occupied level and the lowest empty level of benzvalene are essentially the π–π^* energy levels of the double bond, and the mixing of σ and π atomic orbitals in such molecular orbitals may be called 'hyperconjugation'. Similarly, the two highest occupied and the two lowest empty levels of Dewar benzene can be identified with the two bonding and anti-bonding molecular orbitals resulting from the conjugation of the two double bonds through space.

4. Localization and Relative Stability

The decomposition of total SCF energies into localized bond energies is given in Table 5.

Table 5
Localized Bond Energies (in a.u.)

		Prismane		Benzvalene		Dewar Benzene		Benzene	
K-shells of C	1		−29.291		−29.280		−29.284		−29.307
	2				−29.280		−29.297		
	3				−29.333		−29.297		
	4				−29.294		−29.284		
	5				−29.333		−29.297		
	6				−29.294		−29.297		
σ Bonds C–H	1.1		−2.053		−2.172		−2.068		−2.230
	2.2				−2.172		−2.149		
	3.3				−2.229		−2.149		
	4.4				−2.136		−2.068		
	5.5				−2.229		−2.149		
	6.6				−2.136		−2.149		
σ and π Bonds C–C		1.2	−4.378	1.2	−4.326	1.2	−4.634	1.2	−4.174
		2.3	−4.378	1.2	−4.326	2.3	−4.421	1.2	−4.174
		3.1	−4.378	2.3	−4.749	2.3	−4.142	2.3	−5.145
		1.4	−5.016	3.4	−4.455	3.4	−4.634	3.4	−4.174
		2.5	−5.016	4.5	−4.455	4.5	−4.634	3.4	−4.174
		3.6	−5.016	5.6	−4.455	5.6	−4.421	4.5	−5.145
		4.5	−4.378	6.3	−4.455	5.6	−4.142	5.6	−4.174
		5.6	−4.378	6.4	−4.667	6.1	−4.634	5.6	−4.174
		6.4	−4.378	5.1	−4.749	1.4	−5.334	6.1	−5.145
Total energy			−229.382		−229.523		−229.485		−229.701
σ Bond C=C					−5.412		−5.259		−5.145
π Bond C=C					−3.240		−3.304		−3.203

In unsaturated non-planar molecules, i.e. Dewar benzene or benzvalene, two equivalent sets of quasi-localized orbitals can be constructed for each double bond, namely, two banana bonds, $\tau_1 \tau_2$, or one quasi-σ and one quasi-π bond. The first one is found by applying the localization process to the whole set of occupied (or virtual) SCF molecular orbitals; the second one is deduced from the preceding one by an appropriate two-by-two rotation. In benzvalene, for symmetry reasons, the σ and π components of the double bond are the sum and difference of the τ bonds, and the rotation angle is exactly 45°; in Dewar benzene, the rotation angle leading to σ and π orbitals that satisfy the localization criterion of Boys is practically equal to 45°.

According to the values of Table 5, the destabilization of non-planar benzene isomers with respect to benzene itself is mainly due to a destabilization of the C–H bond system, which is partially counterbalanced by an energy lowering of the C–C bonds. The large energy gap between benzene and its isomers might be peculiar to C–H bonds, and replacement of hydrogens by other substituents might lead to a more favourable situation; indeed, it is known that the hexafluoro and hexakis (trifluoromethyl) derivatives of the benzene isomers are fairly stable [41–42]. Finally, it should be added that there is no strong correlation between localized bond energies and overlap populations. The most striking discrepancy is observed in benzvalene; while the bonds of the three-membered rings contribute to the total energy almost equally, bond C_1–C_4 has a very weak overlap population (0.172) as compared to the other C–C bonds (0.672). Actually, the value of the overlap population of bond C_1–C_4 is related to its particular position in space, rather than to its strength (see Table 6).

Table 6

Relative Energies (in a.u. and Kcal/mole) of Non-Planar Isomers of Benzene

	Total Energy		Energy with Respect to Benzene		
	$E_{SCF}^{(0)}$	$E^{(2)}$	$\Delta E_{SCF}^{(0)}$	$\Delta E^{(2)}$	Total
Benzene	−229.7007	−0.4370			
Dewar benzene	−229.4852	−0.4299	0.2155 (135.2)	0.0071 (4.6)	0.2226 (139.6)
Benzvalene	−229.5230	−0.4551	0.1777 (111.4)	−0.0181 (−11.3)	0.1596 (100.1)
Prismane	−229.3822	−0.4345	0.3185 (199.7)	0.0025 (1.6)	0.3210 (201.3)

The values in Table 6, giving the energy difference between the total energies of benzene and its isomers, are much larger than the experimental isomerization energy into hexamethyl-benzene of the hexamethyl derivatives of Dewar benzene (59.5 Kcal/mole) and prismane (91.2 Kcal/mole) [43]. In order to clarify this question, the stabilizing effect of methyl substituents should be investigated. Furthermore, due to the lack of experimental data on the geometry of benzene isomers, it is not possible to trust entirely the theoretical predictions concerning the relative stability of Dewar benzene and benzvalene. Unlike benzvalene or prismane, it is not to be excluded that a lower total energy might be calculated for Dewar benzene by varying the dihedral angle between the two planes of the molecule.

Appendix

The atomic co-ordinates used in the present calculations are given in Table 7. The exponents and contraction coefficients of the Gaussian basis functions are those of Berthier et al. [4]. Molecular orbital energies are listed in Table 8 according to the symmetry species of the point group C_{2v} (or D_{3h} for prismane and trimethylene-cyclopropane and D_{6h} for benzene). Most calculations were carried out on the CDC 3600 (SCF programs) and IBM 360–75 (perturbation programs) of the CNRS computing centre at Orsay.

Table 7
Coordinates of Atoms (in a.u.)

	Trimethylene-cyclopropane		Dimethylene-cyclobutene		Fulvene		Benzene	
	Y	Z	Y	Z	Y	Z	Y	Z
C_1	0	4.893	−3.276	1.713	0	4.172	0	2.640
C_2	0	2.352	−1.398	0	0	1.631	−2.286	1.320
C_3	−1.358	0	−1.271	−2.785	−2.171	0	−2.286	−1.320
C_4	−3.559	−1.271	1.271	−2.785	−1.358	−2.408	0	−2.640
C_5	3.559	−1.271	1.398	0	1.358	−2.408	2.286	−1.320
C_6	1.358	0	3.276	1.713	2.171	0	2.286	1.320
H_1	1.750	5.966	−2.889	3.729	1.750	5.245	0	4.688
H_2	−1.750	5.966	−5.248	1.143	−1.750	5.245	−4.060	2.344
H_3	−5.363	−0.292	−2.654	−4.301	−4.132	0.607	−4.060	−2.344
H_4	−3.613	−3.322	2.654	−4.301	−2.554	−4.075	0	−4.688
H_5	3.613	−3.322	5.248	1.143	2.554	−4.075	4.060	−2.344
H_6	5.363	−0.292	2.889	3.729	4.132	0.607	4.060	2.344

	Dewar-Benzene			Benzvalene			Prismane		
	X	Y	Z	X	Y	Z	X	Y	Z
C_1	0	1.398	0	0	−1.271	−2.757	0	1.677	1.452
C_2	2.420	1.271	−1.397	0	1.271	−2.757	1.452	−0.838	1.452
C_3	2.420	−1.271	−1.397	0	2.186	0	−1.452	−0.838	1.452
C_4	0	−1.398	0	−1.425	0	1.258	0	1.677	−1.452
C_5	−2.420	−1.271	−1.397	0	−2.186	0	1.452	−0.838	−1.452
C_6	−2.420	1.271	−1.397	1.425	0	1.258	−1.452	−0.838	−1.452
H_1	0	1.492	2.050	0	−2.472	−4.422	0	3.663	1.968
H_2	3.705	2.689	−2.139	0	2.472	−4.422	3.172	−1.832	1.968
H_3	3.705	−2.689	−2.139	0	4.153	0.588	−3.172	−1.832	1.968
H_4	0	−1.492	2.050	−2.136	0	3.193	0	3.663	−1.968
H_5	−3.705	−2.689	−2.139	0	−4.153	0.588	3.172	−1.832	−1.968
H_6	−3.705	2.689	−2.139	2.136	0	3.193	−3.172	−1.832	−1.968

186

Table 8

Orbital Energies (in a.u.)

Symmetry C_{2v}	Trimethylene-cyclopropane	Dimethylene-cyclobutene	Fulvene	Benzene
$A_1 (\sigma)$	$-11.4626\ A'_1$		-11.4390	$-11.4225\ A_{1g}$
	$-11.4614\ E'$	-11.4395	-11.4330	$-11.4221\ E_{1u}$
	$-11.4256\ E'$	-11.4276	-11.4210	$-11.4210\ E_{2g}$
	$-11.4255\ A'_1$	-11.4021	-11.3978	$-11.4205\ B_{1u}$
	$-1.2722\ A'_1$	-1.2798	-1.2596	$-1.2226\ A_{1g}$
	$-1.0627\ E'$	-1.0610	-1.0991	$-1.0836\ E_{1u}$
	$-1.0075\ A'_1$	-0.8846	-0.9278	$-0.8809\ E_{2g}$
	$-0.7634\ E'$	-0.7254	-0.7790	$-0.7654\ A_{1g}$
	$-0.7144\ A'_1$	-0.6938	-0.7469	$-0.6864\ B_{1u}$
	$-0.6727\ E'$	-0.6201	-0.6353	$-0.6533\ E_{1u}$
	$-0.5281\ E'$	-0.5462	-0.5884	$-0.5584\ E_{2g}$
$B_2 (\sigma)$		-11.4389		
	$-11.4614\ E'$	-11.4262	-11.4052	$-11.4221\ E_{1u}$
	$-11.4256\ E'$	-11.4021	-11.3978	$-11.4210\ E_{2g}$
	$-1.0627\ E'$	-1.0646	-1.0374	$-1.0836\ E_{1u}$
	$-0.7634\ E'$	-0.8607	-0.7953	$-0.8809\ E_{2g}$
	$-0.6727\ E'$	-0.7176	-0.6580	$-0.6937\ B_{2u}$
	$-0.6370\ A'_2$	-0.6349	-0.6238	$-0.6533\ E_{1u}$
	$-0.5281\ E'$	-0.5718	-0.5607	$-0.5584\ E_{2g}$
$B_1 (\pi)$	$-0.6370\ A''_2$	-0.6032	-0.5993	$-0.5893\ A_{2u}$
	$-0.4222\ E''$	-0.4050	-0.4348	$-0.4272\ E_{1g}$
$A_2 (\pi)$	$-0.4222\ E''$	-0.4285	-0.3947	$-0.4272\ E_{1g}$
First empty MO	$(-0.0156\ E''_2)$	$(+0.0578\ A_2)$	$(-0.0049\ B_1)$	$(+0.0832\ E_{2g})$

Symmetry C_{2v}	Prismane	Benzvalene	Dewar Benzene
A_1		-11.4465	
	$-11.4079\ A'_1$	-11.4055	-11.4165
	$-11.4072\ E'$	-11.3985	-11.4035
	$-1.4060\ A'_1$	-1.3552	-1.3571
	$-0.9921\ E'$	-1.0967	-0.9609
	$-0.8023\ A'_1$	-0.8284	-0.7730
	$-0.6581\ A'_1$	-0.7825	-0.6770
	$-0.6364\ E'$	-0.6589	-0.4982
	$-0.4707\ E'$	-0.5991	-0.4657
		-0.4272	
B_2		-11.4465	-11.4150
	$-11.4072\ E'$	-11.3970	-11.4028
	$-0.9921\ E'$	-1.0241	-0.9557
	$-0.6364\ E'$	-0.7877	-0.7017
	$-0.4707\ E'$	-0.6340	-0.5655
		-0.5730	
B_1	$-11.4075\ A''_2$		
	$-11.4067\ E''$	-11.4050	-11.4165
	$-1.0658\ A''_2$	-0.8319	-11.5008
	$-0.7397\ E''$	-0.5980	-0.7424
	$-0.6691\ A''_2$	-0.4051	-0.6154
	$-0.4456\ E''$		-0.4410
A_2	$-11.4067\ E''$		-11.4151
	$-0.7397\ E''$		-0.8125
	$-0.4456\ E''$		-0.5523
		-0.5272	
First empty level	$(+0.1860\ A'_2)$	$(+0.1238\ A_2)$	$(+0.0977\ B_1)$

ACKNOWLEDGEMENTS

Part of the study reported above has been performed while one of us (A. Y. Meyer) was staying on a research leave at the Institut de Biologie Physico-Chimique in Paris. We wish to thank Prof. B. Pullman and the Fondation Edmond de Rothschild for a generous research grant.

REFERENCES

1 E. Clementi & D. R. Davis (1966) *J. Comp. Phys.*, 1 : 223.

2 C. C. J. Roothaan (1951) *Rev. Mod. Phys.*, 23 : 69.

3 E. Clementi (1967) *J. Chem. Phys.*, 46 : 4725, 4731, 4737.

4 G. Berthier, L. Praud & J. Serre (1970) in : E. D. Bergmann & B. Pullman (eds.), *Quantum Aspects of Heterocyclic Compounds in Chemistry and Biochemistry (The Jerusalem Symposia on Quantum Chemistry and Biochemistry*, II), The Israel Academy of Sciences and Humanities, p. 40.

5 A. Langseth & B. P. Stoicheff (1956) *Can. J. Phys.*, 34 : 350.

6 N. Norman & B. Post (1961) *Acta Crystallogr.*, 14 : 503.

7 E. A. Dorko, J. L. Hencher & S. H. Bauer (1968) *Tetrahedron*, 24 : 2425.

8 P. S. Epstein (1926) *Phys. Rev.*, 28 : 695.

9 R. K. Nesbet (1955) *Proc. Roy. Soc.*, A 230 : 312.

10 S. F. Boys (1960) *Rev. Mod. Phys.*, 32 : 296.

11 E. Gey, H. Havemann & L. Zulicke (1968) *Theor. Chim. Acta*, 12 : 313.

12 R. H. Pritchard & C. W. Kern (1969) *J. Am. Chem. Soc.*, 91 : 1631.

13 G. Berthier (1953) *J. Chim. Phys.*, 50 : 344.

14 A. Julg (1953) *ibid.*, p. 652.

15 *Ibid.* (1955) 52 : 50.

16 R. D. Brown, F. R. Burden, A. J. Jones & J. E. Kent (1957) *Chem. Comm.*, p. 808.

17 R. D. Brown, F. R. Burden & J. E. Kent (1968) *J. Chem. Phys.*, 49 : 5542.

18 R. J. Buenker & J. D. Peyerimhoff (1969) *Chem. Phys. Lett.*, 3 : 37.

19 K. Siegbahn et al. (1959) in : *ESCA Applied to Free Molecules*, North-Holland Co., Amsterdam.

20 T. Nakajima & A. Pullman (1958) *J. Chim. Phys.*, 55 : 793.

21 B. Mely & A. Pullman (1969) *Theor. Chim. Acta*, 13 : 278.

22 E. Clementi, J. M. Andre, M. C. Andre, D. Klint & D. Hahn (1969) *Acta Phys. Hung.*, 27 : 493.

23 F. Maeder, P. Millie & G. Berthier (1970) *Int. J. Quant. Chem.* 35 : xx.

24 J. H. Day & C. Oestreich (1957) *J. Org. Chem.*, 22 : 214.

25 E. E. Van Tamelen (1965) *Angew. Chem. (Int. Edn)*, 4 : 738.

26 I. G. Bolesov (1968) *Russ. Chem. Rev.*, 37 : 666.

27 G. Huttner & O. S. Mills (1968) *Chem. Comm.*, p. 344.

28 H. C. Volger & H. Hogeveen (1967) *Rec. Trav. Chim. Pays-Bas Belg.*, 86 : 1356.

29 T. Yonezawa, K. Simizu & H. Kato (1968) *Bull. Chem. Soc. Japan*, 41 : 2336.

30 N. C. Baird & M. J. S. Dewar (1969) *J. Am. Chem. Soc.*, 91 : 353.

31 M. Randic & Z. Majersky (1968) *J. Chem. Soc.*, B, p. 1289.

32 C. A. Coulson & W. E. Moffitt (1947) *J. Chem. Phys.*, 13 : 151.

33 P. A. Clark & J. L. Ragle (1966), *SCF-LCAO-MO Calculations with CNDO*, QCPE 100.

34 J. A. Pople & G. A. Segal (1965) *J. Chem. Phys.*, 43 : S 136.

35 R. S. Mulliken (1955) *ibid.*, 23 : 1833.

36 K. Wiberg (1968) *Tetrahedron*, 24 : 1083.

37 C. Trindle (1969) *J. Am. Chem. Soc.*, 91 : 219.

38 C. Trindle & O. Sinanoglu, *ibid.*, p. 853.

39 W. A. Bernett (1967) *J. Chem. Educ.*, 44 : 17.

40 A. L. McClellan (1963) in : *Tables of Experimental Dipole Moments*, Freeman, San Francisco.

41 C. Camaggi, F. Gozzo & G. Cevidalli (1966) *Chem. Comm.*, p. 313.

42 M. G. Barlow, R. N. Haszeldine & R. Hubbard (1969) *ibid.*, p. 202.

43 J.F.M. Oth (1968) *Angew. Chem. (Int. Edn)*, 7 : 646.

Note added in proof:

After our calculations had been completed, an electron diffraction study on hexamethyl-Dewar benzene was published: M. J. Cardillo & S. H. Bauer (1970) *J. Am. Chem. Soc.*, 92 : 2399. In this molecule, the bridge is found to be unusually long (1.63 Å). Other geometrical parameters are: C_1C_2, 1.52 Å; C_2C_3, 1.33Å; fold angle, 124.5°. These values are different from those used by us, but one should bear in mind that at least part of the bond-lengthening and angle-opening is due to the strain imposed by the presence of the six methyl groups. For example, the experimental $CH_3-C_1-C_4$ angle is 115.9°, while a hybridization calculation for Dewar benzene (based on the bond lengths and fold angle measured for its hexamethyl derivative) still indicates the smaller angle of 93°.

Discussion

J. J. C. Mulder:

I would like to ask in the first place if I am clear about your calculation procedure: You minimized the CNDO energies for angles and interatomic distances, then used these in the *ab initio* treatment. Is that right?

A. Y. Meyer:

We did not try any minimization by distances, as the CNDO procedure is known not to be precise in this respect, but we did this for angles. Starting from some set of angles, we effected a CNDO computation, then we used the results to calculate hybridization indices and a new geometry. The process was carried on iteratively to self-consistence.

J. J. C. Mulder:

Do you not distrust your *ab initio* calculations, as these do not contain minimization themselves?

A. Y. Meyer:

At present, it is not feasible to effect calculations for several geometrical structures of these large molecules, let alone a complete energy minimization process, because the computation is very lengthy.

H. Basch (Ford Motor Company, Dearborn, USA):

I have just one comment: Eight hours for the SCF at a minimal basis is very exaggerated, since there are computer programmes available that can do it in a much shorter time. I might quote the SCF time for a molecule like nickel hexafluoride, with seventy-four basis functions. The SCF time, which comprised fifteen cycles, was about an hour.

E. Heilbronner:

I would like to put three questions to Mr Berthier and Mr Meyer jointly. All three have to do with the values of the orbital energies of the top-occupied orbitals that you have calculated.
The first question is: On the slide which Mr Berthier has shown there were two π orbitals of E_{2g} and A_{1u} symmetry and a σ orbital in between. Now, there is still some doubt as to whether the second π orbital is really at -12.3 eV, as deduced by some authors from photo-electron spectroscopic data. There are results which seem to suggest that it should lie at much lower energies. I would like to ask you if you have any comments on that?
The second one is: Do you know what the orbital gap will be in Dewar benzene between the a_1 and the b_2 orbitals, which are the two top π orbitals according to your calculations?
The last question is: Is the top orbital of prismane essentially the E'' combination of the Walsh orbitals of the three-membered rings, or do other carbon-hydrogen σ orbitals come on top?

A. Y. Meyer:

The orbital symmetry and energies are tabulated in the appendix of our joint paper. Defining the axes and symmetry planes as in the figures, we have, for Dewar benzene: $\phi_{+2}(A_2, \varepsilon = 0.1417$ a.u.$)$, $\phi_{+1}(B_1, \varepsilon = 0.0977$ a.u.$)$, $\phi_{-1}(B_2, \varepsilon = -0.4410$ a.u.$)$, $\phi_{-2}(A_1, \varepsilon = -0.4657$ a.u.$)$. For prismane

189

we find: ϕ_{+2} (A_2'', $\varepsilon = 0.2357$ a.u.), ϕ_{+1} (A_2', $\varepsilon = 0.1860$ a.u.), ϕ_{-1} (E'', $\varepsilon = -0.4456$ a.u.), ϕ_{-2} (E', $\varepsilon = -0.4707$ a.u.).

Mrs A. Pullman:

I have another point concerning the *ab initio* calculation in these basis sets. We all know perfectly well that Clementi's set is not the best nowadays. But there is one very good thing Clementi has said, namely, that it is very important to calculate a number of molecules with the same basis set. If you calculate one molecule with a given basis set and another molecule with another basis set, and another one with still another basis set, you end up with a lot of calculations that are not related to each other and do not permit any conclusions. Here is a good argument in favour of continuing with this type of basis set for the time being — at least for people who have at their disposal the programme, the knowledge of the basis set and of background. We have understood quite a number of things by looking at calculations that have already been made with this particular type of basis set.

B. Pullman:

I do not think I have much to add. I would just like to say that these types of calculations, anyway, have the importance of giving experimentalists suggestions as to what type of experiment might be important to do. And, whatever the bond lengths are that you will find by minimizing, the essential results, say, on dipole moment, will probably be preserved to some extent. And these are stimulating.

J. J. C. Mulder:

I would like to ask you: Have you any estimate of how far from the Hartley-Fox limit you are in your SCF calculations?

G. Berthier:

Although our basis set is a minimal one, the total SCF energy obtained for benzene is not too far from the best value published until now. If the computer time were not so large (3 to 5 hours on CDC 3600 for the whole SCF process in the case of planar C_6H_6 molecules; 8 hours for non-planar isomers), it would be possible to recalculate these compounds with basis sets, including all the improvements you want. Frankly, I do not think that the relative stability of benzene isomers could be very much modified at the level of the SCF approximation. Instead of trying to reach the Hartley-Fox limit—a typical non-observable—we have preferred to go beyond the SCF approximation immediately. Of course, the correction term $E^{(2)}$ we calculate is only the correlation energy that can be recovered in our limited basis set.

J. J. C. Mulder:

I would like to say that I realize that the computing-time question is a very important one, but as long as we are certain that we cannot do complete configuration-interaction treatments of molecules of any kind, it is still worthwhile to know how far we can get with only one set of determinants for these closed-shell type molecules.

H. Basch (Ford Motor Company, Dearborn, USA):

I think I can say something about that. Recent experience has shown that when one compares molecules of the types Mr Berthier is studying, even using as small basis sets as he is using, one gets very good results regarding the kinds of properties that Mr Berthier is comparing. Thus, the question of how far he is from the Hartley-Fox limit or how much correlation there is, is in a sense irrelevant.

Discussion

Mrs A. Pullman:

At the beginning you said that the method used is largely empirical, in spite of the name *ab initio* generally used. Although I understood perfectly what you meant, I think that the statement may be misleading for non-specialists, who usually think of 'empirical' methods as methods in which some parameters are fitted for agreement with experiments, a feature that does not exist in the all-electron SCF-LCAO calculation that you used.

My second comment is a warning of the same sort especially directed at our colleagues the experimentalists—it refers to the equivalence between Clementi's contracted basis set and a Slater minimal basis set. It is true as far as the number of functions is concerned, but not as far as accuracy in results is concerned.

As concerns the IS-orbital energies and their connection with the charges, I would like to mention that I have tried to use a relation similar to the relation established many years ago for lone-pair ionization potentials by Mr Nakajima and myself: using SCF charges, it does not work. While it might have been hoped that CNDO charges would work better, they do not, at least on a general scale.

My last point concerns the irritating question of the dipole moment of fulvene, which appears nowadays to be much smaller than was thought before, so that the 1 Debye value found by theory is about twice the experimental value. A very simple possibility for the disagreement, with experiment, of all methods could be the neglect of configuration-interaction effects. Indeed, when doubly excited configurations (σ and π) are included after a CNDO-SCF calculation, we have shown (C. Giessner-Prettre & A. Pullman, *Theoretica Chimica Acta*, in press) that not only does a lowering of the ground-state energy occur, but also a non-negligible decrease of the dipole moment is obtained. Thus, we have run the same type of calculation for fulvene and found the following results in Debye units (CNDO/2 stands for the usual parametrization and DBJ for the Del Bene-Jaffé parametrization):

	SCF	Doubly Excited Configurations Included
CNDO	0.84	0.76
DBJ	0.74	0.59

It would be extremely interesting to introduce configuration-interaction on the *ab initio* results. It may be mentioned that in the case of formic acid Peyerimhoff and Buenker found that doubly excited configurations lowered the SCF-calculated moment from 1.815 to 1.372 Debyes; see S. D. Peyerimhoff & R. I. Buenker (1969) *J. Chem. Phys.*, 50:1846. This confirms the existence of the effect.

G. Berthier:

I have mentioned the fact that *K*-shell ionization potentials may be related to the total charge distribution rather than to the charge of an individual carbon atom, because the primitive relationship postulated by ESCA spectroscopists has been recently replaced by an empirical equation of that type.

In the frame of the second-order perturbation theory, it is not possible to estimate any correction term for the dipole moment given by the SCF method, because the matrix elements with singly excited configurations vanish by virtue of Brillouin's theorem. Higher-order perturbation or true configuration-interaction calculations should be carried out.

H. Basch:

I have a question which may be relevant. I just wondered if Professor Berthier has looked at one more isomer of benzene, and that is hypothetical planar cyclohexatriene. It has been suggested on and off, for many years, that perhaps the hexagonal benzene is not the most stable structure of benzene, but that some planar cyclohexatriene may be more stable, and that the benzene structure may represent some sort of small barrier that is easily penetrated, perhaps of the order of the zero-point energy of one particular normal mode of vibration, which would be that of a carbon–carbon stretch.

191

Theoretical and Experimental Studies of Double-Bond Fixation in some Cyclic Unsaturated Hydrocarbons

by A. SKANCKE

Department of Chemistry, University of Oslo

I. Introduction

CYCLIC HYDROCARBONS built up of sp^2 π-hybridized carbon atoms are known to be of two types: aromatics and pseudo-aromatics. The molecules belonging to the first group are recognized by nearly uniform charge distributions, whereas the molecules of the other group exhibit a significant variation in electronic densities. This leads to characteristic differences for a series of measurable properties that may be used in classifying molecules in one of these groups. Among these are molecular structure, chemical shift, electronic transitions, dipole moment and diamagnetic susceptibility.

In the present work, I have tried to combine a theoretical study of bond distances with available results of experimental structure determinations for a few unsaturated cyclic hydrocarbons that are of importance in this context. The theoretical approach used will be presented in the following section.

II. Theoretical Calculations

1. *Computational Procedure*

The Pariser-Parr-Pople method has been applied in the study of the π-electron systems. This method is based upon the Roothan-Hall equations, which, in turn, are obtained by combining the Hartree-Fock equations with the LCAO approximation:

$$\phi_i = \sum_{\mu} c_{i\mu} \kappa_{\mu}. \tag{1}$$

A simplification of the Roothan-Hall equations was introduced by Pariser and Parr [1–2] and by Pople [3] by neglecting complicated two-electron integrals, except for the pure Coulomb terms. This is the so-called ZDO (zero differential overlap) approximation. Using standard notation, this approximation yields the following integrals:

$$(\mu \,|\, v) = S_{\mu v} = \delta_{\mu v}; \tag{2}$$

$$(\mu \,|\, H^{core} \,|\, \mu) = \alpha_{\mu} \neq 0; \tag{3}$$

$$(\mu \,|\, H^{core} \,|\, v) = \beta_{\mu v} \neq 0; \qquad \mu \text{ and } v \text{ neighbours}; \tag{4}$$

$$(\mu \,|\, H^{core} \,|\, v) = 0; \qquad \mu \text{ and } v \text{ non-neighbours}; \tag{5}$$

192

Studies of Double-Bond Fixation in some Cyclic Unsaturated Hydrocarbons

$$(\mu\alpha \,|\, \nu\beta) \;=\; \gamma_{\mu\nu}\delta_{\mu\alpha}\delta_{\nu\beta}. \tag{6}$$

The integrals presented above can be evaluated numerically by using different schemes of parametrization. In this way rather different — although internally consistent — parameter sets may be obtained. It is, however, considered to be beyond the scope of this presentation to discuss the different methods applied.

In the following, the particular scheme applied in our calculation will be briefly discussed. This scheme is evaluated in accordance with a general theoretical analysis of the ZDO approximation by Fischer-Hjalmars [4].

The diagonal element of the core operator, α_μ, is decomposed, as suggested by Goeppert-Mayer and Sklar [5], in the following way:

$$\alpha_\mu \;=\; W_\mu - \sum_{\nu \neq \mu} \gamma_{\mu\nu}. \tag{7}$$

This relation is valid for the case where each atom contributes one electron to the π-electron system. The parameter W_μ, which is usually considered to be a pure atomic quantity, is made dependent on the surroundings of atom μ through the relation

$$W_\mu \;=\; W_0 + \sum_{\nu} q_\nu \Delta W_\mu(R_{\mu\nu}). \tag{8}$$

Here W_0 is the ionization potential of the assumed planar methyl radical. The constant q_μ is zero if the neighbour atom ν is a hydrogen atom, and unity if it is an sp^2 π-hybridized carbon atom. Deviation from a chosen standard distance is accounted for by assuming the linear relation

$$\Delta W_\mu(R_{\mu\nu}) \;=\; \Delta W_0 + \delta(R_{\mu\nu} - R_0), \tag{9}$$

where the subscript zero refers to values for the benzene molecule. Similar relations are assumed for the parameters β and γ:

$$\beta_{\mu\nu} \;=\; \beta_0 + \delta_\beta(R_{\mu\nu} - R_0), \tag{10}$$

$$\gamma_{\mu\nu} \;=\; \gamma_0 + \delta_\gamma(R_{\mu\nu} - R_0). \tag{11}$$

For non-neighbouring atoms, the Coulomb integrals were evaluated by using a method due to Roos [6].

The different semi-empirical parameters to be determined within this scheme — altogether eight — have been evaluated numerically by invoking observed values for molecular ionization potentials and electronic transitions for some chosen reference molecules. For details, see Roos and Skancke [7].

2. Results and Discussion

The scheme outlined above has been applied in a study of the π-electron structure of the ground states of the molecules fulvene (I), 3, 4-dimethylenecyclobutene (II) and 1, 2-dimethylenebenzocyclobutene (III) [8]. The aim of this investigation was to see whether a study of this kind could be used for elucidating the possible aromatic character of the above-mentioned compounds.

The concept of aromaticity is by no means an unambiguous one. Several criteria, both of direct and indirect nature, have been suggested as a measure of aromatic character. Here

193

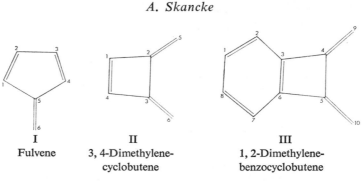

I	II	III
Fulvene	3, 4-Dimethylene- cyclobutene	1, 2-Dimethylene- benzocyclobutene

Labelling of molecules and numbering of atoms

we have applied a very simple, and easily interpretable, measure of aromaticity, namely, the distribution of π electrons in the molecular ground states as described by the calculated one-particle density matrix.

The molecules I and II are cyclic isomers of benzene, whereas molecule III is isoelectronic with naphthalene. For molecule I an experimental UV spectrum is interpreted to give a ground-state structure of nearly pure alternating single and double bonds [9]. Furthermore, an observed dipole moment of about 1 D indicates a substantial polarity in this molecule [10–11]. A recent electron diffraction study of the closely related dimethylfulvene by Chiang and Bauer [Chiang, personal communication] shows a pronounced single-bond–double-bond structure of that molecule.

For molecule II, a vapour phase UV spectrum has been reported [12], and an NMR spectrum of this molecule has given indications of olefinic character [12]. For molecule III, the experimental information is both scarce and uncertain, but a UV spectrum in ethanol solution is available [13].

In the computational scheme, the semi-empirical parameters have been expressed as functions of internuclear distances, and a point was made in investigating the effect of the assumed nuclear configuration on the resulting calculated populations.

Two different starting models were applied, one assuming all bonds to be equal to 1.40 Å, the other assuming pure 'single-bond–double-bond' character, with lengths equal to 1.47 Å and 1.34 Å, respectively. The resulting bond orders, $p_{\mu\nu}$, were used to calculate corresponding distances from the formula

$$R_{\mu\nu} = 1.517 - 0.18\,p_{\mu\nu}. \tag{12}$$

The bond distances thus obtained were applied in a re-evaluation of the parameter values, and a new cycle was performed. The iterative procedure was repeated until convergence was reached. The results are given in Table 1. As is shown in the table, three iterations were performed for molecules I and III, whereas two were sufficient for molecule II. The table shows quite clearly that in the case of these molecules, the final bond distances obtained are independent of the starting model. Column A_1 demonstrates that one cycle is sufficient to obtain a structure with strongly alternating character from an 'aromatic' starting model. The obtained bond distances indicate strongly that molecules I and II are olefinic in character, while molecule III consists of a normal, aromatic hexagonal ring and an alternating four-membered ring system quite similar to the one in molecule II. A further comparison of bond distances demonstrates that the external bonds, C_5–C_6 in I, C_2–C_5 in II and C_4–C_9 in III, are nearly identical. The same is true for the 'single' bonds adjacent to this

194

external bond. There is a difference of about 0.05 Å between the bond C_1–C_4 in II and the bond C_3–C_6 in III, but the larger length of the latter bond may be understood by considering the conjugative effect of the benzene ring.

Table 1

Bond Distances (in Å) Obtained from the Different Cycles in the Iterative Process Described

Molecule	Bond	A_1	A_2	A_3	Exp.	B_3	B_2	B_1
I	1–2	1.367	1.359	1.357	1.340 ± 0.006*	1.357	1.357	1.355
	2–3	1.435	1.446	1.450	1.462 ± 0.009	1.450	1.450	1.456
	1–5	1.453	1.463	1.466	1.476 ± 0.008	1.464	1.464	1.467
	5–6	1.361	1.356	1.354	1.347 ± 0.010	1.355	1.355	1.353
II	1–2	1.459	1.467		1.488 ± 0.009		1.468	1.469
	1–4	1.362	1.355		1.357 ± 0.005		1.354	1.353
	2–5	1.353	1.351		1.335 ± 0.003		1.350	1.350
	2–3	1.470	1.476		1.516 ± 0.020		1.476	1.476
III	1–2	1.398	1.398	1.398		1.394	1.388	1.374
	1–8	1.398	1.397	1.397		1.401	1.408	1.424
	2–3	1.400	1.399	1.399		1.404	1.408	1.426
	3–6	1.412	1.410	1.409		1.405	1.399	1.386
	3–4	1.465	1.470	1.471		1.471	1.470	1.471
	4–5	1.465	1.472	1.472		1.473	1.473	1.473
	4–9	1.353	1.351	1.351		1.351	1.351	1.350

* Experimental Values for dimethylfulvene

A=start with all bonds equal to 1.40 Å; B=start with 'single' bonds of 1.47 Å and 'double' bonds of 1.34 Å

The electronically excited states of the molecules were studied using configurational mixing, including all the singly-excited configurations. Calculated spectra and corresponding observables are given in Table 2. The agreement between calculated and observed data is satisfactory, perhaps with the exception of molecule III. In view of the good agreement between experiments and calculations for the spectra of molecules I and II, it is tempting to explain at least some of the discrepancies by experimental uncertainties for molecule III. However, it must also be emphasized that the semi-empirical parameters on which these calculations were based were obtained from model molecules not containing strained systems, such as four-membered rings. As a consequence, minor discrepancies should be interpreted with some reservations.

III. Experimental Structure Determinations

For molecule I, the bond lengths from the calculations are compared to the results of dimethylfulvene by Chiang and Bauer (see above). As seen from Table 1, the agreement between calculation and experiment is very good, all distances being well within the error limit (taken as three times the standard deviation). However, one interesting observation is made in the comparison: It seems that replacement of the methylene hydrogens in fulvene by methyl groups leads to a structure that is even more pronouncedly olefinic. This is in contrast to the true aromatics, where replacement of a hydrogen by a substituent is known to have very little effect on the carbon skeleton.

Table 2

Calculated and Observed Electronic Spectra

(frequencies in cm^{-1})

Molecule	Calc.	Obs.	$f_{calc.}$	$f_{obs.}$	Pol. $_{calc.}$
I	26,600	27,600	0.05	0.01	x
	40,800	41,300	0.54	0.34	y
	54,450		0.13		y
II	37,800⎫		0.12		y
	38,600⎭	41,600	0.01	$\log E = 4.2$	x
	48,700	47,000	0.77	$\log E = 4.8$	x
	51,000	48,800	0.89	$\log E = 5.0$	y
	63,900		0.10		y
III	35,100	30,400	0.04	$\log E = 4.2$	x
	40,100	36,600	0.05	$\log E = 3.3$	y
	44,600		0.35		x
	44,750		0.13		y
	45,300	43,500	1.33	$\log E = 4.8$	x
	52,300		0.14		y

A sample of molecule II was obtained and its molecular structure investigated by means of the electron diffraction method [14]. The aim of this investigation was to test the validity of the conclusions drawn from the SCF calculations. The final results are given in Table 1, which also includes the results from the theoretical calculations. The table shows four different bond distances, although the deviation of distance C_2-C_3 is rather large, and the observed difference between the distances C_1-C_2 and C_2-C_3 is within the error limit. Comparing with the results from the calculation, only distance C_2-C_5 is slightly outside the range of the error limit.

The main conclusion reached in the theoretical work, a pronounced single-bond–double-bond alteration, is in complete agreement with experimental findings. Furthermore, it is worth noticing that the predicted differences between the two single and the two double bonds, respectively, are qualitatively confirmed by the experimental results.

REFERENCES

1 R. Pariser & R. G. Parr (1953) *J. Chem. Phys.*, 21 : 466.

2 *Ibid.*, p. 767.

3 J. A. Pople (1953) *Trans. Faraday Soc.*, 49 : 1375.

4 I. Fischer-Hjalmars (1965) *J. Chem. Phys.*, 42 : 1962.

5 M Goeppert-Mayer & A. L. Sklar (1938) *ibid.*, 6 : 645.

6 B. Roos (1965) *Acta Chem. Scand.*, 19 : 1715.

7 B. Roos & P. N. Skancke (1967) *ibid.*, 21 : 233.

8 A. Skancke & P. N. Skancke (1968) *ibid.*, 22 : 175.

9 P. A. Straub, D. Meuche & E. Heilbronner (1966) *Helv. Chim. Acta*, 49 : 517.

10 G. W. Wheland & D. E. Mann (1949) *J. Chem. Phys.*, 17 : 264.

11 R. Ferreira (1965) *Theor. Chim. Acta*, 3 : 147.

12 M. L. Heffernan & A. J. Jones (1966) *Chem. Comm.*, p. 120.

13 M. P. Cava, R. J. Pohl & M. J. Mitchell (1963) *J. Am. Chem. Soc.*, 85 : 2080.

14 A. Skancke (1968) *Acta Chem. Scand.*, 22 : 3239.

Discussion

E. D. Bergmann:

Have you also calculated the dipole moments for these compounds?

Mrs A. Skancke:

Yes, the calculated values for molecules I, II and III are 0.57 D, 1.35 D and 1.23 D, respectively.

E. D. Bergmann:

It is most gratifiying that for the first time a calculation has given a value for the dipole moment which agrees with experiment. However, is it not surprising that this same method gives double the experimental value for 1,2-dimethylene-cyclobutene?

Mrs A. Skancke:

But the σ part of the dipole moment is not taken into account, and this might go in the same or in the opposite direction. So, it is not really surprising that the total calculated dipole moment may be different from the experimental one.

E. D. Bergmann:

No. II is higher than the experimental one. It is No. I, I think, which interests us, because of the discussion we had yesterday. It is very much nearer to the experimental value than anything which has been calculated, except for Mrs Pullman's very accurate determination of this. But for No. II, 0.86 is, I think, the experimental value which the Americans have found.

Mrs A. Pullman:

1. It is quite remarkable that you obtain such a low value of the μ_π for fulvene. Since the σ moment is very small, your value is on the whole very good. The discrepancy for the square molecule, however, remains puzzling.

2. I was surprised by the fact that in identifying the spectrum of the square isomer, you seem to disregard the first calculated transition at 37800 cm^{-1}, which has a calculated intensity of 0.12. On what grounds do you neglect this calculated transition?

Mrs A. Skancke:

There were two close lines at 37800 and 38600, and I should have put a bracket on the slide in order to show that the observed line is the mean of the two calculated values.

Electronic Structure of 1,2-, 1,3- and 1,4-Dehydrobenzenes

by J. SERRE, L. PRAUD *and* P. MILLIÉ

Laboratoire de Chimie de l'École Normale Supérieure, Paris

THE 1,2-BENZYNE, C_6H_4, is a benzene molecule in which two neighbouring hydrogens are removed. There are two other possible isomers, the 1,3-benzyne and the 1,4-benzyne, corresponding, respectively, to the abstraction of hydrogens in the *meta* or in the *para* position.

Physical evidence for the existence of 1,2-dehydrobenzene in the gas phase is found by means of time-resolved mass spectrometry [1]. By this technique, the rate of disappearance of dehydrobenzene can be determined. The corresponding rate constant is very large ($k \simeq 10^9 l \cdot mole^{-1} \cdot s^{-1}$) [2]; a similar value is obtained by flash photolysis experiments [3].

It has been shown by time-resolved mass spectrometry, again, that the products from benzenediazonium-3-carboxylate decomposition may contain 1,3-benzyne [4], and that this molecule is probably in a singlet state. Another C_6H_4 species [5] has been observed in the decomposition of benzenediazonium-4-carboxylate, and is probably 1,4-benzyne. The latter isomer is a very long-lived species, since the corresponding peak in the mass spectrometer persists for as long as two minutes.

In order to obtain an idea on the relative stability of the three isomers and to learn something about the nature of the bond between the two dehydrogenated carbons in each of these molecules, a non-empirical LCAO-MO-SCF calculation, using a basis of Gaussian orbitals, was carried out. In such a calculation the molecular orbitals are linear combinations of contracted Gaussian functions, identical to the ones chosen by Clementi for the carbons and the hydrogens of pyridine [6]. In this case, the dimension of the basis is 34. The chosen geometry is the same for the three molecules — the benzene geometry ($C{=}C : 1.397 \text{Å}$; $C{-}H : 1.084 \text{Å}$). In the case of 1,2-benzyne, it is known that the C_1–C_2 bond is probably similar to the one in the first excited state of acetylene [7], and the value of 1.397Å is not too far from the one determined by spectroscopy in the latter molecule. This choice of geometry enables us to compare the energies of the three benzynes and that of benzene, since a similar computation on benzene had been done with the same basis [8].

The energies of the SCF molecular orbitals occupied in the ground state are given in Table 1 for 1,2-benzyne, and in Table 2 for 1,3-benzyne and 1,4-benzyne. The energy of the first virtual orbital is also indicated in each case; this orbital is of σ type. In Fig. 1 the levels of the molecular orbitals of benzene and those of the three benzynes are compared. The

198

Table 1
SCF Energies of 1,2-Benzyne

σ	π
−11.4509	
−11.4509	
−11.4397	
−11.4384	
−11.4349	
−11.4339	
−1.2313	
−1.0903	
−1.0781	
−0.8786	
−0.8610	
−0.7463	
−0.6939	
−0.6743	
−0.6236	
	−0.5974
−0.5948	
	−0.4404
	−0.4275
−0.4126	
−0.0335 *	

* Energy of the first virtual orbital

Table 2
SCF Energies of 1,3-Benzyne and 1,4-Benzyne

1, 3-Benzyne			1, 4-Benzyne	
σ	π		σ	π
−11.4509			−11.4408	
−11.4459			−11.4406	
−11.4456			−11.4395	
−11.4436			−11.4391	
−11.4336			−11.4385	
−11.4335			−11.4382	
−1.2425			−1.2406	
−1.0961			−1.0973	
−1.0882			−1.0781	
−0.8848			−0.8870	
−0.8596			−0.8611	
−0.7579			−0.7203	
−0.6936			−0.6998	
−0.6775			−0.6791	
−0.6262			−0.6659	
	−0.6028			−0.6028
−0.5971			−0.5658	
	−0.4417			−0.4423
	−0.4363			−0.4341
−0.3619			−0.2871	
−0.0865*			−0.1616 *	

* Energy of the first virtual orbital

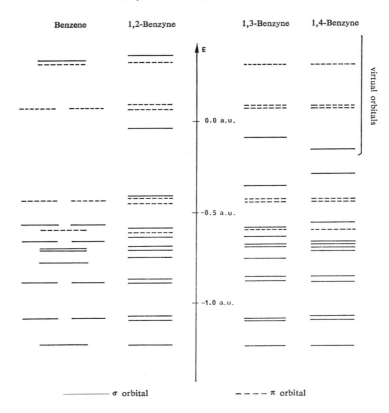

Fig. 1
SCF orbital energies

energy variations of the π orbitals are small. The occupied π orbitals become slightly more stable from benzene to the benzynes, and, of course, the degeneracy is lifted. The energies of the occupied σ orbitals, with the exception of the highest ones, are only slightly modified in comparison with benzene. The situation is completely different for the highest occupied orbitals. The effect of hydrogen abstraction is to destabilize strongly the σ orbitals that are localized mainly on the corresponding carbons. This effect varies for each benzyne and is particularly important in 1,4-benzyne.

The orbitals were localized by the procedure of Foster and Boys [9]. The values given in Figs. 2–3 include the kinetic and nuclear attraction energy I_i, the orbital energy e_i of the i-th localized orbital, and also a part of the total nuclear repulsion energy, R_i, so that the sum over all the electron pairs i, $\sum_i (e_i + I_i + R_i)$, is equal to the total energy E_T of the molecule.

In 1,2-benzyne, by comparison with benzene, it is found that the two σ bonds, close to the 1,2-bond, are destabilized. If the coefficients of the molecular orbitals localized between the two dehydrogenated carbons are considered, the more stable one ($e = -5.24$ a.u.) is found to be built mainly with the $2s$, $2p_y$ and $2p_z$ orbitals of carbons 1 and 2. This orbital is bonding for the $2p_z$ part and anti-bonding for the $2p_y$ one. Therefore, this orbital can be considered as constructed of two $2p$ orbitals pointing to each other. The isodensity curves

200

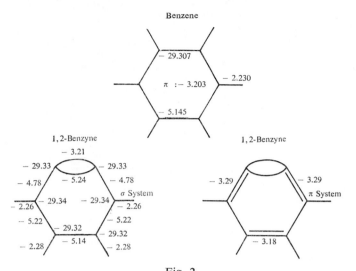

Fig. 2
Energies of the localized bonds in benzene,
the σ system of 1, 2-benzyne and the π system of 1, 2-benzyne

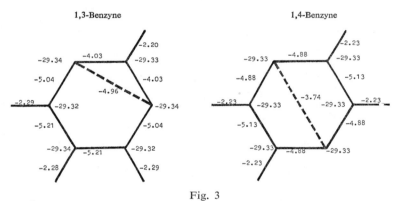

Fig. 3
Energies of the localized bonds in the σ system of 1, 3-benzyne and 1, 4-benzyne

of this orbital are given in Fig. 4. The other molecular orbital ($e = -3.21$ a.u.) contains $1s$, $2s$ and $2p_z$ orbitals, belonging to the same carbons. It is bonding for the $2p_z$ component. This orbital can be depicted as being obtained by the overlap of two $2p$ orbitals with parallel axes. In some ways, it is a π-type orbital, except that the orbital axes are in the molecular plane. The isodensity curves of this orbital are given in Fig. 5.

In 1,3-benzyne the σ orbital between the two carbons C_1 and C_3 is found to consist mainly of the $2p_y$ orbitals of these carbons. Therefore, it can be considered as being localized very close to the line C_1C_3. The σ bonds between carbons C_1 and C_2 and between carbons C_2 and C_3 are distinctly less stable.

201

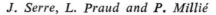

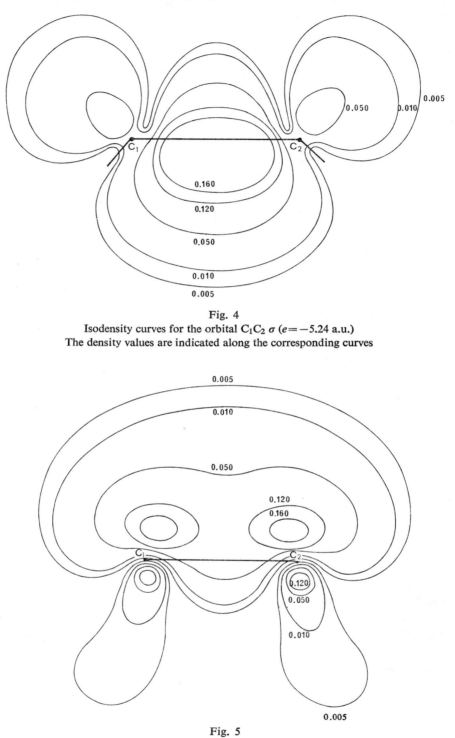

Fig. 4

Isodensity curves for the orbital $C_1C_2\ \sigma\ (e = -5.24$ a.u.)

The density values are indicated along the corresponding curves

Fig. 5

Isodensity curves for the orbital $C_1C_2\ p\ (e = -3.21$ a.u.)

The density values are indicated along the corresponding curves

In 1,4-benzyne the σ orbital between the two carbons C_1 and C_4 is also localized very close to the line C_1C_4, but this orbital is much less stable than the other σ orbitals.

In the π system of 1,2-benzyne, the π bonds may be localized, as indicated in Fig. 2. In fact, the other set of localized orbitals is also found with an adapted starting point in the Boys procedure.

By adding the gross atomic population of the two atoms of each CH group, the σ charge diagrams given in Figs. 6–7 are obtained. They show an excess of σ charges on the two dehydrogenated carbons in 1,2- and 1,3-benzyne. In 1,4-benzyne the reverse is true. In all three molecules there is a lack of π electrons on the dehydrogenated carbons. In 1,2-benzyne

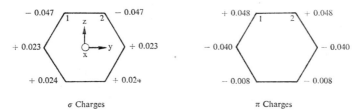

σ Charges   π Charges

Fig. 6

σ and π charges of 1, 2-benzyne

The σ charges are obtained by adding the σ charges of a carbon and of its adjoining hydrogen

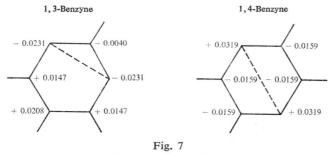

Fig. 7

σ Gross atomic populations

this lack compensates for the excess of σ electrons. In 1,3- and 1,4-benzyne the gross atomic populations are positive (Fig. 8). The dipole moments are rather large: 1.02 D for

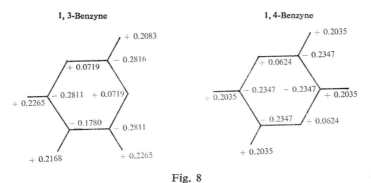

Fig. 8

Gross atomic populations

203

1,2-benzyne and 1,5 D for 1,3-benzyne. In the two molecules they are directed along O_z and correspond to a negative charge transfer from the centre of the ring towards the line passing through the two dehydrogenated carbons.

A second-order perturbation calculation has been done for benzene and for the three benzynes, starting with the localized SCF molecular orbitals. The unperturbed Hamiltonian, H_o, is the one used by Epstein [10] and Nesbet [11]; it is defined with respect to the chosen basis as the diagonal part of the Hamiltonian, H:

$$\langle \Psi_l^0 | H_o | \Psi_k^0 \rangle = \delta_{kl} \langle \Psi_l^0 | H | \Psi_k^0 \rangle,$$

where the functions Ψ_l^0, the eigenfunctions of H_o and of S^2, are Slater determinants or linear combinations of Slater determinants built with the same basis of molecular orbitals.

With such a definition of H_o, the perturbation is the non-diagonal part of H:

if $H = H_o + V$. $$\langle \Psi_l^0 | V | \Psi_k^0 \rangle = (1 - \delta_{kl}) \langle \Psi_l^0 | H | \Psi_k^0 \rangle,$$

With such a zero-order Hamiltonian, the first-order correction to the energy is equal to zero:

$$E^{(1)} = \langle \Psi_o^0 | V | \Psi_o^0 \rangle = 0;$$

the second-order correction $E^{(2)}$ is

$$E^{(2)} = - \sum_{l \neq 0} \frac{(\langle \Psi_l^0 | H | \Psi_o^0 \rangle)^2}{E_l^0 - E_o^0},$$

where the Ψ_l^0 are all the possible diexcited functions. For benzene, a minimal basis set yields 27,720 terms to be included in the sum over l. For the three benzynes, there are 22,050 terms. The results for these last molecules are given in Fig. 9. It may be noticed that the largest energy lowering is in 1,4-benzyne. According to the second-order perturbation calculation, 1,4-benzyne is more stable than 1,3-benzyne, but less stable than 1,2-benzyne.

In 1,4-benzyne the largest contribution to $E^{(2)}$ comes from the diexcitation involving the $C_1C_4\sigma$ bond, namely, $(C_1C_4\sigma, C_1C_4\sigma) \rightarrow (C_1C_4\sigma^*, C_1C_4\sigma^*)$, and, in this particular case, it is rather large in comparison with the total amount (0.104 a.u.).

In 1,2-benzyne the two first contributions come from diexcitations involving the two bonds between the dehydrogenated carbons, and in 1,3-benzyne the first contribution comes from the diexcitation $(C_1H_1, C_1H_1 \rightarrow C_1H_1^*, C_1H_1^*)$. In these last two molecules the contributions are only a few per cent of the total second-order energy.

From these different results it is possible to estimate the energy necessary for abstracting two hydrogens from benzene in the *ortho, meta* or *para* position. The hydrogen energy, E_H, is the one obtained with the same Gaussian basis (0.9996 a.u.). The results are summarized in Table 3. It is clear that the abstraction energy is too large in 1,3-benzyne; this molecule should be the least stable of the three molecules.

Electronic Structure of 1,2-, 1,3- and 1,4-Dehydrobenzenes

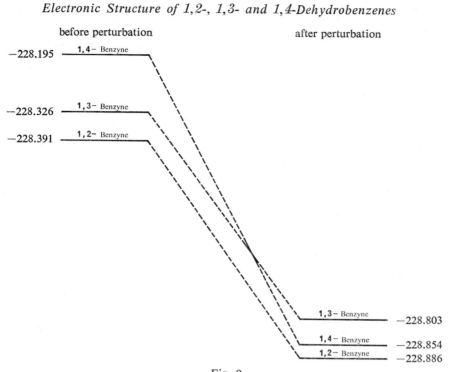

Fig. 9
Total energies (in a.u.)

Table 3

Two Hydrogen Abstraction Energies
(in Kcal mole^{-1})

	$E_{benzyne} + 2 E_H - E_{benzene}$		
	$E_{SCF}^{(0)}$	$E^{(2)}$	E_{Total}
1, 2-Benzyne	193	−36	157
1, 3-Benzyne	235	−25	210
1, 4-Benzyne	317	−139	178

REFERENCES

1 R. S Berry, J. Clardy & M. E Schafer (1964) *J. Am. Chem. Soc.*, 86 : 2738.

2 M. E. Schafer & R. S. Berry (1965) *ibid.*, 87 : 4497.

3 G. Porter & J. I. Strinfeld (1968) *J. Chem. Soc.* (A), p. 877.

4 R. S. Berry, J. Clardy & M. E. Schafer (1965) *Tetrahedron Lett.*, p. 1011.

5 *Ibid.*, p. 1003.

6 E. Clementi, H. Clementi & D. R. Davis (1967) *J. Chem. Phys.*, 46 : 4731.

7 G. W. King & C. K. Ingold (1953) *J. Chem. Soc.*, p. 2725.

8 L. Praud, P. Millie & G. Berthier (1968) *Theor. Chim. Acta* (Berlin), 11 : 169.

9 J. M. Foster & S. F. Boys (1960) *Rev. Mod. Phys.*, 32 : 300.

10 P. S. Epstein (1926) *Phys. Rev.*, 28 : 695.

11 R. K. Nesbet (1955) *Proc. Roy. Soc.* (London), A 230 : 312, 322.

Discussion

E. D. Bergmann:

I do not remember if anybody has measured the half-life time of benzyne. Would you think it is possible to make any measurements during the period of the existence of the molecule, because some of your data are certainly worth trying to verify — I think the dipole moment, or some other constants that one might be able to measure.

Mrs J. Serre:

R. S. Berry and co-workers studied the decomposition of benzenediazonium-4 (or 3)-carboxylate by time-resolved mass spectroscopy; R. S. Berry, J. Clardy & M. E. Schafer (1965) *Tetrahedron Letters*, p. 1003; R. S. Berry, J. Clardy & M. E. Schafer (1965) *ibid.*, p. 1011. They found that the peak at m/e = 76, which very likely corresponds to 1, 4-dehydrobenzene in the decomposition of the first carboxylate, persists for as long as two minutes and that the peak at 76, which must correspond to 1, 3-dehydrobenzene in the decomposition of the second molecule, drops in intensity about 400 μsec. after photolysis.

E. Heilbronner:

Have you done any calculations on the abstraction of one hydrogen from benzene, to give phenyl?

Mrs J. Serre:

The fact is that there are symmetry properties that we will lose if we just look at C_6H_5. Moreover, this last system is an incomplete shell system, and we would have to choose an open-shell method convenient for this problem.

E. Heilbronner:

I am asking this because, according to the calculations which Mr Meyer has reported, the values must be too high by a factor of at least 2 to 3. I have also a question that is perhaps a little bit tongue-in-cheek. If we take the 1, 2-, 1, 3- and 1, 4-benzynes, which of these three is aromatic and which is not?

Mrs J. Serre (added in Nov. 1970):

In the calculations reported at this Symposium the geometry taken for Dewar benzene corresponded to a too short C–C bond: 1.48 Å, instead of 1.56 Å, which has recently been found for the hexamethyl derivative; M. J. Cardillo & S. H. Bauer (1970) *J. Amer. Chem. Soc.*, 92:2399. The results of the computation, repeated with this new geometry, are in good agreement with the experimental values.

G. Berthier:

In conformation problems, the predictions of the SCF theory are not generally modified by the inclusion of second-order correction terms. For instance, this is the case for rotation barriers of ethane-like compounds. The inversion of the total energy calculated for 1, 3-benzyne and 1, 4-benzyne is the only example known until now.

206

Discussion

Mrs J. Serre:

The inversion is very likely due to the small difference between the first highest occupied σ level and the lowest unoccupied σ^* level in 1,4-benzyne; these two orbitals roughly correspond to the 1–4 bond. In fact, the triplet of this molecule should be very close to the ground state.

B. Binsch:

I wonder what you mean by the energy of hydrogen abstraction in your calculations of the energy difference between benzyne and benzene. Does the energy refer to the formation of two hydrogen atoms or one hydrogen molecule?

Mrs J. Serre:

It refers to two hydrogen atoms, which were computed with the same basis, of course.

B. Binsch:

Has the difference in zero-point vibration energy between benzene and benzyne been taken into account?

Mrs J. Serre:

No, we did not take account of this difference in the computations.

B. Binsch:

Would it not be more accurate, or more appropriate, to take the molecule with its zero-point vibration energy, rather than to do what you are doing? That could make some difference, because you are talking about several carbon-hydrogen stretches, which have 1000 wave numbers each.

H. Basch:

Do you have any evidence from your calculations that would indicate that the 1,4-dehydrobenzene might be non-planar in its equilibrium configuration?

Mrs J. Serre:

No. We wanted only to compare it with benzene. It is very clear that the difference between the two σ orbitals is too small.

E. D. Bergmann:

As to 1,4-benzyne, perhaps one should try to dehydrogenate 2,3,5,6-tetramethyl-Dewar benzene in the 1,4-position:

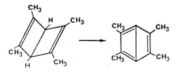

207

The Indenylium Ion

A Study in Anti-Aromaticity

by M. RABINOVITZ

Department of Organic Chemistry, the Hebrew University of Jerusalem

AMONG THE VARIOUS MODIFICATIONS of the general notion of aromaticity, the concept of anti-aromaticity has recently been put forward [1–2]. Hückel's rule of aromaticity, which has been widely used for the explanation of the 'aromatic' character in cyclic conjugated systems and for the prediction of new aromatic systems, implied the 'magic' number of $4n + 2$ π electrons and emphasized the closed-shell configuration strongly stabilized by resonance.

In contrast to the aromatic compounds, a series of non-aromatic cyclic conjugated systems having $4n$ π electrons has been defined as 'pseudo-aromatic'. The elusive cyclobutadiene ($n = 1$) is one of the most discussed of these. Cyclobutadiene is not only an unstable system, but it appears both theoretically and experimentally that the system is destabilized by the conjugation of the two double bonds [3]. Such systems, 'in which electron delocalization considerably raises the energy', have been called anti-aromatic by Breslow [1–2]. A representative case — the cyclopropenyl anion — has been investigated extensively [4]. In simple Hückel theory, the resonance energy of this ion is calculated to be zero; in more sophisticated theories it is negative.

This destabilization, due to conjugation, is exemplified by the comparison of the base-catalysed hydrogen exchange rates of triphenylcyclopropene and triphenylmethane or of cyclopropenes and cyclopropanes. Breslow and co-workers have shown that deuterium exchange in cyclopropane derivatives is at least several hundred times faster than in cyclopropene derivatives, whilst ring strain effects are comparable in both anions. The slower exchange rate in cyclopropenes indicates that the conjugation of the negative charge with a double bond destabilizes the anion of cyclopropene, which thus proves to be anti-aromatic.

The case of the cyclopentadienyl cation — another anti-aromatic system — seems to be similar [5–6]. The pK_{R^+}'s of a number of relevant carbinols have been measured; e.g., the carbonium ions II–IV have been compared. The data show that the 9-phenylfluorenyl cation (II) [7] is less stable than the closely-related triphenylmethyl cation (I), although the former is certainly more planar than the latter and should be more highly conjugated, and more stable, were it not for the effect of anti-aromaticity.

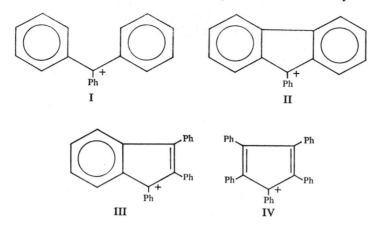

I II III IV

The planar cycloheptatrienyl anion with eight π electrons was expected to be anti-aromatic [8]. Indeed, the anion of heptaphenylcycloheptatriene (V) was found to be relatively unstable; the removal of the proton from the parent hydrocarbon was possible only with very strong bases.

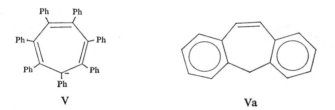

V Va

Similar difficulties have also been encountered in the proton abstraction from 2,3, 6,7-dibenzo-2,4,6-cycloheptatriene (Va) with butyl lithium [9]; they can be ascribed to the same effect. It has been known, of course, that annelation in this case (as in that of fluorene) obliterates to some extent the properties of the fundamental odd-membered ring systems.

It was predicted that any cyclic system with $4n\,\pi$ electrons and three-fold or greater symmetry should be capable of existing as a triplet diradical with the spins of the two electrons unpaired [10]. Examples are the following:

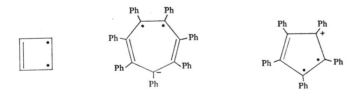

In fact, the heptaphenylcycloheptatrienyl anion (eight π electrons in the ring and a potential seven-fold symmetry) did not appear as a triplet, but only a singlet was detected [8]. On the other hand, in the case of the pentaphenylcyclopentadienyl cation Breslow and co-workers [10] easily detected the triplet state in equilibrium with the singlet state.

209

We selected for our investigation the indenylium cation (as III). We hoped that the two species, A and B, would be distinguishable by physical methods.

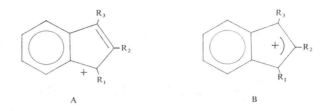

A   B

If the indenylium ion is a classical ion (A), R_1 and R_3 should be different from R_2 (both R_1 and R_3 are substituents on an allylic carbonium ion), while anti-aromaticity in B would demand that not only R_1 and R_3, but also R_2 should show similar properties.

It did not seem advisable to study the unsubstituted indene (R=H), because the hydrogen atoms are so near to the unsaturated system that they would probably be affected, in an undefined manner, by inductive (and perhaps steric) effects. Furthermore, indenol itself and its cation are none too stable. We therefore used the triarylindenols, in which the aryl groups were phenyl and p-tolyl (the methyl groups serving as an excellent indicator) in various combinations [11]; we have also studied some cases in which the methyl group was linked directly to C_2 or C_3, the other substituents being aryl groups.

The required indenols were prepared by the reaction of disubstituted indenones with aryl-magnesium halides* (Table 1). The NMR experiments were carried out on a 100 MHz spectrometer (Varian HA – 100 D) at −70° in methylene chloride as solvent, and consisted in a comparison of the carbinols and the corresponding cations generated by means of boron trifluoride. The solutions of the salts of these carbonium ions were of deep red colour and were fairly stable from −70° to 0°C; the NMR spectrum did not change with time. Addition of water to the purple solutions of, e.g., triphenylindenol and boron trifluoride caused decolouration, and the original carbinol was recovered, so that obviously no structural change had taken place in the course of the reaction.**

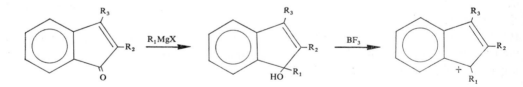

The NMR absorptions of the aromatic protons were very complex and difficult to interpret. However, the methyl hydrogen atoms offered a useful means of answering our question, both when they were attached to the *para*-position of the aryl groups and when they were linked directly to carbon atoms of the five-membered ring. Table 1 summarizes the NMR

---

* All carbinols were analysed and the results were within the limits of the analytical error. The solid compounds were recrystallized, the oils were chromatographed on silica and eluted with benzene-petroleum ether.

** Unsymmetrical carbinols ($R_1 \neq R_3$) when reacted with BF_3 and decomposed with water gave a mixture of carbinols.

Table 1

1, 2, 3-Trisubstituted Indenols [12–13]

NMR Absorption of the Ring Protons of the Substituents *

No.	Substituents	m.p. (solvent)	R_1 (cps)	R_2 (cps)	R_3 (cps)
1	$R_1 = R_2 = R_3 =$ Phenyl	126° (CH$_3$NO$_2$)	745	708	728
2	$R_1 = R_3 = p$-Tolyl $R_2 =$ Phenyl	164° (CH$_3$COOH)	747	707	726
3	$R_2 = p$-Tolyl $R_1 = R_3 =$ Phenyl	156° (CH$_3$NO$_2$)	747	694	724
4	$R_3 = p$-Tolyl $R_1 = R_2 =$ Phenyl	174° (CH$_3$NO$_2$)	750	709	727
5	$R_1 = p$-Tolyl $R_2 = R_3 =$ Phenyl	142° (CH$_3$COOH)	744	708	723
6	$R_1 = R_2 = p$-Tolyl $R_3 =$ Phenyl	—**	745	690	723
7	$R_2 = R_3 = p$-Tolyl $R_1 =$ Phenyl	158° (CH$_3$NO$_2$)	745	691	720
8	$R_1 = R_2 = R_3 = p$-Tolyl	156° (CH$_3$NO$_2$)	747	690	721
9	$R_1 = p$-Tolyl; $R_2 =$ Me $R_3 =$ Phenyl	124° (CH$_3$NO$_2$)	746	(176)	727
10	$R_1 = p$-Tolyl; $R_3 =$ Me $R_2 =$ Phenyl	—**	744	709	(200)

* Centre of absorption

** The compound did not crystallize; it was chromatographed and analysed

spectral data of the indenols, and Table 2 the absorption of the methyl group in the various carbonium ions.

The differences in the chemical shift of the methyl hydrogen atoms between the carbinol and the carbonium ion are identical in position 1 (and 3) and in position 2. If the carbonium ion is only an allylic ion, the chemical shifts of the methyl hydrogen atoms at positions 1 and 2, relative to the corresponding carbinols, should be different. As this is not the case, the results indicate that the carbonium ion is not simply an allylic ion.

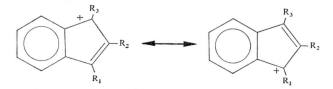

The anti-aromaticity of the system is established by comparison of the NMR spectra of the indenols and the corresponding cations, in which the methyl groups are — directly or indirectly — linked to carbon atoms 2 and 1 (or 3).

When the substituents are *p*-tolyl, there is a constant chemical shift difference to low field of the order of 16–18 cps; when the substituents are methyl, the shift increases to 40–50 cps to low field [12–13]. This indicates that the charge is evenly distributed between atoms 1, 2 and 3 of the five-membered ring (which would not be expected for a 'normal' allylic ion). A more detailed NMR analysis is complicated by the existence in the substituents of 12–15

211

Table 2

Methyl Proton Absorption of 1, 2, 3-Trisubstituted Indenols
and the Corresponding Carbonium Ions

No.	Substituent	Indenol			Carbonium Ion			Δ cps		
		R_1	R_2	R_3	R_1	R_2	R_3	R_1	R_2	R_3
2	$R_1 = R_3 = p$-Tolyl R_2 = Phenyl	234	—	238	252	—	252	18	—	14
3	$R_2 = p$-Tolyl $R_1 = R_3$ = Phenyl	—	220	—	—	238	—	—	18	—
4	$R_3 = p$-Tolyl $R_1 = R_2$ = Phenyl	—	—	236	—	—	252	—	—	16
5	$R_1 = p$-Tolyl $R_2 = R_3$ = Phenyl	233	—	—	250	—	—	17	—	—
6	$R_1 = R_2 = p$-Tolyl R_3 = Phenyl	233	221	—	251	239	—	18	18	—
7	$R_2 = R_3 = p$-Tolyl R_1 = Phenyl	—	220	237	—	239	252	—	19	15
8	$R_1 = R_2 = R_3 = $ p-Tolyl	233	219	236	251	237	251	18	18	15
9	$R_1 = p$-Tolyl; R_2 = Me R_3 = Phenyl	231	176	—	250	228	—	19	52	—
10	$R_1 = p$-Tolyl; R_3 = Me R_2 = Phenyl	232	—	200	250	—	240	18	—	40

hydrogen atoms, the absorptions of which are superimposed on the spectrum of the AA'BB' four-spin system at positions 4, 5, 6 and 7.

On the strength of these results, we resumed the study of Breslow's pentaphenylcyclopentadienyl cation. Our results are not quite consistent with those reported by Breslow [5–6]. We observed the formation of a deep-blue solution of the pentaphenylcyclopentadienyl cation at −70°; it was impossible to identify any absorption of the aryl protons due to the paramagnetism of the solution. At lower temperature, there appeared an intense and well-defined peak, which disappeared completely after a short while without a visible change in the colour of the solution. The peak lies by 50 cps in lower field, compared with the absorption of the aryl protons in the carbinol. It is thus obvious that the situation is more complex than Breslow assumed: there exists a precursor of the blue species, described by Breslow, which is not paramagnetic and therefore gives an NMR spectrum.

REFERENCES

1 R. Breslow (1968) *Angew. Chem.* (*Int. Edn*), 7 : 565.
2 Idem (1965) *Chem. Eng. News*, 43 : 90.
3 M. P. Cava & M. J. Mitchell (1967) *Cyclobutadiene and Related Compounds*, Academic Press, New York.
4 R. Breslow, J. Brown & J. J. Gajewski (1967) *J. Am. Chem. Soc.*, 89 : 4383.
5 R. Breslow & H. W. Chang (1961) *ibid.*, 83 : 3727.
6 R. Breslow, H. W. Chang & W. A. Yager (1963) *ibid.*, 85 : 2033.
7 E. A. Chandross & C. F. Sheley Jr (1968) *ibid.*, 90 : 4345.
8 R. Breslow & H. W. Chang (1965) *ibid.*, 87 : 2200.
9 E. D. Bergmann, D. Ginsburg, Y. Hirshberg, M. Mayot, A. Pullman & B. Pullman (1951) *Bull. Soc. Chim. France*, 18 : 697.
10 R. Breslow, H. W. Chang, R. Hill & E. Wasserman (1967) *J. Am. Chem. Soc.*, 89 : 1112.
11 C. F. Koelsch (1934) *ibid.*, 56 : 1337.
12 N. C. Deno (1963) *ibid.*, 85 : 2991.
13 G. A. Olah (1964) *ibid.*, 86 : 5682.

Discussion

D. J. Bertelli:

I am curious whether or not you looked at the NMR spectra of your carbonium ions in terms of the coupling constants in the six-membered ring, because, if I understand you correctly, you are trying to evaluate the perturbation that this kind of system has on the benzene ring. We know that as one takes a benzene ring of that nature and perturbs it, one should get a divergence of the vicinal coupling constants in the six-membered ring. I wonder if you had looked at this possibility.

M. Rabinovitz:

The aromatic spectrum was very complex. We did not actually look into it, because we have a very good probe to measure the charges produced in the five-membered ring by having sharp singlets, as shown for positions 1, 2 and 3. We did not try to evaluate the perturbation in the benzene ring.

D. J. Bertelli:

Well, I would question whether or not arguments based on chemical shifts are really that good in terms of the charge distributions, because even if the ion looks like B (supra, p. 210), it is not clear that the charge distribution should be equivalent at R_1 and R_2. How do you know that this should even be true? Did you try MO calculation?

M. Rabinovitz:

I think that if you take a series of the same compounds, not changing the skeleton, and measure the same substituent, you could gain at least qualitative knowledge.

D. J. Bertelli:

You have a correlation, but does the correlation mean anything? In other words, if you change R_3 and R_2 etc. in a systematic manner, you get a systematic result, but that systematic result does not really reflect what the charge densities might be on those various carbon atoms.

M. Rabinovitz:

What our results clearly show is that the indenylium cation is not a regular allylic ion, a property that we attribute to its anti-aromaticity. The charge distribution that we can study is limited to positions 1, 2 and 3 only. In regular cyclic allylic ions the methyl substituents show a difference of more than 0.7 ppm between their absorptions in position 1, 3 and 2, while we observe a difference of only 0.1 ppm. As to the analysis of the spectra in order to observe any perturbations at the benzene ring, it is difficult to analyse the spectra of the indenylium ions. These spectra were run in dilute solutions at low temperatures and high gain. Even the symmetrical ions, e.g. 1 and 8 in Table 2, have 12–15 aromatic protons of the phenyl substituents in the same region as the four AAB′B′ or ABCD protons to be analysed. This is why we could not study whether there is a perturbation of the benzene rings or not.

D. J. Bertelli:

Forgive me for throwing in some of my own results here, but I think it is a good idea because of a series of compounds that we looked at recently.

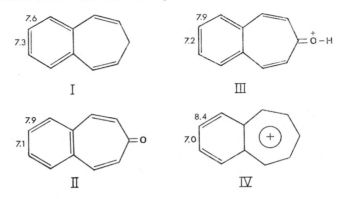

If we take a look at the NMR spectra of the six-membered ring in this series, we get very much what I think you are looking for in the sense that if we would look at the vicinal coupling constants in I, they differ by about 0.3 cycles per second. If we go to the tropone, this may go to 0.8; if we go to the conjugate cation — I am not exactly sure of these numbers, they are approximate — we have about 0.7; if we go to the tropenium cation, about 1.4. So I think, to a certain extent, these coupling constants are a reasonably good indication of the perturbation of the second ring upon the π system of the benzene. And this is why I suggest that your systems look quite adaptable to this kind of analysis.

M. Cais:

But, as you yourself pointed out, this perturbation must not necessarily be due to a difference in charge distribution; there can be other factors involved.

E. D. Bergmann:

On the other hand, the influence of the seven-membered ring on your six-membered ring should get an entirely different order of magnitude than the one which you have here, because you are not getting a condensed ring system in the R_1, R_2 and R_3. So, the effect of the odd-membered ring on your benzene ring might well be different from the effect of the odd-membered ring in our case.

New Developments
in the Chemistry of Push-Pull Cyclobutadienes

by R. GOMPPER *and* G. SEYBOLD

Institut für Organische Chemie der Universität München

THERE HAS BEEN a remarkable increase in the number of papers on cyclobutadiene chemistry during the last years, and an excellent summary has been published [1].

Molecular orbital theory predicts that cyclobutadiene should be a highly unstable species, due to the fact that it is an anti-Hückel $4n$ π system. This prediction is in accord with the known chemistry of cyclobutadiene and its alkyl-, aryl- and halogen-substituted derivatives. In spite of this discouraging prospect, chemists have made numerous attempts to synthesize cyclobutadienes in which the instability of the anti-Hückel system would be decreased by introducing some special structural or electronic elements into the cyclobutadiene molecule. Roberts [2] suggested that push-pull substituents might be sufficient in stabilizing cyclobutadiene, and the synthesis of stable push-pull *o*-quinodimethanes [3] in our laboratory prompted us to attempt a synthesis of cyclobutadiene with two pairs of electron-releasing and electron-attracting substituents. The HMO diagram, resulting from

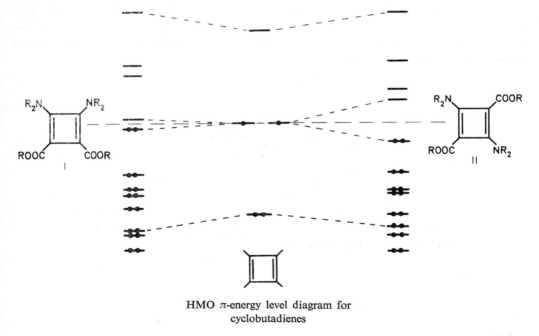

HMO π-energy level diagram for
cyclobutadienes

first-order perturbation calculations, shows that 'cross push-pull substituted' cyclobutadiene II should be more stable than the corresponding isomer I.

In order to synthesize II, many futile experiments had to be carried out before the following procedure [4], based on Viehe's work on ynamines [5], could be successfully applied.

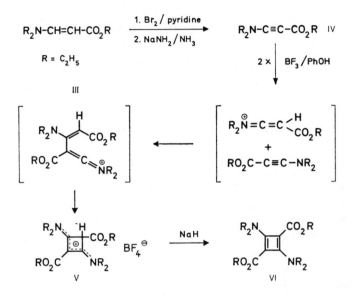

Ethyl β-diethylaminoacrylate (III) was brominated to ethyl-α-bromo-β-diethylamino-acrylate, which, after treatment with sodamide, gave ethyl diethylaminopropiolate (IV) in fair yield. In the presence of a mixture of boron trifluoride and phenol, the propiolate IV 'dimerizes' to the cyclobutenium salt V. Salt V is a surprisingly strong acid, as was shown by H–D exchange. With sodium hydride in benzene or tetrahydrofuran, V forms its conjugate base, the pale-yellow cyclobutadiene VI, almost quantitatively. Some properties of VI, which is the first stable monocyclic cyclobutadiene that has been made, are listed in the following table.

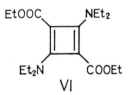

VI

mp:	50–52° C
IR (in CCl_4):	1,675 cm^{-1}, 1,620 cm^{-1}, 1,555 cm^{-1}, 1,480 cm^{-1}
UV (in THF):	295 nm, log $\varepsilon = 4.18$
MS:	338 (100%) 309 (65%) 293 (30%) 249 (40%) 72 (40%) 29 (40%)
NMR (in C_6D_6):	$\tau = 5.84$ (O–CH$_2$–), $\tau = 6.08$ (N–CH$_2$–), m; $\tau = 8.93$, t
Mol. wt. (osm.):	f. 339, calc. 338.

Two properties of the push-pull cyclobutadiene **VI** are especially remarkable: its thermal stability and its sensitivity towards water and other protonic solvents. Thus, the main problem in handling **VI** is to avoid exposure to moisture extremely carefully.

A *square* structure for the 'crossed' isomer **II** is inferred from calculated bond orders (Scheme 1). Recent calculations of R. Weiß and J. N. Murrell [private communication] favour a rectangular structure for **II** as well as for **VI**, due to second-order bond fixation. All attempts to synthesize the benzocyclobutadiene **VII** have failed so far. Further

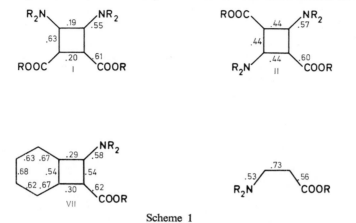

Scheme 1

evidence for the reliability of predictions based on these HMO calculations is obtained by comparing calculated C=O bond orders with $\gamma_{C=O}$ of typical carbonyl compounds (Fig. 1).

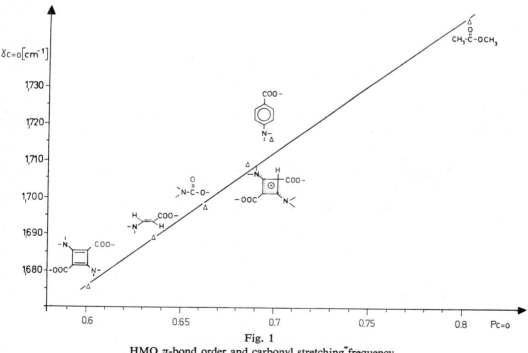

Fig. 1

HMO π-bond order and carbonyl stretching frequency

Finally, there is a rather good correlation between the experimental long-wavelength absorptions and the HMO transition energies calculated (Fig. 2).

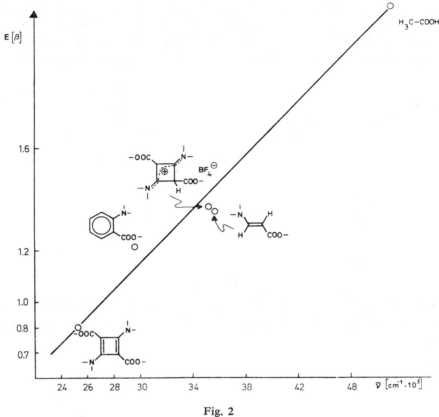

Fig. 2
Long-wavelength absorption and HMO transition energies

The simple VB theory is less suitable to explain the different stabilities of the cyclobutadienes I and II (Scheme 2).

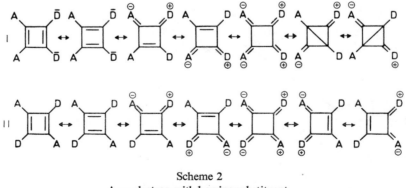

Scheme 2
A = electron-withdrawing substituent
D = electron-donating substituent

Formally, both compounds possess the same number of mesomeric structures. However, these structures are of different energies. The large number of favourable dipolar resonance structures may be used as an argument that VI should be a strong nucleophile, as well as a strong electrophile. The following observations proved this assumption to be correct: a characteristic feature of VI, in particular, is its extremely high reactivity towards electrophiles. Scheme 3 summarizes reactions of VI with simple electrophiles. Protonation leads to the cyclobutenium salt V, which is used as a starting material for the synthesis of VI (see above); this provides one form of evidence for the structure of VI. Similar cyclobutenium salts are formed with methyl iodide, bromine, acetyl bromide and tropylium perchlorate, as can be inferred from their IR and NMR spectra. The bromine adduct VIII eliminates bromine on heating.

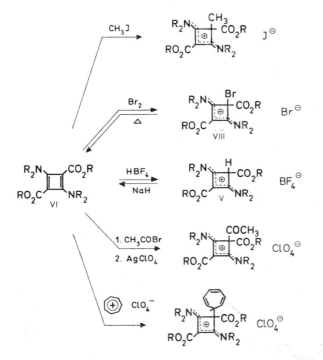

Scheme 3

Reaction of VI with water (cf. Scheme 4) gives rise to the formation of IX and X, whose structures are proposed by Neuenschwander [6]. The mechanism of the production of IX became apparent from the corresponding reaction of VI with methanol, which afforded the substituted butadiene XII. Obviously, addition of methanol to VI gives the intermediate push-pull substituted cyclobutene XI, which undergoes rapid ring opening. The salt V is reduced by $NaBH_4$ in acetonitrile to the butadiene XIII. IR and NMR spectra of XII and XIII are almost identical.

No reactions of other known cyclobutadienes with electrophiles have been reported which would make comparison of reactivities of the various cyclobutadienes possible. However, since cycloaddition reactions have been studied for a number of cyclobutadienes, we were anxious to learn about the reactivity of VI in cycloaddition reactions. By using acetylenes

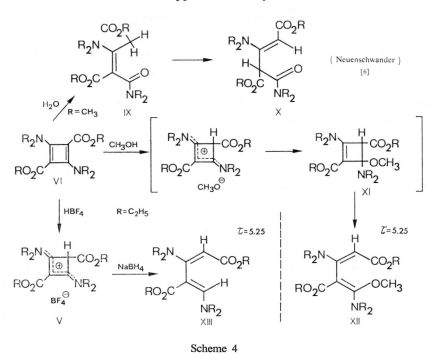

Scheme 4

as dienophiles, we might expect the formation of various Dewar benzenes. However, even at $-80°C$ only the benzene derivatives XIV and XV were isolated.

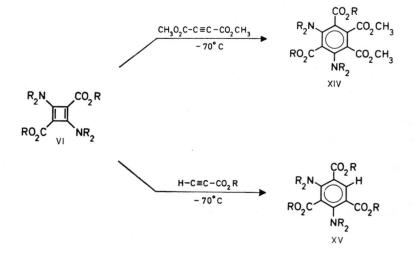

This gives rise to two questions, illustrated by Scheme 5: First, why are push-pull substituted Dewar benzenes (XVI) assumed to be intermediates in the formation of XIV and XV, which are thermally unstable? Second, do cycloaddition reactions of acetylenes

220

to VI occur *via* a concerted reaction mechanism, or do they proceed in a two-step reaction with dipoles XVII as intermediates?

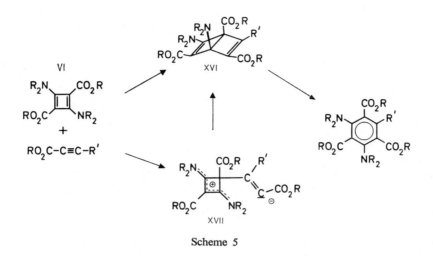

Scheme 5

The first question may be answered by means of a simple HMO model (Fig. 3). Fig. 3 shows the calculated energy profile diagrams for the ring-opening reactions of the unsubstituted and a push-pull substituted Dewar benzene. The considerably lower activation energy of the push-pull system accounts for the fact that it is impossible to isolate compounds of this type. Furthermore, according to Fig. 3, valence isomerization of XVI involves a polar transition state. Partial positive and negative charges arise at positions 1

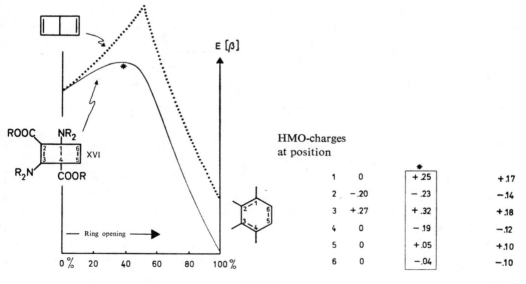

HMO-charges at position

position		*	
1	0	+ .25	+ .17
2	− .20	− .23	− .14
3	+ .27	+ .32	+ .18
4	0	− .19	− .12
5	0	+ .05	+ .10
6	0	− .04	− .10

Fig. 3

221

and 4 by going along the reaction coordinate from XVI to the transition state, and then disappear during the further reaction. In a formal manner this is shown in Scheme 6.

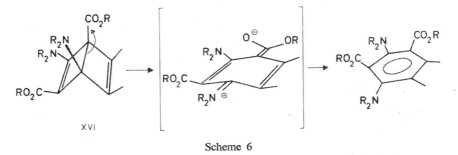

Scheme 6

Heterolytic fission of the strained C_1–C_4 bond is facilitated, since the developing positive and negative charges can be stabilized by the electron-releasing and electron-attracting substituents. The energy barrier for the ring opening, as expected from Woodward-Hoffmann rules, is thus decreased.

The second question with regard to the mechanism of the cycloaddition may be answered as well.

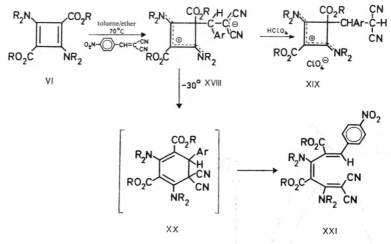

Cyclobutadiene VI reacts with *p*-nitrobenzylidenemalononitrile at $-70°$C to form an orange-red crystalline product which, upon protonation with perchloric acid, is converted into the stable colourless perchlorate XIX. Structure XIX is confirmed by the infra-red spectrum. Thus, the orange-red intermediate should be the dipolar compound XVIII. At $-30°$C, XVIII is transformed into an orange-yellow product, which was shown to be the hexatriene XXI by its NMR spectrum. XXI is obviously formed *via* the cyclohexadiene derivative XX.

These results allow to conclude that cycloaddition reactions to push-pull substituted cyclobutadienes generally proceed through dipolar intermediates. A further support for this view is the fact that VI reacts with TCNE to form the stable dipolar molecule XXII.

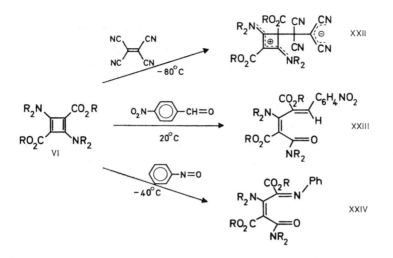

From the reactions of VI with *p*-nitrobenzaldehyde and nitrosobenzene, however, only the open-chain products, XXIII and XXIV, could be isolated, very probably because of the instability of the primarily formed cycloadducts.

The only cycloaddition reactions of VI that result in the formation of isolable cyclohexa-dienes (XXV and XXVI) are those with N-phenylmaleimide and 4-phenyl-1,2,4-triazolin-3,5-dione. The structure of XXVI has not yet been proved completely; however, since XXVI is colourless, a 9-membered ring structure is very unlikely.

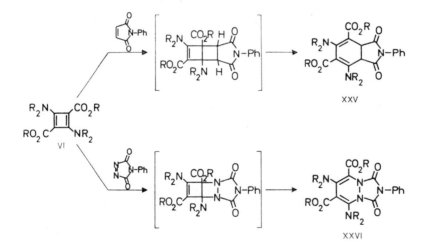

Cycloadditions of heterocumulenes to VI proceed as shown in the following scheme, 6-membered rings being the only products. Structural evidence is obtained mainly from spectral data.

223

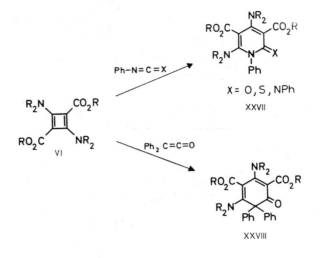

The typical reaction sequence, cycloaddition followed by valence isomerization, is also observed for addition reactions of methylene-cyclopropenes to VI. The dicyanomethylene-cyclopropene XXVII and the related quinocyclopropene derivative XXVIII afford the methylene-cycloheptatrienes XXIX and XXX in good yields. XXIX shows an intensely yellow colour, XXX a deep blue colour.

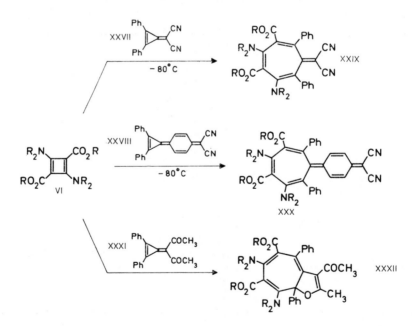

Diacetylmethylene-diphenylcyclopropene (XXXI) and VI, however, form a colourless 1:1 adduct. From IR and NMR data, the furocycloheptane structure XXXII is tentatively assigned to this adduct. Its formation may be explained as follows:

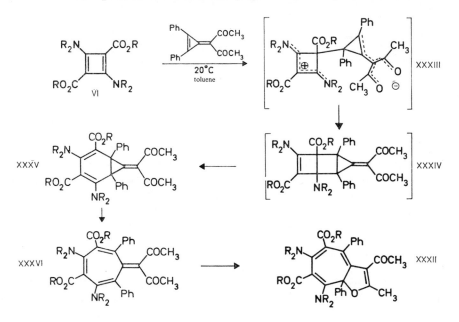

XXXI and VI react to form the dipolar molecule **XXXIII**, which cyclizes to **XXXIV**. This intermediate subsequently isomerizes to the norcaradiene **XXXV**, which rearranges to the methylene-cycloheptatriene **XXXVI**. Reactions of VI with the dicyanomethylene-cyclopropenes, **XXVII** and **XXVIII**, stop at this stage. However, because of its strong polar nature, **XXXVI** is transformed into **XXXII**. Partial negative charges on the oxygen atoms of **XXXVI** and a formal positive charge in the 7-membered ring account for this result.

The physical properties of VI are in agreement with those expected for a 'real' cyclobutadiene. Its reactions are characterized by the facile formation of polar transition states as well as of polar intermediates possessing cyclobutenium structure.

REFERENCES

1 M. P. Cava & M. J. Mitchell (1967) *Cyclobutadiene and Related Compounds*, Academic Press, New York–London.

2 J. D. Roberts (1958) *Special Publication, No. 12*, The Chemical Society, London, p. 111.

3 R. Gompper, E. Kutter & H. Kast (1967) *Angew. Chem.*, 79 : 147.

4 R. Gompper & G. Seybold (1968) *ibid.*, 80 : 804.

5 H. G. Viehe (1967) *ibid.*, 79 : 744.

6 M. Neuenschwander & A. Niederhauser (1970) *Helv. Chim. Acta*, 53 : 519.

Discussion

J. J. C. Mulder:

I would like to make the following remark: You have formulated dipolar intermediates in your cycloaddition reactions. It could very well be that, if you had not done this, but had formulated this as a concerted reaction, you would have ended up with a reaction which is anti-Woodward-Hoffmann. At least in the case where you have no evidence for the appearance of a Dewar benzene, i.e., where you get directly benzene in a thermal reaction, this could be an anti-Woodward-Hoffmann case.

R. Gompper:

I think in cases where we have highly polar systems you do not need the Woodward-Hoffmann rule. You have always stabilization of polar or dipolar intermediates. We actually isolated such dipolar intermediates, and we think that in all these cases of push-pull systems, or polar systems, it is easier to do without the Woodward-Hoffmann rules and to postulate non-concerted reactions.

J. J. C. Mulder:

I concur wholeheartedly with that, although it is possible to show by calculation that if you have a suitably substituted system of this kind, the reaction should go anti-Woodward-Hoffmann.

M. Cais:

I would like to ask two questions — the first with regard to this square *v.* rectangular structure. Would you care to comment on the proposition that cyclobutadiene iron tricarbonyl has a square structure, whereas cyclobutadiene itself has been suggested by Dewar to be a rectangular structure?

The second question is: Are you quite certain that in your cyclobutenium ion there is no species where the charge has gone on to one of the nitrogens?

R. Gompper:

It is easy to see from the spectrum that the charge sits on the carbon. If you make some HMO calculation, you find that all the bonds in the ring are of the same order, and this is not so surprising, as you have a highly symmetric compound. So, it is most likely that the molecule is square. This fits the NMR data.

D. J. Bertelli:

I would like to comment on your NMR data, because I don't think they really indicate what you said. If I follow you correctly, you are trying to say that the equivalence of the alkyl groups on the nitrogen indicates that the ground state is square, but this is really meaningless, because a rapid valence bond isomerization would give you exactly the same result.

R. Gompper:

I think we have only a hint, no more; anyhow, less than absolute evidence.

226

Bond Distortions and Related Phenomena in Non-Benzenoid Aromatic Hydrocarbons

by T. NAKAJIMA, A. TOYOTA *and* H. YAMAGUCHI

Department of Chemistry, Tohoku University, Sendai, Japan

I. INTRODUCTION

RECENTLY, Binsch et al. [1–5] have developed an elegant general theory of treating the problem of double-bond fixation in conjugated π-electron systems. This theory allows a sharp distinction to be made between first-order bond fixation, from which bond-length–bond-order relationships may be derived, but which leaves the full symmetry of the molecule unaffected, and second-order bond fixation, which may result in a symmetry reduction. Information about second-order bond fixation is obtained by examining the eigenvalues and eigenvectors of the bond–bond polarizability matrix. The most favourable second-order bond distortion is given by the eigenvector belonging to the largest eigenvalue λ_{max}. If λ_{max} becomes greater than a certain critical value, second-order effects in the π-electron energy overcome the σ-bond compression energy, and the molecule will, in general, lose its original full symmetry.

On the basis of this theory, Binsch et al. have examined the second-order effects on bond lengths and stability of non-alternant π-electron systems and proposed a new aromaticity criterion that is entirely based on double-bond fixation.

However, as they themselves admit, the theory of Binsch et al. gives only the type of second-order bond distortions — i.e., the normalized components of the displacement co-ordinate corresponding to the energetically most favourable distortion — and does not provide information about actual magnitudes of distortions or equilibrium bond distances at which the nuclei of the real molecule will settle.

In this paper, we present a general method of predicting the energetically most favourable geometrical structure — i.e., equilibrium bond distances of conjugated hydrocarbons — without presuming the type of bond distortion (that is, alternation) [6].

II. METHOD OF CALCULATION

The method of calculation used is the self-consistent formalism of the Pariser-Parr-Pople method, in which bond lengths and, consequently, the resonance and Coulomb repulsion integrals are allowed to vary with bond order at each iteration until self-consistency is reached [7]. The C–C bond lengths r are correlated with bond orders p by the aid of the formula $r(\text{Å}) = 1.520 - 0.186p$.

The Coulomb repulsion integrals are calculated using the Mataga-Nishimoto formula.

The resonance integral is assumed to be in the form $\beta = Be^{-ar}$, the value of a being taken as 1.7Å^{-1} [7].

As the starting geometrical structures for iterative calculation, we adopt various distorted ones in which bond lengths are distorted, so that the set of displacement vectors may form a basis for an irreducible representation of the full symmetry group of the molecule. In case of pentalene (I), for example, there are 3, 2, 2 and 2 distinct bond distortions belonging to a_g, b_{1g}, b_{2u} and b_{3u}, respectively. If self-consistency is achieved at two or more different nuclear arrangements, the total energies should be compared with each other in order to determine which one is most favourable. The total energy is assumed to be the sum of the π-electron energy and the σ-electron energy, the latter being calculated by using the harmonic oscillator model, with the force constant equal to 714 Kcal/Å [8].

III. Results and Discussion

1. *Neutral Non-Alternant Hydrocarbons*

The equilibrium bond lengths at the energetically most favourable geometrical structure of the molecules examined (I–XVI; see figure) are listed in Table 1 and compared with available experimental data.

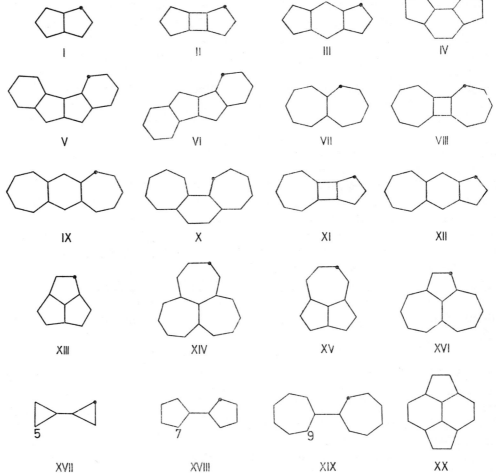

Table 1
Bond Lengths of Non-Alternant Hydrocarbons*

Molecule	Bond	Bond length (Å)	Molecule	Bond	Bond length (Å)	Molecule	Bond	Bond length (Å)
I	1–2	1.356		5–16	1.409		10–11	1.435
	1–8	1.469		6–7	1.364		11–12	1.404
	2–3	1.450		7–15	1.455	XII	1–2	1.371
	3–4	1.370		15–16	1.467		1–14	1.441
	4–8	1.446	VII	1–2	1.357		2–3	1.428
II	1–2	1.383		1–12	1.460		3–4	1.382
	1–10	1.419		2–3	1.450		4–5	1.435
	2–3	1.420		3–4	1.362		4–14	1.462
	3–4	1.381		4–5	1.445		5–6	1.376
	4–5	1.431		5–6	1.371		6–7	1.443
	4–10	1.488		6–12	1.453		6–12	1.460
	9–10	1.391	VIII	1–2	1.367		7–8	1.369
III	1–2	1.424		1–14	1.442		8–9	1.434
	1–12	1.392		2–3	1.440		9–10	1.373
	2–3	1.372		3–4	1.366		10–11	1.429
	3–4	1.447		4–5	1.442		11–12	1.382
	4–5	1.381		5–6	1.366		12–13	1.436
	4–12	1.444		6–7	1.460		13–14	1.376
	11–12	1.435		6–14	1.471	XIII	1–2	1.358
IV	1–2	1.449		13–14	1.381		1–9	1.468
	1–12	1.371	IX	1–2	1.430		2–3	1.474
	2–3	1.357		1–16	1.385		2–10	1.464
	3–4	1.463		2–3	1.371		3–4	1.351
	4–5	1.368		3–4	1.439		4–5	1.463
	4–12	1.454		4–5	1.364		5–10	1.370
	5–6	1.442		5–6	1.452	XIV	1–2	1.450
	11–12	1.458		6–7	1.375		1–15	1.360
V	1–2	1.366		6–16	1.449		2–3	1.368
	1–16	1.440		15–16	1.440		3–4	1.461
	2–3	1.437	X	1–2	1.443		3–16	1.456
	3–4	1.364		1–16	1.376		4–5	1.356
	4–5	1.446		2–3	1.363		5–6	1.454
	5–6	1.375		3–4	1.449		6–7	1.358
	5–16	1.449		4–5	1.358		7–8	1.453
	6–7	1.459		5–6	1.457		8–16	1.379
	7–8	1.376		6–7	1.375	XV	1–2	1.404
	7–15	1.439		6–16	1.450		1–11	1.397
	8–9	1.451		7–8	1.432		2–3	1.397
	9-10	1.404		15–16	1.451		3–4	1.435
	9–14	1.413	XI	1–2	1.374		3–12	1.445
	10–11	1.396		1–12	1.434		4–5	1.371
	11–12	1.398		2–3	1.426		5–6	1.437
	12–13	1.397		3–4	1.380		6–12	1.395
	13–14	1.400		4–5	1.463	XVI	1–2	1.399 (1.412)**
	14–15	1.462		4–12	1.448		1–13	1.400 (1.393)
	15–16	1.390		5–6	1.381		2–3	1.433 (1.437)
VI	1–2	1.397		5–11	1.446		2–14	1.442 (1.472)
	1–16	1.399		6–7	1.427		3–4	1.374 (1.380)
	2–3	1.397		7–8	1.376		4–5	1.429 (1.434)
	3–4	1.396		8–9	1.429		5–6	1.373 (1.367)
	4–5	1.402		9–10	1.374		6–7	1.435 (1.446)
	5–6	1.457					7–14	1.400 (1.429)

* Atomic positions are numbered consecutively, starting from the dotted position in the figure and proceeding in a clockwise fashion along the periphery.

** X-ray bond lengths for 3, 5, 9, 11-tetramethylaceheptylene [9]

It should be noted that in all the molecules, except for IV, VI, X, XV and XVI, two different self-consistent nuclear arrangements, one belonging to the full symmetry group of the molecule and the other belonging to a reduced symmetry group, are obtained. In pentalene (I), for example, the starting bond distortions, belonging to a_g, b_{2u} and b_{3u} representations, all converge into the unique self-consistent set of bond lengths belonging to the molecular symmetry group D_{2h}, and distortions belonging to b_{1g} converge into another set of bond lengths belonging to C_{2h}. In such a case, the nuclear arrangement belonging to the lower molecular symmetry group should, in principle, be energetically favoured as compared with that belonging to the full molecular symmetry group. The stabilization energies which favour the lower-symmetry nuclear arrangement for I, II, III, V, VII, VIII, IX, XI, XII, XIII and XIV are predicted to be 8.4, 0.6, 2.4, 7.2, 12.1, 5.3, 5.9, 2.5, 6.6, 2.7 and 8.6 Kcal/ mole, respectively.

It may thus be stated that molecules IV, VI, X, XV and XVI exhibit only the first-order bond distortions and show no higher-order effects which result in a symmetry reduction, while all the remaining molecules, excepting XIII and XIV, which suffer the pseudo-Jahn-Teller effect, suffer more or less higher-order bond distortions, the extent of which is estimated from the stabilization energy defined above. In all the latter molecules, the first-order bond lengths of the peripheral bonds are nearly equalized, and the double-bond fixations of these molecules are due essentially to the higher-order effects.

Of the molecules examined, pentalene (I) and heptalene (VII) show a marked tendency to the higher-order double-bond fixation. In both cases, there exists strong bond alternation in the peripheral carbon skeleton. This prediction is in agreement with the previous theoretical investigations [8, 10–13] and available experimental facts [14]. The *s*-indacene (III), recently prepared by Hafner's group [15], and its seven-membered analogue IX exhibit moderate double-bond fixations, higher-order effects being stronger in IX than in III. As for III, this conclusion is in agreement with results obtained previously by a different method [16] and with experimental information. It is interesting to note that molecule XII, whose perimeter is composed of $4n + 2$ carbon atoms, shows double-bond fixation to a greater extent than III or IX.

In *as*-indacene (IV) and its seven-membered analogue X the higher-order effects are absent. This is in contradiction with the earlier results obtained by Binsch et al. [3] on the one-electron model, but in agreement with their later results in the SCF scheme [5]. In these molecules, two different self-consistent geometrical structures, corresponding to the two possible Kekulé-type structures, are obtained. The differences in total energies between the two structures are 4.3 and 3.0 Kcal/mole for IV and X, respectively.

Of particular interest is dibenzopentalene (V), which suffers strong higher-order effects. In one of the six-membered rings of this molecule, bond lengths are smoothed out as in benzene, while in the other ring a strong double-bond fixation exists. This conclusion is not in agreement with the prediction due to Binsch et al. [3, 5], which indicates that the two six-membered rings are equivalent and that both of them exhibit moderate double-bond fixation.

In the so-called 'bowtiene' (II), predicted to be non-aromatic by Binsch et al., higher-order effects turned out to be of an intermediate magnitude. The double-bond fixations in the periphery of this molecule are of approximately the same order of magnitude as in naphthalene; the molecule, therefore, provides an example in which a clear-cut distinction between aromatic and non-aromatic character cannot be made. On the other hand, the seven-membered analogue of bowtiene (VIII) exhibits strong higher-order double-bond fixation, and it should be non-aromatic.

In all the molecules examined, higher-order effects in molecules containing seven-membered rings are stronger than in the five-membered analogues.

2. *Dianions of Non-Alternant Hydrocarbons*

The dianions of I, III, IV and VII have been examined. In these dianions, higher-order double-bond fixations are absent, and bond lengths are largely smoothed out. Dianions of I, III and IV have been prepared, and experimental information [17–21] is in accordance with these conclusions.

3. *Anions and Cations of Fulvalenes*

Recently, the cation and anion radicals of heptafulvalene have been prepared, and their ESR spectra have been studied by Sevilla et al. [22]. The hyperfine spectrum of the cation radical is consistent with three sets of four protons. On the other hand, the low-temperature spectrum of the anion radical is consistent with two groups of two protons. Comparison of these splittings with those of the higher-temperature spectra reveals that the unpaired spin in the heptafulvalene anion is essentially localized on one of the rings, and, therefore, that the molecular symmetry group of the anion radical should be C_{2v}.

The nuclear arrangements of fulvalene radicals are calculated using the open-shell SCF-MO method [23], in conjunction with the variable bond-length technique.

In anion radicals of triafulvalene (XVII) and heptafulvalene (XIX) and in the cation radical of pentafulvalene (XVIII), the starting distortions, belonging to a_g, b_{1g} and b_{2u} irreducible representations of the point group D_{2h}, all converge into the unique self-consistent set of bond lengths belonging to the symmetry group D_{2h}, and distortions belonging to b_{3u} converge into another set of bond lengths belonging to the point group C_{2v}, the stabilization energies being predicted to be 60.2, 9.4 and 17.8 Kcal/mole, respectively. The calculated bond lengths and spin densities listed in Table 2 indicate that in one of the rings of these radicals there exists a marked double-bond fixation, while in the other ring bond lengths are nearly equalized, the unpaired electron residing essentially on the latter. The cation radicals of XVII and XIX and the anion radical of XVIII, on the other hand, suffer no symmetry reduction and show moderate double bond fixation.

In Table 2, the hyperfine splittings, calculated using McConnell's relationship, with $|Q| = 26\,G$, are presented and compared with experimental values. In the heptafulvalene anion and cation radicals, the theoretical values are in fairly good agreement with experimental data.

Table 2
Bond Lengths, Spin Densities and Proton Hyperfine Splittings

Molecule	Bond	Bond Length (Å)	Atom	Spin Density	Splitting Constant (G)	
					Theor.	Exp.
XVII$^+$	1–2	1.374	1	0.120	3.12	
	1–3	1.435	3	0.260		
	3–4	1.418				
XVII$^-$	1–2	1.436	1	0.500	13.0	
	1–3	1.486	3	0		
	3–4	1.365	4	0		
	4–6	1.460	6	0.000		
	5–6	1.353				
XVIII$^+$	1–2	1.416	1	0.379	9.85	
	1–5	1.441	2	0.120	3.12	
	2–3	1.396	5	0		
	5–6	1.388	6	0		
	6–10	1.461	9	0.000		
	8–9	1.477	10	0.000		
	9–10	1.353				
XVIII$^-$	1–2	1.378	1	0.065	1.69	
	1–5	1.429	2	0.097	2.52	
	2–3	1.428	5	0.186		
	5–6	1.421				
XIX$^+$	1–2	1.379	1	0.043	1.10	0.075
	1–7	1.427	2	0.082	2.13	2.90
	2–3	1.426	3	0.059	1.53	1.72
	3–4	1.380	7	0.133		
	7–8	1.422				
XIX$^-$	1–2	1.400	1	0.275	7.15	8.22
	1–7	1.437	2	0.040	1.04	
	2–3	1.399	3	0.189	4.91	5.02
	3–4	1.422	7	0		
	7–8	1.396	8	0		
	8–14	1.449	12	0.000		
	11–12	1.354	13	0.000		
	12–13	1.426	14	0.000		
	13–14	1.360				

4. Diamagnetic Susceptibilities

The diamagnetic susceptibility of a ring π-electron system, attributable to induced ring currents in its π-electron network, is one of the important quantities indicative of π-electron delocalization. In this paper, diamagnetic susceptibilities are calculated using the London-Hoarau method [24], in the framework of the Wheland-Mann-type SCF-MO approximation. The resonance integral is assumed to be in the form $\beta = Be^{-ar}$, the value of a being determined so as to reproduce the bond lengths and the electron densities obtained above for non-alternant hydrocarbons.

In Table 3, theoretical diamagnetic susceptibilities for some of the non-alternant hydrocarbons are presented (in units of $\Delta K_{benzene}$) and compared with the experimental exaltations [25]. The predicted values are in good agreement with experimental data, except for the case of aceheptylene (XV). In pentalene (I), s-indacene (III), heptalene (VII) and pyracylene (XX), the magnetic susceptibilities are predicted to be paramagnetic, and in I, III and VII double-bond fixation results in a large diminution of paramagnetism. It may be concluded that the magnetically-induced ring currents decrease sensitively with double-bond fixation, and in I, III, VII and XX they are extremely impeded. A part of the evidence for these conclusions is given by the NMR spectra of III [15], VII [14] and XX [26], which show the absence of benzenoid ring current effects.

Table 3
Theoretical and Experimental
Diamagnetic Susceptibility Exaltations

Molecule	$\Delta K/\Delta K_{benzene}$	$\Lambda/\Lambda_{benzene}$
Fulvene	0.10	0.08
Fulvalene	0.075	
Heptafulvene	0.16	
Heptafulvalene	0.18	0.15
Azulene	2.04	2.16
Pentalene (I), D_{2h}	−1.35	
Pentalene (I), C_{2h}	−0.41	
s-Indacene (III), D_{2h}	−0.88	
s-Indacene (III), C_{2h}	−0.38	
Heptalene (VII), D_{2h}	−4.41	
Heptalene (VII), C_{2h}	−0.61	−0.45
Aceazulylene (XV)	2.14	2.18
Aceheptylene (XVI)	1.72	0.0
Dibenzopentalene (VI)	1.10	1.00
Pyracylene (XX)	−0.40	

IV. CONCLUSION

We have discussed the aromaticity of conjugated π-electron systems from the viewpoint of double-bond fixation. Double-bond fixation is assumed to be a sensitive indicator of aromaticity, aromaticity being reduced with double-bond fixation.

This aromaticity criterion differs from the conventional one, in which aromaticity is correlated with the global π-electronic properties, such as the delocalization energy or the diamagnetic ring current. In molecules V, VI, XV and XVI, for example, the peripheral C–C skeleton is composed of two distinguishable parts: one in which bond lengths are smoothed out and the other in which a strong double-bond fixation exists. That is, there are two distinct regions, one aromatic and the other polyolefinic, in the peripheries of these molecules. Furthermore, in the heptafulvalene and triafulvalene anions and in the pentafulvalene cation, one of the rings should be aromatic, while the other ring should be non-aromatic from the viewpoint of double-bond fixation. These molecules should exhibit either aromatic or polyolefinic characteristics, as the case may be. A clear-cut distinction between aromatic and non-aromatic character is impossible.

T. Nakajima, A. Toyota and H. Yamaguchi

REFERENCES

1 G. Binsch, E. Heilbronner & J. N. Murrell (1966) *Molec. Phys.*, 11 : 305.
2 G. Binsch & E. Heilbronner (1968) in : *Structural Chemistry and Molecular Biology* (eds. A. Rich & N. Davidson), Freeman, San Francisco, p. 815.
3 G. Binsch & E. Heilbronner (1968) *Tetrahedron*, 24 : 1215.
4 G. Binsch, I. Tamir & R. D. Hill (1969) *J. Am. Chem. Soc.*, 91 : 2446.
5 G. Binsch & I. Tamir (1969) *J. Am. Chem. Soc.*, 91 : 2450.
6 T. Nakajima & A. Toyota (1969) *Chem. Phys. Lett.*, 3 : 272.
7 H. Yamaguchi, T. Nakajima & T. L. Kunii (1968) *Theor. Chim. Acta*, 12 : 349.
8 L. C. Snyder (1962) *J. Phys. Chem.*, 66 : 2299.
9 E. Carstensen-Oeser & G. Habenmehl (1968) *Angew. Chem. (Int. Edn)*, 7 : 543.
10 P. C. den Boer-Veenendaal & D. H. W. den Boer (1961) *Molec. Phys.*, 4 : 33.
11 T. Nakajima, Y. Yaguchi, R. Kaeriyama, & Y. Nemoto (1964) *Bull. Chem. Soc. Japan*, 37 : 272.
12 P. C. den Boer-Veenendaal, J. A. Vliegenthart & D. H. W. den Boer (1962) *Tetrahedron*, 18 : 1325.
13 T. Nakajima & S. Katagiri (1964) *Molec. Phys.*, 7 : 149.
14 H. J. Dauben Jr & D. J. Bertelli (1961) *J. Am. Chem. Soc.*, 83 : 4659.
15 K. Hafner, K. H. Häfner, C. König, M. Kreuder, G. Ploss, G. Schulz, E. Sturm & K. H. Vöpel (1963) *Angew. Chem.*, 75 : 35; *Angew. Chem. (Int. Edn)*, 2 : 123.
16 T. Nakajima, T. Saijo & H. Yamaguchi (1964) *Tetrahedron*, 20 : 2119.
17 T. J. Katz & M. Rosenberger (1962) *J. Am. Chem. Soc.*, 84 : 865.
18 J. T. Katz, M. Rosenberger & R. K. O'Hara (1964) *ibid.*, 86 : 249.
19 K. Hafner (1963) *Angew. Chem.*, 75 : 1041; (1964) *Angew. Chem. (Int. Edn)*, 3 : 165.
20 T. J. Katz & J. Schulman (1964) *J. Am. Chem. Soc.*, 86 : 3169.
21 T. J. Katz, V. Balogh & J. Schulman (1968) *ibid.*, 90 : 734.
22 M. O. Sevilla, S. H. Flajser, G. Vincow & H. J. Dauben Jr (1969) *ibid.*, 91 : 4139.
23 H. C. Longuet-Higgins & J. A. Pople (1955) *Proc. Phys. Soc.*, A 68 : 591.
24 J. Hoarau (1956) *Ann. Chim.* (Paris), 1 : 544.
25 H. J. Dauben Jr, J. D. Wilson & J. L. Laity (1969) *J. Am. Chem. Soc.*, 91 : 1991.
26 B. M. Trost & G. M. Bright (1967) *ibid.*, 89 : 4244.

234

Discussion

E. Heilbronner:

I would like to make three comments and ask one question. The first comment is on the method which Mr Binsch and I developed and which you talked about. One of the points of this method is that you get a parameter, namely, the eigenvalue λ of the polarizability matrix, which we had used at the time as a 'criterion' for aromaticity. As far as I am concerned, this was a case of juvenile delinquency, of which I am now ashamed. I was probably under the influence of Mr Binsch!

One other point, which I believe is the most important result of our work, is that first-order double bond fixation depends on the slope of the resonance integral β with respect to the interatomic distance R. The direction of distortion, that is, the irreducible representation we are going to fall into when we do have distortion, is determined by the second derivative of β with respect to R.

Finally, the amount of double-bond fixation also depends, if we believe our statement, on the third derivative of β. This result, which I believe to be generally true (and I think that the results of M.J.S. Dewar have shown this to be the case), raises an important problem. You have to know very precisely what the value of the third derivative of β is, and you have to know what the third-power term is in the Morse potential you are going to assume for the σ bond.

Now, my question is: What sort of potential have you assumed for the σ bond? Have you taken into account the fact that this potential will change with the type of σ bond that you have, going from five- or three-membered rings to six- and seven-membered rings? In this context it is also important to realize that, in the type of formula you use for β, the first, the second and the third derivatives all depend on one single parameter, namely, on a. This means that these quantities are no longer independent.

SCF-MO-CI Calculations on the Electronic Structures of Non-Benzenoid Aromatic Hydrocarbons

by H. KURODA,* T. OHTA* *and* T. L. KUNII**

Department of Chemistry * *and Computer Centre,* ** *University of Tokyo*

I. Introduction

THE SEMI-EMPIRICAL self-consistent field molecular-orbital (SCF-MO) method within the Pariser-Parr-Pople (PPP) formalism has been successfully applied to the study of the electronic spectra of aromatic hydrocarbons. In the case of non-benzenoid aromatic hydro-carbons, however, we encounter a considerable difficulty. This is known to be due to the alternation of C–C bond lengths within their π-conjugated systems [1–3]. We have to determine two-centre integrals by taking into account the bond alternation effect, whereas the exact molecular geometry is known for few of these hydrocarbons.

Dewar and Gleicher [4] have proposed a method in which the two-centre integrals between bonded atoms are adjusted at each iteration of the SCF-MO calculation according to the calculated bond order. This method — called 'the variable β, γ procedure' or 'the variable bond-length procedure' — has been applied by several authors to the calculation of the electronic structure of non-benzenoid aromatic hydrocarbons [4–8]. Although this method is most suited for these calculations, we should realize that the results obtainable by this method are strongly dependent on the assumption concerning the relation between the two-centre integrals and the bond length.

In the present paper, we shall describe the results of our critical examination of several problems associated with the application of the variable β, γ procedure for non-benzenoid aromatic hydrocarbons, with emphasis on the choice of the semi-empirical parameters.

II. Procedure of Calculation

1. *Molecular Geometry*

The initial geometries of the molecules were constructed by assuming an equal bond length, 1.395 Å, for all C–C bonds within the π-conjugated system. The five-, six- and seven-membered rings were taken as the regular pentagon, hexagon and heptagon, respectively. The molecular geometries thus obtained are illustrated in Fig. 1.

2. *SCF-MO Calculation*

The valence-state ionization potential of the carbon atom was taken as 11.16 eV, according to Hinze and Jaffé [9], and the one-centre repulsion integral was taken as 11.13 eV. For the

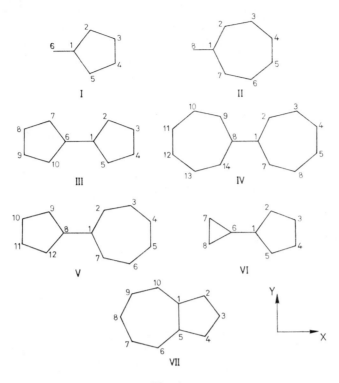

Fig. 1
Molecular geometries of non-benzenoid aromatic hydrocarbons

two-centre resonance integral, $\beta_{\mu\nu}$, and the two-centre repulsion integral, $\gamma_{\mu\nu}$, we have examined various approximations, which we shall describe later.

At each iteration of the SCF-MO calculation, the bond length, $l_{\mu\nu}$, was adjusted according to the calculated bond order, $p_{\mu\nu}$, by assuming the following relation:

$$l_{\mu\nu} = 1.517 - 0.180 p_{\mu\nu}, \tag{1}$$

and the values of the two-centre integrals between nearest-neighbour atoms were re-estimated according to the calculated bond length or the bond order. The integrals between non-nearest-neighbour atoms as determined from the initial geometry were kept unchanged.

The iteration was repeated until a sufficient self-consistency was achieved. In most cases, satisfactory results were obtainable with ten iterations.

3. *Configuration Interaction*

We performed the calculation of configuration interaction (CI) by taking into account the forty lowest singly-excited configurations. No doubly-excited configurations were considered.

All calculations were performed on HITAC 5020 E in the Computer Centre of the University of Tokyo.

237

III. Effect of the Bond-Length Dependence of the Two-Centre Integrals

A number of different approximations have been proposed concerning the two-centre resonance integral, $\beta_{\mu\nu}$. Among these, we shall take up the following seven methods:

(A) Mulliken's magic formula [10]:

$$\beta_{\mu\nu} = -\frac{(I_\mu + I_\nu)}{2}\frac{S_{\mu\nu}}{1 + S_{\mu\nu}};\tag{2}$$

(B) Nishimoto-Forster's formula [11]:

$$\beta_{\mu\nu} = A_0 + A_1 p_{\mu\nu}\tag{3}$$

$$(A_0 = -1.90\text{eV}; \; A_1 = -0.51\,\text{eV});$$

(C) Overlap approximation:

$$\beta_{\mu\nu} = (\beta_0/S_0)S_{\mu\nu};\tag{4}$$

(D) Wolfsberg-Helmholtz's approximation [12]:

$$\beta_{\mu\nu} = KS_{\mu\nu}(H_{\mu\mu} + H_{\nu\nu});\tag{5}$$

(E) Pariser-Parr's formula [13], with $a = 2.58$ [14]:

$$\beta_{\mu\nu} = \beta_0 \exp[a(l_0 - l_{\mu\nu})];\tag{6}$$

(F) Ohno's formula [15]:

$$\beta_{\mu\nu} = S_{\mu\nu}\left(K\frac{\alpha'_\mu + \alpha'_\nu}{2} + \frac{\alpha''_\mu + \alpha''_\nu}{2}\right),\tag{7}$$

$$\alpha'_\mu = \langle\chi_\mu| -\tfrac{1}{2}\varDelta + v_\mu|\chi_\mu\rangle,$$

$$\alpha''_\mu = \sum_{\nu=\mu} \langle\chi_\mu|v_\nu|\chi_\mu\rangle;$$

(G) Kon's formula [16]:

$$\beta_{\mu\nu} = -17.464\, l_{\mu\nu}^{-6}.\tag{8}$$

It should be noticed that the dependence on the interatomic distance varies appreciably among these approximations.

For the two-centre repulsion integral, $\gamma_{\mu\nu}$, we have used two methods: (I) the Pariser-Parr method [13], and (II) the Nishimoto-Mataga method [17]. Thus, we have examined fourteen different approximations obtained by the combination of the seven methods of $\beta_{\mu\nu}$ with the two methods of $\gamma_{\mu\nu}$. In this paper, we shall indicate the method used by the combination of an alphabetic letter, representing $\beta_{\mu\nu}$, and a Roman numeral, representing $\gamma_{\mu\nu}$. Thus, the representation by the method A – II means that the Mulliken magic formula is used for $\beta_{\mu\nu}$, and the Nishimoto-Mataga method is used for $\gamma_{\mu\nu}$.

The predicted transition energies are markedly dependent on the methods used for the two-centre integrals. Although we have performed calculations on the seven non-alternant aromatic hydrocarbons shown in Fig. 1, as well as on several alternant hydrocarbons, we will show here only the results obtained for heptafulvene, fulvalene and azulene

(Figs. 2–4), since the general feature of our results can be well demonstrated by these examples.

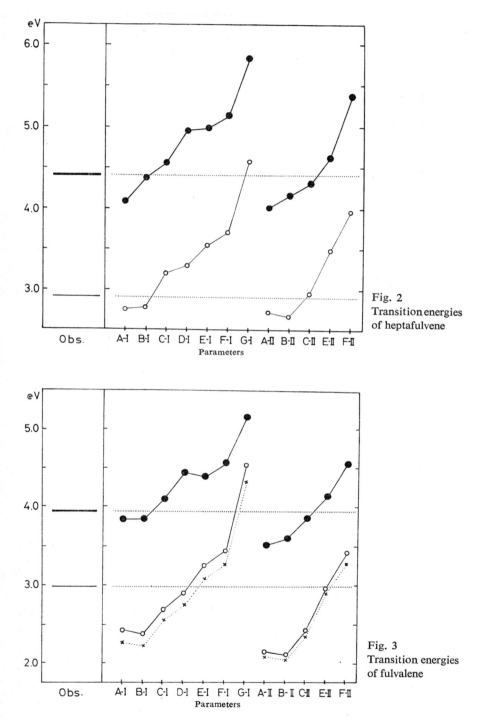

Fig. 2
Transition energies
of heptafulvene

Fig. 3
Transition energies
of fulvalene

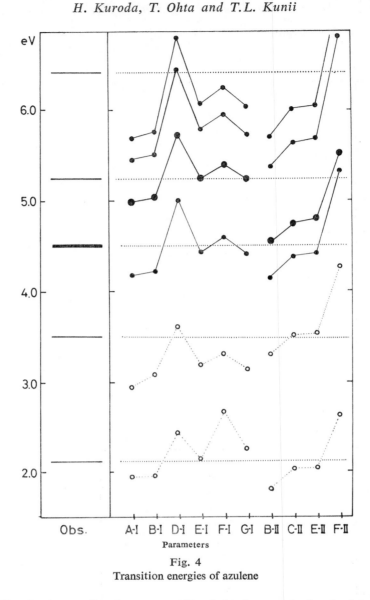

Fig. 4

Transition energies of azulene

As seen in Fig. 2, the predicted energy, ΔE, of the lowest singlet-singlet transition of heptafulvene changes markedly, depending on the method used for the two-centre integrals, the width of the variation being as large as 2 eV. The energy change of the next transition parallels the change of the predicted energy of the lowest singlet excited state.

The results obtained for fulvalene are shown in Fig. 3. Here the dependence on the methods used for the two-centre integrals is almost the same as in heptafulvene. We have found that the situation is almost the same for other non-alternant hydrocarbons as well, except for azulene. The behaviour of the latter is found to be quite different, as shown in Fig. 4. The predicted transition energies are not strongly affected by the choice of the methods for the two-centre integrals. This type of behaviour is generally also found in alternant hydrocarbons.

240

The bond orders of fulvalene, calculated by the use of various methods, are compared in Table 1. Although all of the methods can predict the presence of a marked bond-alternation, the extent of alternation depends appreciably on the method that is used. In order to make a quantitative comparison, we shall introduce a quantity, Δ, defined by Equation 9, and call it 'the degree of bond-order fluctuation' (DBF):

$$\Delta = 2\left[\sum_{\text{bond}} (p_{\mu\nu} - \bar{p})/n\right]^{\frac{1}{2}}, \tag{9}$$

where n is the number of bonds in the π-conjugated system, $\bar{p}$ is the mean value of the bond orders between bonded atoms — i.e., $\bar{p} = (\sum_{\text{bond}} p_{\mu\nu})/n$ — and the summation is to be made over all bonds. The values of DBF are given in the last column of Table 1.

Table 1

Bond Orders of Fulvalene

Method	Bond Orders				DBF*
	P (1–2)	P (2–3)	P (3–4)	P (1–6)	
A – I	0.2997	0.8924	0.3484	0.8215	0.5532
B – I	0.3072	0.8854	0.3607	0.8125	0.5460
C – I	0.3030	0.8887	0.3555	0.8173	0.5616
D – I	0.3100	0.8866	0.3546	0.8093	0.5520
E – I	0.2734	0.9111	0.3178	0.8518	0.6102
F – I	0.2722	0.9128	0.3137	0.8532	0.6128
G – I	0.2170	0.9453	0.2493	0.9067	0.7062
A – II	0.3250	0.8090	0.3588	0.7926	0.5258
B – II	0.3288	0.8760	0.3628	0.7865	0.5186
C – II	0.3209	0.8831	0.3518	0.7986	0.5352
E – II	0.3417	0.8654	0.3768	0.7701	0.5824
F – II	0.3086	0.8984	0.3316	0.8133	0.5616

* Degree of bond-order fluctuation defined by Equation 9

As we compare the values of DBF with the changes in the transition energies, we notice that there is a correlation between them: a method that gives a larger DBF usually gives a higher transition energy. Naturally, the calculated values depend not only on the method used for obtaining $\beta_{\mu\nu}$, but also on the method used for $\gamma_{\mu\nu}$. However, all of the results of our calculation indicate that it is the relation between $\beta_{\mu\nu}$ and $l_{\mu\nu}$, and particularly the slope of the $\beta_{\mu\nu} - l_{\mu\nu}$ curve, that most strongly affects the predictions about the transition energies and DBF.

It is well established that the azulene molecule possesses an aromatic character and has no alternation of bond distances. In agreement with such experimental results, the calculated value of DBF is small, irrespective of the choice of the method used for calculating the two-centre integrals. The calculated values of DBF are 0.2026, 0.2110, 0.2194 and 0.2258 for the methods B – II, C – II, E – II and F – II, respectively. A similar behaviour is found for alternant hydrocarbons.

From the results described above, it may be concluded that although the variable β, γ procedure provides a good way to take into account the bond-alternation effect, we have to determine carefully the dependence of the two-centre resonance integral on the bond length in order to obtain a reasonable result. Although it is claimed that the methods given in Equations 2–8 are applicable to calculations on alternant hydrocarbons by the ordinary procedure of PPP, it should be realized that some of these methods give very unsatisfactory results when used in the variable β, γ calculation on non-alternant hydrocarbons.

If the Pariser-Parr method is used for $\gamma_{\mu\nu}$, a satisfactory agreement is obtainable between prediction and experiment with regard to the energies of the singlet excited states when we use either the Mulliken magic formula (Equation 2) or the Nishimoto-Forster approximation (Equation 3). On the other hand, a relatively good result is obtained by the overlap approximation (Equation 4) when the Nishimoto-Mataga method is used for $\gamma_{\mu\nu}$.

It is not easy to say which $\gamma_{\mu\nu}$ procedure is preferable for obtaining a good over-all agreement between prediction and experiment. However, in the present study we shall use Nishimoto-Mataga's method to determine the best function of $\beta_{\mu\nu}$. For this purpose, we have adopted the Pariser-Parr-type formula:

$$\beta_{\mu\nu} = -2.38 \exp[a(1.397 - l_{\mu\nu})] \tag{10}$$

and treated the value of a in the above equation as an adjustable parameter. In performing the calculations on the seven hydrocarbons shown in Fig. 1, we have assigned various values to a, in the range of 1.5 to 2.6, and compared the obtained results with the observed spectra. An example is shown in Fig. 5.

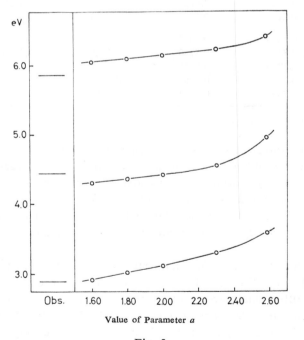

Fig. 5

Dependence of the predicted transition energies of heptafulvene on the value of a in the Pariser-Parr formula for $\beta_{\mu\nu}$

Calculations on the Electronic Structures of Non-Benzenoid Aromatic Hydrocarbon

In this way, we have found that a good over-all agreement is obtained when a is taken to equal $1.80\ A^{-1}$. Thus, we have decided to adopt the following equations for the two-centre integrals:

$$\beta_{\mu\nu} = -2.38\exp[1.80\,(1.397 - l_{\mu\nu})]; \quad [\text{eV}] \tag{11}$$

$$\gamma_{\mu\nu} = 14.3897/(1.294 + l_{\mu\nu}). \quad [\text{eV}] \tag{12}$$

The results calculated by this method are tabulated in Table 4.

Although we can obtain satisfactory results for the transition energies, we encounter difficulty in predicting the dipole moments. The dipole moments obtained by the present method are given in Table 2. It can be seen that the predicted values are always larger than the observed values. This is especially true for azulene, where the predicted value is more than twice the observed value.

Table 2

Calculated and Experimental Dipole Moments

Compound	Dipole Moment (D)		
	Calc.	Exp.	Ref.
Fulvene	1.32	1.1	[18]
Heptafulvene	0.93	0.7	[24]
Sesquifulvalene	5.59		
Calicene	6.70		
Azulene	2.82	1.08	[25]

The dependence of the predicted dipole moment on the value assigned to a is shown in Table 3. It can be observed that the predicted value increases with a decrease in a. Therefore, we have to assign a considerably large value to a to obtain an agreement between the predicted and observed dipole moment. However, this is not allowed if a good prediction of the spectrum is desired. Apparently it is impossible to find the parameter that simultaneously satisfies the requirements of predicting the spectrum and the dipole moment. This might be one of the defects inherent in the approximations adopted in the PPP method.

Table 3

Dependence of the Predicted Dipole Moment
on the Parameter a

Assumed Value of a	Predicted Dipole Moment [D]	
	Heptafulvalene	Azulene
1.60	1.00	2.87
1.80	0.93	2.82
2.00	0.86	2.79
2.30	0.77	2.73
2.58	0.74	2.69

Table 4

Transition Energies (ΔE) and Oscillator Strengths (f)
of Non-Benzenoid Aromatic Hydrocarbons

Compound	Calc.		Obs.		Ref.
	ΔE (eV)	f	ΔE (eV)	f	
Fulvene	3.37	0.035 (y)	3.32	0.012	[18]
	5.09	0.608 (x)	5.13	0.32	
	6.87	0.287 (x)			
	7.41	0.266 (y)			
Heptafulvene	3.01	0.045 (y)	2.91	0.02	[19]
	4.37	0.486 (x)	4.43	0.3	
	6.09	0.097 (y)	5.83		
	6.16	1.282 (x)			
	6.47	0.003 (y)			
Fulvalene	2.45	0			[20]
	2.49	0.016 (y)	2.98		
	3.90	1.177 (x)	3.95	0.4	
	5.58	0			
	6.26	0			
	6.48	0			
Heptafulvalene	2.19	0	Tail		[20]
	2.24	0.015 (y)			
	3.17	1.313 (x)	3.42	0.38	
	4.95	0			
	5.07	0			
	5.31	0			
	5.42	0.258 (y)			
	5.80	1.353 (x)			
	5.89	0.012 (y)			
Sesquifulvalene	2.74	0.009 (y)			[21]*
	2.81	0.033 (y)	(3.05 log ε = 4.38)		
	3.18	1.118 (x)			
	3.69	0.002 (x)	(4.24 log ε = 4.40)		
	5.14	0.279 (x)	(5.53 log ε = 4.20)		
	5.40	0.003 (y)			
	5.48	0.021 (x)			
Calicene	3.56	0.035 (y)			[22]**
	3.94	0.038 (y)			
	4.17	0.926 (x)	(4.13 log ε = 4.64)		
	4.45	0.002 (x)			
	6.58	0.264 (x)	($\geqq$ 6.2)		
Azulene	2.06	0.025 (y)	2.13	0.045	[23]
	3.35	0.005 (x)	3.50	0.08	
	4.42	0.137 (y) ⎫	4.52	1.10	
	4.79	1.880 (x) ⎭			
	5.70	0.414 (y)	5.24	0.38	
	6.05	0.010 (x)			
	6.42	0.196 (y)	6.42	0.65	
	6.74	0.394 (y)			
	7.05	0.732 (y)			

* Spectrum of benzylsesquifulvalene   ** Spectrum of tetrachloro-di-n-propyl-calicene

Calculations on the Electronic Structures of Non-Benzenoid Aromatic Hydrocarbons

IV. Bond Alternation and Molecular Symmetry

Although the effect of bond alternation is automatically taken into account in the variable β, γ procedure, there is one important restriction that should be kept in mind, namely, that the molecular symmetry defined by the initial molecular geometry is usually maintained throughout the calculation. Sometimes this restriction brings about a situation in which the iterations of the SCF-MO calculation do not converge to the value corresponding to the minimum total π-electron energy.

We encounter such a situation in the case of heptalene. If we construct the initial molecular geometry of heptalene by assuming that the seven-membered ring is a regular heptagon, it possesses a symmetry of D_{2h}. This symmetry is kept unchanged throughout the calculation if we apply the ordinary variable β, γ procedure, although it must be lowered in order to show bond alternation. In other words, bond alternation can never be taken into account in such a calculation.

To remove the above-mentioned restriction on the ordinary variable β, γ procedure, we have tried two different methods: (i) we constructed the initial molecular geometry as usual, assuming equal C–C bond length, but modified the β values in the Hückel molecular orbital (HMO) calculation so that they were alternately 0.95β and 1.05β; (ii) we modified the C–C bond lengths in the initial geometry so that they were alternately 1.43Å and 1.36Å. We shall refer to the former procedure as the β-alternation method and to the latter as the l-alternation method, thus distinguishing them from the ordinary procedures.

The extent of the alternation in HMO β or in the initial bond length assumed in these procedures has no vital effect on the final result of the SCF-MO calculation. Hence, these parameters may be assigned arbitrarily. We have confirmed that in the cases of naphthalene and other alternant aromatic hydrocarbons the final results of the calculation always converge to the value that is essentially equal to that obtained by the ordinary procedure. The same situation is found also in the case of azulene. In the case of heptalene, however, a significant change appears if we apply the procedures described above, as can be seen in Table 5. A marked bond alternation is predicted by the β-alternation method, as well as by the l-alternation method, but not by the ordinary procedure. In general, the results given by the β-alternation method and those given by the l-alternation method can be regarded as essentially the same, even though there are minor differences. Subsequently, we shall use the β-alternation method preferentially, since it is the more convenient of the two.

We calculated the total π-electron energy, E_π, and the total π-bond energy, E, which is defined by Equation 13:

$$E = E_\pi + \sum_{\mu < \nu} q_\mu q_\nu \gamma_{\mu\nu}, \tag{13}$$

where q_μ is the charge of the μ-th atomic core, being equal to unity for carbon atoms. The calculated values of E_π and E indicate that a structure with bond alternation is more stable than the symmetrical structure obtained by the ordinary procedure.

245

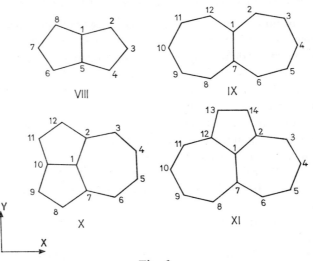

Fig. 6

Molecular geometries of pentalene (VIII), heptalene (IX), cyclopent [c, d] azulene (X)
and cyclopenta [e, f] heptalene (XI)

Table 5

Bond Orders, $P(i–j)$, and
Total π-Bond Energy, E, of Heptalene

Bonds	Bond Orders		
	Method 1*	Method 2**	Method 3 $\triangledown$
1–2, 7–8	0.5197	0.7780	0.7973
2–3, 6–7	0.6819	0.4334	0.4110
3–4, 9–10	0.6189	0.8328	0.8478
4–5, 10–11	0.6189	0.3957	0.3787
5–6, 11–12	0.6819	0.8636	0.8734
6–7, 1–12	0.5197	0.3317	0.3241
1–7	0.5211	0.3904	0.3695
DBF	0.1362	0.4424	0.4806
E_π (eV)$^{\triangledown\triangledown}$	−384.09	−384.77	−384.88
E (eV)	−151.22	−152.11	−152.30

* The ordinary procedure of the variable-β, γ method
** The β-alternation method
$\triangledown$ The l-alternation method
$\triangledown\triangledown$ Total π-electron energy

The predicted transitions are shown in Fig. 7. The transition energies predicted for the symmetrical structure are very low compared to the experimental values. The agreement is much better in the results obtained by the β-alternation method. This fact again supports the structure of lower symmetry.

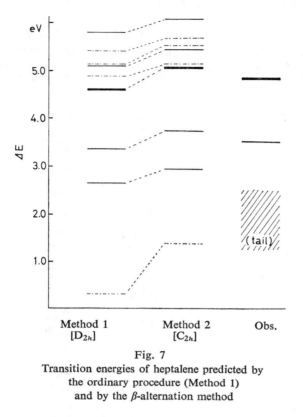

Fig. 7
Transition energies of heptalene predicted by
the ordinary procedure (Method 1)
and by the β-alternation method

We have examined pentalene, cyclopenta [e, f] heptalene and cyclopent [c, d] azulene, using the same technique. The calculated results are given in Tables 6–7.

The β-alternation method predicts the presence of a strong bond alternation in pentalene. Such a structure can be regarded as being more stable than the symmetrical structure, as far as the total π-electron energy is concerned. It is interesting, however, that a converse conclusion may be derived if we look at the total π-bond energy. This is an exceptional case, where an inconsistency is found between the total π-bond energy and the total π-electron energy.

According to the present calculation, compounds XI and X are predicted to have no significant bond alternation. However, the latter has been reported to be non-aromatic from studies on its chemical reactivity. There seems to be a possibility that cyclopent [c, d] azulene does not have a planar structure, and, if this is the case, the conclusion of our calculation is not directly applicable.

Table 6

Bond Orders, $P(i–j)$, and Total π-Bond Energies, E, of Pentalene, Cyclopenta [e, f] heptalene and Cyclopent [c, d] azulene

	Bond	Method 2*	Method 1**
Pentalene	1–2	0.8060	
	4–5	0.2771	0.5086
	5–6	0.8060	
	1–8	0.2771	
	2–3	0.3788	
	3–4	0.8795	0.6572
	6–7	0.3788	
	7–8	0.8795	
	1–5	0.4002	0.5589
	DBF	0.5060	0.1408
	E_π (eV)	−217.50	−216.94
	E (eV)	−102.21	−102.68
Cyclopenta [e, f] heptalene	1–2	0.4559	0.4216
	1–12	0.3941	
	2–3	0.4488	0.4727
	11–12	0.5060	
	3–4	0.7994	0.7798
	10–11	0.7517	
	4–5	0.4701	0.4591
	9–10	0.5286	
	5–6	0.8065	0.7872
	8–9	0.7603	
	6–7	0.4347	0.4591
	7–8	0.4903	
	2–14	0.6339	0.6405
	12–13	0.6320	
	13–14	0.6607	0.6531
	1–7	0.6348	0.6414
	DBF	0.2702	0.2730
	E_π (eV)	−492.43	−492.44
	E (eV)	−178.80	−178.80
Cyclopent [c, d] azulene	1–2	0.4269	0.4085
	1–7	0.3944	
	2–3	0.6475	0.6528
	6–7	0.6468	
	3–4	0.6364	0.6301
	5–6	0.6364	
	4–5	0.6648	0.6710
	7–8	0.4791	0.4603
	2–12	0.4486	
	8–9	0.7838	0.7985
	11–12	0.8078	
	9–10	0.4664	0.4490
	10–11	0.4357	
	1–10	0.6652	0.6686
	DBF	0.2632	0.2668
	E_π (eV)	−395.75	−395.76
	E (eV)	−154.40	−154.40

* The ordinary procedure   ** The β-alternation method

Table 7

Transition Energies (ΔE) and Oscillator Strength (f)

	Method 2*		Method 1**		Experiment		Ref.
	ΔE (eV)	f	ΔE (eV)	f	ΔE (eV)	f	
Heptalene	1.38	0	0.26	0	(1.3	2.5, tail)	[26]
	2.94	0.348	2.67	0.543 (x)			
	3.74	0.175	3.36	0.243 (y)	3.52	0.15	
	5.08	1.334	4.62	1.864 (x)	4.84	0.52	
	5.15	0	4.87	0			
	5.45	0.855	5.11	0.209 (y)			
	5.54	0	5.16	0			
	5.67	0	5.42	0			
Pentalene	1.58	0	0.35	0			
	3.77	0.300	3.59	0.505 (x)			
	4.74	0.183	4.50	0.294 (y)			
	6.30	0	5.93	1.159 (x)			
	6.55	1.261	6.21	0			
Cyclopenta [e, f] heptalene	1.49	0.012	1.51	0.012 (x)	1.16 (1.67), 1.21 (1.65), 1.38 (2.06), 1.57 (2.11)		[27]▽
	2.90	0.014	2.90	0.014 (y)	2.74 (2.30), 2.95 (2.81)		
	3.46	0.309	3.47	0.310 (x)	3.16 (4.17), 3.21 (4.15) 3.34 (4.09)		
	3.80	0.148	3.81	0.140 (y)	3.98 (4.00), 4.15 (4.10)		
	4.47	0.040	4.47	0.022 (x)			
	4.59	0.501	4.59	0.515 (y)	4.71		
	4.97	1.499	4.96	1.497 (x)	4.90		
	5.47	0.444	5.48	0.435 (y)			
	5.61	0.235	5.62	0.245 (x)			
	5.77	0.214	5.78	0.190 (y)			
Cyclopent [c, d] azulene	1.65	0.003	1.67	0.003 (y)	2.61 (2.94)		[27]▽▽
	2.91	0.0006	2.91	0.0008 (x)	3.23 (7.33), 3.31 (3.11), 3.40 (3.21)		
	3.66	0.081	3.66	0.078 (x)	3.65 (4.80)		
	3.86	0.121	3.86	0.116 (y)	3.88 (3.82)		
	4.84	0.409	4.83	0.410 (y)	4.82 (4.65)		
	4.97	1.598	4.97	1.616 (x)			
	5.67	0.593	5.67	0.577 (y)			

 * The β-alternation method   ▽ Spectrum of the 2, 4-dimethyl derivative
** The ordinary method   ▽▽ Spectrum of the 4, 6-dimethyl derivative

V. Bond Alternation and Aromaticity

The prediction of aromaticity has long been a problem of chemical interest in the theoretical studies on the cyclic π-conjugated system. The concept of delocalization energy, based on the HMO method, has often been used for this purpose. As is well known, however, this method fails to give a correct prediction, particularly in the case of non-alternant hydrocarbons. A much more refined definition of the stabilization energy associated with the delocalization of π electrons has been given by Dewar and his collaborators on the basis of the SCF-MO calculation [4]. The resonance energy thus defined has been considered to provide a criterion for aromaticity. However, it should be remembered that the resonance energy depends not only on the extent of delocalization of π electrons, but also on the size of the conjugated system. We believe that the degree of bond alternation can provide a more direct measure of the localization of the π bond.

The values of DBF evaluated by the present method are given in Table 8. In a monocyclic system, DBF should be zero if there is a complete delocalization of π bonds, and it should be equal to unity if π bonds are completely localized. Although the situation is somewhat complicated in a polycyclic system, a larger value of DBF means a higher localization of π bonds. It can be seen from Table 8 that the value of DBF is less than 0.3 for a typical aromatic hydrocarbon, whereas it is in the range of 0.4 to 0.6 in non-benzenoid aromatic hydrocarbons, except for azulene. Although it may not be necessary to draw any clear-cut limit distinguishing between aromatic and non-aromatic compounds, we may expect that a molecule will exhibit aromatic character if DBF is smaller than 0.3, and it will be non-aromatic if DBF is above 0.4. We have found that this criterion of aromaticity can be successfully applied to the cyclic polyenes.

Table 8
Degree of Bond-Order Fluctuation (DBF)

Compound	DBF	Compound	DBF
Benzene	0	Cyclo-butadiene	1.0000
Naphthalene	0.2226	Butadiene	0.6100
Anthracene	0.2516		
Pyrene	0.2478		
Azulene	0.21184	Calicene	0.3870
Cyclopent [c, d] azulene	0.2632	Sesquifulvalene	0.3982
Cyclopenta [e, f] heptalene	0.2730	Heptalene	0.4424
		Heptafulvalene	0.4890
		Pentalene	0.5060
		Heptafulvene	0.5328
		Fulvalene	0.5404
		Fulvene	0.5570

VI. EFFECTS OF INCLUDING NON-NEAREST-NEIGHBOUR β TERMS

The necessity of including non-nearest-neighbour β terms has been pointed out by several authors [20, 28]. Although we can obtain satisfactory results for the prediction of the spectra of hydrocarbons by this method without including the non-nearest-neighbour β terms, there are defects which cannot be removed simply by the improvement of the semi-empirical parameters. As we have already pointed out, such an example can be found in the discrepancy between the predicted and observed dipole moments. Thus, we have examined the effect of including the non-nearest-neighbour β terms in the SCF-MO-CI calculation by the variable-β, γ procedure.

We have carried out the calculation using the formula of $\beta_{\mu\nu}$ given in Equation 11. The calculation has also been done by using the Katagiri-Sandorfy formula (Equation 14) with the parameter $C_{\mu\nu} = 14.5$ eV, which we have recently determined from a comparison of the predicted and observed spectra of naphthalene:

$$\beta_{\mu\nu} = -\frac{S_{\mu\nu}}{4}[C_\mu + C_\nu + \gamma_{\mu\mu} + \gamma_{\nu\nu} - 2\gamma_{\mu\nu}]. \tag{14}$$

We shall denote the calculation based on Equation 11 as Method PP, and that based on Equation 14 as Method KS.

The energies of the singlet excited state predicted by the calculation including the non-nearest-neighbour β terms are listed in Table 9. As we compare these results with the ones obtained by the calculation neglecting the non-nearest-neighbour β terms, we notice that the lower-energy levels are not appreciably affected by the inclusion of the non-nearest-

Table 9

Transitions Calculated by Including Non-Nearest-Neighbour β Terms

Compound	Method PP		Method KS	
	ΔE (eV)	f	ΔE (eV)	f
Fulvene	3.26	0.023 (y)	2.87	0.012 (y)
	5.06	0.277 (x)	5.33	0.253 (x)
	6.09	0.227 (y)	6.15	0.245 (y)
	6.59	1.321 (x)	6.76	1.452 (x)
Heptafulvene	2.92	0.046 (y)	2.87	0.049 (y)
	4.15	0.524 (x)	4.20	0.586 (x)
	5.53	0.152 (x)	5.78	0.023 (y)
	5.72	0.022 (y)	5.87	0.160 (x)
	6.33	0.018 (y)	6.55	0.123 (y)
	6.40	1.098 (x)	6.59	0.982 (x)
Fulvalene	2.69	0.014 (y)	2.65	0.014 (y)
	2.78	0	2.72	0
	4.01	1.000 (x)	4.09	1.085 (x)
	5.44	0	5.68	0
	5.71	0	5.92	0
	6.03	0.370 (y)	6.22	0.380 (y)
Heptafulvalene	2.06	0.012 (y)	1.95	0.011 (y)
	2.22	0	2.12	0
	3.01	1.173 (x)	3.02	1.260 (x)
	4.35	0	4.46	0
	4.71	0	4.92	0
	4.95	0	5.09	0
	5.09	0.013 (y)	5.19	0.024 (y)
	5.20	0.077 (x)	5.33	0.069 (x)
	5.25	0	5.37	0
	5.36	0.352 (y)	5.54	0.397 (y)
Sesquifulvalene	2.62	0.020 (y)	2.58	0.020 (y)
	2.79	0.015 (y)	2.77	0.015 (y)
	3.07	0.974 (x)	3.12	1.048 (x)
	3.88	0.045 (x)	3.74	0.057 (x)
	4.73	0.000 (x)	4.89	0.001 (x)
	5.43	0.016 (x)	5.64	0.013 (x)
	5.51	0.019 (y)	5.66	0.028 (y)
Calicene	3.36	0.023 (y)	3.41	0.025 (y)
	3.63	0.044 (y)	3.68	0.044 (y)
	4.06	0.747 (x)	4.12	0.016 (x)
	4.20	0.013 (x)	4.19	0.802 (x)
	5.90	0.636 (x)	6.10	0.678 (x)
	5.96	0.070 (y)	6.16	0.080 (y)
Azulene	1.87	0.020 (y)	1.96	0.021 (y)
	3.30	0.029 (x)	3.44	0.037 (x)
	4.15	0.177 (y)	4.36	0.183 (y)
	4.68	1.962 (x)	4.85	2.045 (x)
	5.23	0.003 (y)	5.46	0.001 (y)
	5.55	0.018 (x)	5.81	0.021 (x)
	5.77	0.011 (y)	6.00	0.014 (y)
	6.20	1.171 (y)	6.45	1.237 (y)
	6.44	0.004 (x)	6.73	0.007 (x)
	6.54	0.246 (x)	6.83	0.263 (x)

neighbour β terms, but that significant changes appear in the higher-energy levels. Unfortunately, however, there is little experimental data concerning the higher-energy levels of these molecules. Thus, at present, we cannot draw any definite conclusion regarding the merits of the inclusion of all β terms in predicting the spectra of non-benzenoid aromatic hydrocarbons.

The bond orders of heptafulvene and azulene are given in Table 10. The bond alternation is usually predicted to be somewhat small by Method KS, as compared to that predicted by Method PP. In both methods, the inclusion of the non-nearest-neighbour β terms causes an increase in DBF.

Table 10

Effect of Including Non-Nearest-Neighbour β Terms
on the Predicted Bond Orders

	Bond	Method PP		Method KS	
		Excl.	Incl.	Excl.	Incl.
Heptafulvene	1–2, 7–1	0.3167	0.3058	0.3282	0.3267
	2–3, 6–7	0.8795	0.8865	0.8698	0.8777
	3–4, 5–6	0.3790	0.3660	0.3948	0.3764
	4–5	0.8638	0.8729	0.8527	0.8647
	1–8	0.8918	0.9004	0.8832	0.8829
	DBF	0.5328	0.5505	0.5098	0.5255
Azulene	1–2, 4–5	0.6172	0.6296	0.6161	0.6285
	2–3, 3–4	0.6573	0.6492	0.6572	0.6481
	5–6, 1–10	0.6127	0.6163	0.6117	0.6166
	6–7, 9–10	0.6560	0.6514	0.6560	0.6507
	7–8, 8–9	0.6456	0.6471	0.6454	0.6469
	1–5	0.2748	0.2547	0.2820	0.2630
	DBF	0.2118	0.2224	0.2076	0.2172

The inclusion of the non-nearest-neighbour β terms causes an appreciable change in the charge distribution, as seen in Table 11. In the case of heptafulvene, the charge at the

Table 11

Effect of Including Non-Nearest-Neighbour β Terms
on the Charge Distribution

	Atom	Method PP		Method KS	
		Excl.	Incl.	Excl.	Incl.
Heptafulvene	1	−0.0084	+0.0158	−0.0090	+0.0514
	2, 7	+0.0233	+0.0174	+0.0266	+0.0330
	3, 6	+0.0035	+0.0002	+0.0032	−0.0093
	4, 5	+0.0099	+0.0058	+0.0111	+0.0080
	8	−0.0650	−0.0628	−0.0741	−0.1147
Azulene	1, 5	−0.0153	−0.0060	−0.0157	−0.0051
	2, 4	−0.0945	−0.0823	−0.0975	−0.0861
	3	−0.0037	+0.0150	−0.0043	+0.0173
	6, 10	+0.0948	+0.0764	+0.0975	+0.0771
	7, 9	−0.0618	−0.0223	−0.0168	−0.0221
	8	+0.0672	+0.0533	+0.0693	+0.0548

1-position changes from negative to positive. In the case of azulene, the negative charges at the bridge positions, 1 and 5, decrease considerably when all β terms are included. It has been reported by Flurry and Bell [28–29] that the charges at these positions change from negative to positive if all β terms are included. In the present calculation, however, they remain negative even if all β terms are included.

As a result of the changes in the charge distribution, a marked change appears in the prediction of the dipole moment, as shown in Table 12.

Table 12

Dipole Moments of Heptafulvene and Azulene

(in Debye units)

	Method PP		Method KS		Exp.
	Excl.	Incl.	Excl.	Incl.	
Heptafulvene	0.93	0.70	1.05	1.00	0.7
Azulene	2.82	1.95	2.93	2.01	1.08

It should be noticed that the predicted dipole moment is appreciably reduced by the inclusion of all β terms. The predicted value shows good agreement with the experimental one in the case of heptafulvene. In azulene, the predicted value is still too large compared to the experimental value, but the discrepancy between them has been significantly improved by the inclusion of all β terms in comparison with the result obtained by neglecting the non-nearest-neighbour β terms. This effect is of particular interest, since we have seen that, within the method of neglecting the non-nearest-neighbour β terms, it is not possible to find a function of $\beta_{\mu\nu}$ that satisfies both the requirement for predicting the spectra of non-benzenoid aromatic hydrocarbons and that for predicting their dipole moments.

REFERENCES

1 T. Nakajima & S. Katagiri (1963) *Molec. Phys.*, 6 : 149.

2 M. Asgar Ali & C. A. Coulson (1961) *ibid.*, 4 : 65.

3 P. C. den Boer-Veenendaal & D.H.W. den Boer, *ibid.*, p. 33.

4 M. J. S. Dewar & G. J. Gleicher (1965) *J. Am. Chem. Soc.*, 87 : 685.

5 H. Kuroda & T. Kunii (1967) *Theor. Chim. Acta* (Berlin), 7 : 220.

6 J. E. Bloor, B. R. Gilson & N. Brearley (1967) *ibid.*, 8 : 35.

7 J. Jurg & P. François (1967) *ibid.*, 7 : 249.

8 H. Yamaguchi, T. Nakajima & T. L. Kunii (1969) *ibid.*, 12 : 349.

9 J. Hinze & H. H. Jaffé (1962) *J. Am. Chem. Soc.*, 84 : 540.

10 R. S. Mulliken (1949) *J. Chim. Phys.*, 46 : 497.

11 K. Nishimoto & L. S. Forster (1965) *Theor. Chim. Acta* (Berlin), 3 : 407.

12 M. Worfsberg & L. Hermoholtz (1952) *J. Chem. Phys.*, 20 : 837.

13 R. Pariser & R. G. Parr (1953) *ibid.*, 21 : 767.

14 T. Nakajima (1965) *Bull. Chem. Soc. Japan*, 38 : 83.

15 K. Ohno, Y. Tanabe & F. Sasaki (1963) *Theor. Chim. Acta* (Berlin), 1 : 378.

16 H. Kon (1955) *Bull. Chem. Soc. Japan*, 28 : 275.

17 N. Mataga & K. Nishimoto (1957) *Z. Phys. Chem.*, 13 : 140.

18 J. Thiec & J. Wiemann (1956) *Bull. Soc. Chim. France*, p. 177.

19 W. von E. Doering & D. W. Wiley (1960) *Tetrahedron*, 11 : 183.

20 W. von E. Doering (1959) *Theoretical Organic Chemistry* (*Kekulé Symposium*), Butterworths, London, p. 35.

21 H. Prinzbach & W. Rosswog (1961) *Angew. Chem.*, 73 : 543.

22 Y. Kitahara, I. Urata, M. Ueno, K. Sato & H. Watanabe (1966) *Chem. Comm.*, p. 180.

23 R. Pariser (1956) *J. Chem. Phys.*, 25 : 1112.

24 Y. Kurita, S. Seto, T. Nozoe & M. Kubo (1953) *Bull. Chem. Soc. Japan*, 26 : 272.

25 A. G. Anderson Jr & B. M. Steckler (1959) *J. Am. Chem. Soc.*, 81 : 4941.

26 H. J. Dauben Jr & D. J. Bartelli (1961) *ibid.*, 83 : 4659.

27 K. Hafner & J. Schneider (1959) *Ann.*, 623 : 37.

28 R. L. Flurry Jr & J. J. Bell (1967) *J. Am. Chem. Soc.*, 89 : 525.

29 R. L. Flurry Jr, E. W. Stout & J. J. Bell (1967) *Theor. Chim. Acta* (Berlin), 8 : 203.

Discussion

B. Binsch:

As you know, we — that is, Mr Heilbronner and myself — have come to essentially the same conclusion that you have reached, using a different method, and I am very glad to see that we agree on this point, namely, that there should be bond alternation in molecules such as pentalene and heptalene. But I would like to call your attention to one essential difference, namely, that in our calculations we obtain the transformation properties of the energetically most favourable distortion directly, whereas in your calculation you have to assume a certain transformation, a certain symmetry of the species as a starting point, and if you do not choose the right one you will get a minimum, but that minimum may not be the deepest minimum for the molecule.

H. Kuroda:

Indeed, as you have pointed out, the results of our calculation depend somewhat on the symmetry of the assumed initial molecular geometry. We cannot exclude the possibility of the presence of more than one minimum. But we have not encountered such a case in the hydrocarbons taken up in the present study.

E. Heilbronner:

In your last slide you infer that only benzene has all π bonds of equal length. Could you say briefly how much of a bond alternation is predicted for [14] annulene? This is the only case where we know a derivative with a planar π system, namely, 15, 16-dimethyldihydropyrene, whose structure has been determined. In this molecule all the π bonds are, within the limits of error, practically equal.

H. Kuroda:

0.4 and 0.6 alternate in a [14] annulene, and the standard deviation is 0.2.

E. Heilbronner:

Do you also infer that you have bond alternation in [18] annulene?

H. Kuroda:

According to the crystal structure analysis of [18] annulene, there is no bond alternation, but there is bond distortion. But, if we take the molecular geometry found in the crystal, we cannot predict exactly the absorption spectrum of [18] annulene. If we make the calculation with constant bond length, either without or with bond distortion, we never get the results that correspond to the absorption spectrum of the solution. In solution, there could be some puckering of the ring, and that might be the origin of the deviation.

E. Heilbronner:

What do you get for the energy of the first excited state, that is, the B_{2u} state of the [18] annulene? Experimentally, the first transition is at 1.3 eV. This is terribly low, and it creates a problem, which you have hinted at, but which we should perhaps discuss later.

Discussion

H. Kuroda:

I have not noticed that you have found a low-energy transition in [18]annulene. According to our calculation, the first transition is predicted to be at 1.50 eV if we take a structure without bond alternation, and at 2.25 eV if we consider bond alternation. Higher energy transitions are not strongly dependent on the bond alternation effect. Therefore, your spectrum seems to suggest that the [18]annulene molecule has, even in solution, a molecular geometry similar to that found in the crystal.

E. D. Bergmann:

To what extent has the planarity or non-planarity of the cycloheptatriene ring an effect on your data for the dipole moments? Do you think it changes them a great deal, or only a little? Do you think that the difference between the observed and the calculated dipole moment may be accounted for by deviation from planarity?

H. Kuroda:

I do not think so.

G. Wagnière:

I have not quite understood whether you adjust the β's to the bond lengths in every iteration of an SCF calculation, or whether you do a complete SCF calculation with given β's, and then adjust the β's and start all over again. In other words, do you iterate with respect to energy and bond lengths simultaneously, or one after another? If the former is the case, do you not possibly run into convergence difficulties?

H. Kuroda:

We obtain a bond order, and using this we estimate the β and γ, the nearest neighbour β and the nearest neighbour γ. We iterate with respect to energy and bond length at the same time. In some cases it takes a lot of iteration, but in most cases, less than twenty are needed for satisfactory convergence.

Mrs A. Pullman:

May I ask you what values you used for the one-centre Coulomb integrals $\gamma_{\mu\mu}$ and for the one-centre core integrals α or $H_{\mu\mu}$?

H. Kuroda:

We used the following values:

$$H_{\mu\mu} = 11.16 \text{ eV}; \ \gamma_{\mu\mu} = 11.13 \text{ eV}.$$

New Aspects in Fulvene Chemistry
Cyclic Cross-Conjugated 10-π-Electron Systems

by K. HAFNER

Institut für Organische Chemie der Technischen Hochschule Darmstadt

NUMEROUS non-alternant π-electron systems contain the element of cross-conjugation [1–3]. In bicyclic and linear annelated polycyclic conjugated systems, this cross-conjugation is not associated with branching of the π-electron system. On the other hand, such branching will occur in pericondensed tri- and polycyclic compounds. The participation of the element of cross-conjugation in the π-electron systems of such polycycles affects their properties and results in characteristic differences in bonding character and reactivity, compared with monocyclic conjugated compounds with the same number of π electrons [4]. Thus, azulene (I) shows the characteristic reactions of both of the fulvenoid structural elements contained in the bicyclic system. Addition of nucleophiles takes place at the pentafulvene part, producing an anionic 6-π-electron system (II), whereas addition of electrophiles takes place at the heptafulvene part, producing a tropylium cation (III) [5–7]. In a similar way, the heptafulvene system of 2H-benz-

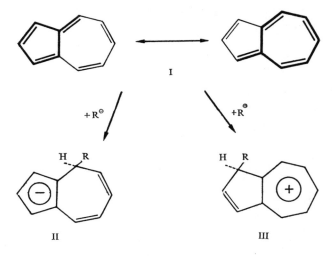

This research was supported in part by the Deutsche Forschungsgemeinschaft, in part by the Fonds der Chemischen Industrie, Verband der Chemischen Industrie, and in part by Prof. Dr K. Ziegler, Studiengesellschaft Kohle m.b.H.

256

azulene (IV) is responsible for the easy and reversible protonation in the 1-position [8]. Just as the reactivity of I is understandable in terms of a formal combination of penta- and heptafulvene, so may qualitative interpretations be made of the chemical properties of tri- and tetracyclic conjugated π-electron systems [1–3] containing these structural elements. Like benzene in the benzenoid series, penta- and heptafulvenes are topologically the basic units found in many non-benzenoid π-electron systems.

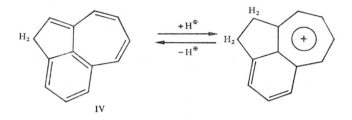

IV

The same may be expected for other cross-conjugated systems. In contrast to tria-, penta- and heptafulvenes, and with the exception of the bridged homologues of sesquifulvalene studied by Prinzbach [9–10], fulvenes with larger π-electron systems are still unknown [11]. Such systems are of interest in connection with the attempted synthesis of higher homologues of azulene, e.g., bicyclo[7.1.0]decapentaene (V) or bicyclo[7.5.0]tetradecaheptaene (VI).

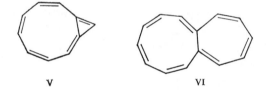

V   VI

Information about configuration, bond character and chemical properties of the nonafulvene system VII should give some indications as to the properties expected for the hypothetical 'azulenoid' bicyclic compounds V and VI. With this in mind, we have started in the Darmstadt laboratory an investigation of cyclic cross-conjugated 10-π-electron systems.

a   b
VII

Inspection of molecular models shows that nonafulvenes should be non-planar. As is known for pentafulvenes [12–14], they should, therefore, show high bond alternation. On the other hand, it was hoped that they would also show high polarizability, like the pentafulvenes [15]. The known tendency of cyclononatetraene to undergo irreversible valence isomerization to *cis*-3a,7a-dihydroindene above 0°C [16] suggested that similar behaviour should be expected for nonafulvenes. It was expected, therefore, that the monocyclic cross-conjugated 10-π-electron system VII would be thermally rather unstable.

257

For these reasons it seemed promising to study first a 10-π-electron-containing fulvene which would be planar due to a bridging σ bond. Such a system is found in the 2,3-benzo-fulvene (*iso*-benzofulvene) VIII, which is, of course, also *o*-quinoid. The reactivity of this truly planar bicyclic compound should be much affected by the tendency to form the energetically favoured benzenoid sextet. On the other hand, a high polarizability of the cross-conjugated system, in the sense of the dipolar resonance form VIIIb, might be expected. Furthermore, it was suggested that, as in the case of the pentafulvenes [17], the presence of electron-donating groups at the exocyclic C atom should increase the resonance stabilization of the 10-π-electron system.

a VIII b

As Heilbronner [18] has shown in the case of cyclodecapentaene, naphthalene and azulene, the influence of a bridging bond on the energy levels of the molecular orbitals can be determined by a simple quantum-mechanical calculation. In principle, this method may be extended to monocyclic cross-conjugated systems. Calculations for the two bridged nonaful-venes [K. Pfeiffer, unpublished results], the well-known 1,2-benzofulvene [11] and the unknown 2,3-benzofulvene VIII, show, for example, that in the UV spectra the longest-wavelength absorption of the latter should display a bathochromic shift, and that of the former a hypsochromic shift, compared with nonafulvene VII. Furthermore, in accor-dance with calculations by Sadlej [19], high reactivity of the 2,3-benzofulvene VIII in the 1- and 4-positions is expected. A decrease of the high electron density at C_1 and C_4 should result in an increase of thermal and electronic stability, as well as in a bathochromic shift of the longest-wavelength absorption.

iso-Benzofulvenes cannot be prepared by methods analogous to those used for penta-fulvenes [20], since isoindene (2H-indene), unlike cyclopentadiene, cannot be isolated [21–22]. However, indene-2-(N,N-dimethylformimmonium) perchlorate (IX) proved to be a suitable and easily accessible starting material for a synthesis of the 2,3-benzofulvene system. This salt is formed in 23% yield, together with 4% of the 6-dimethylamino-1,2-benzo-

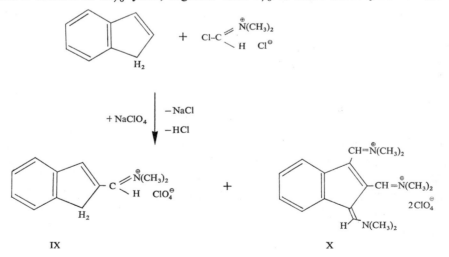

IX   X

fulvene-3, 4-bis (N, N-dimethylamino-formimmonium) perchlorate (X) by reaction of indene with chloro-N, N-dimethyliminium chloride in the presence of sodium perchlorate [23]. Hydrolysis of IX yields 35% of the temperature- and light-sensitive indene-2-aldehyde XI, which has previously been prepared by Arnold [24]. Reduction of IX with lithium aluminium hydride gives the stable 2-(N, N-dimethylaminomethyl)indene XII in 85% yield. This can easily be transformed to the quaternary ammonium hydroxide XIII [25].

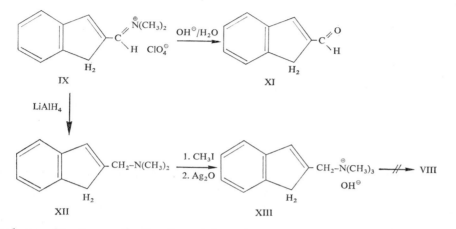

By analogy with the synthesis of *o*-xylylene by pyrolysis of *o*-methylbenzyltrimethyl-ammonium hydroxide [26], the hydroxide XIII should lead to the 2,3-benzofulvene VIII. In fact, only high-molecular weight products were formed in all attempts to isolate VIII, which is probably due to the extreme thermal instability of the 2,3-benzofulvene system. However, just as pentafulvene is stabilized by electron-donating groups at the exocyclic C atom [17], so 2,3-benzofulvene (VIII) might be. The immonium salt IX is a likely starting material for the preparation of 6-(N, N-dimethylamino)-2,3-benzofulvene (XIV). The bicyclic cross-conjugated system could be obtained simply by deprotonation. Indeed, the formation of XIV can be followed spectrophotometrically after addition of N, N-diiso-propyl-N-ethylamine to IX at room temperature. XIV exists only for a few seconds and shows UV-absorption maxima at 364 and 376 nm [25], which are, in accordance with MO calculations, at longer wavelengths than those of the corresponding 1,2-benzofulvene derivative (277, 282 and 350 nm [27]). The formation of XIV was confirmed by the isolation of the 1,3-adduct XV with N-phenylmaleimide. The structure of XV follows from spectroscopic and analytical data [25].

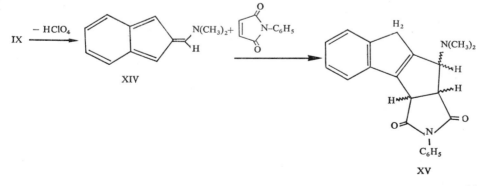

259

The extremely low stability of XIV must be due, among other things, to the quantum-mechanically predicted high reactivity in the 1- and 4-positions. As with the iso-electronic *iso*-benzothiophene, *iso*-benzofuran and *iso*-indole [28–29], suitable substituents in these positions should confer enhanced stability.

Indeed, this turned out to be the case. 6-(N,N-dimethylamino)-1,4-diphenyl-2,3-benzo-fulvene (XVII) was obtained in 26% yield as deep blue prisms, sensitive to oxygen and moisture, by Vilsmeier formylation of 1,3-diphenylindene, followed by deprotonation of the immonium salt XVI [23].

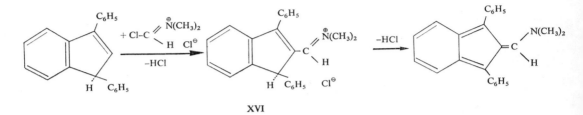

XVI

The structure of XVII accords with elemental analysis and NMR spectra (Fig. 1). The latter show, in addition to a multiplet for 10 aromatic protons centred at $\tau = 2.63$ and a four-proton multiplet centred at $\tau = 3.25$, a singlet for the exocyclic proton at $\tau = 2.01$ and two broad signals at $\tau = 6.89$ and 7.36 for the N-methyl protons. That two signals are seen for the methyl protons is due to hindrance of rotation around the C–N bond. The large difference in chemical shift is the result of a shielding effect of one of the out-of-plane phenyl groups. At elevated temperatures, the restriction of rotation is reduced, and the two signals coalesce. The UV spectrum of XVII in tetrahydrofuran shows a broad maximum at 593 nm (log $\varepsilon = 3.55$) and further fine-structured absorptions at shorter wavelengths — 403 (4.42), 339 (4.39), 260 (4.17) and 250 (4.17) nm (log ε).

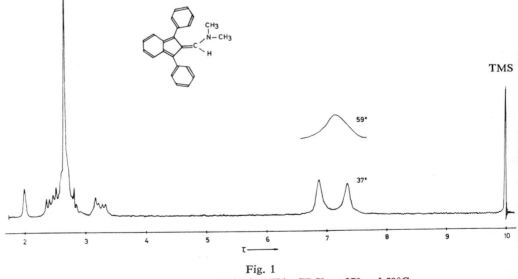

Fig. 1
NMR spectrum (60 MHz) of XVII in CDCl₃ at 37° and 59°C

The chemical behaviour of the 1,4-diphenyl-*iso*-benzofulvene derivative XVII is typical of an *o*-quinoid cross-conjugated system. Whilst 6-(N,N-dimethylamino) pentafulvene reacts only with 2N sodium hydroxide to give sodium formylcyclopentadienide [30] by an addition-elimination mechanism, XVII is hydrolysed by water alone to give 88% of 1,3-

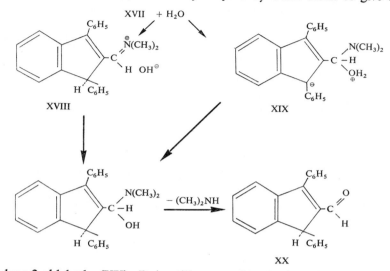

diphenylindene-2-aldehyde (XX). It is still uncertain whether the primary step in this reaction is a protonation in either the 1- or the 4-position to give the immonium hydroxide XVIII, or a nucleophilic attack of water at the exocyclic C atom to give the indenyl anion XIX. Furthermore, the 1,4-diphenyl-*iso*-benzofulvene derivative XVII shows a high reactivity towards dienophiles, which was similarly observed for XIV. Cycloaddition with N-phenylmaleimide takes place in the way of a Diels-Alder reaction at the 1- and 4-posi-

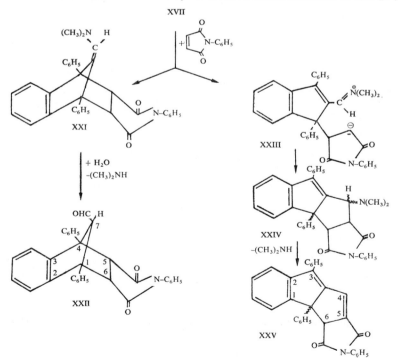

tions, but also as a 1,3-dipolar addition at the 1- and 6-positions. One of the two reaction products — the 1,4-adduct XXI — was hydrolysed during work-up, giving the aldehyde XXII. The other product is probably formed *via* Michael addition to the electron-rich 1-position of XVII, to give the primary adduct XXIII, which then undergoes ring closure to XXIV. This then, under the reaction conditions, loses dimethylamine to give the conjugated 1,2-benzo-6,6a-dihydropentalene derivative XXV. A thermally initiated synchronous dipolar *cis* addition of N-phenylmaleimide is not to be expected in view of the Woodward-Hoffmann rules [31]. The structures of the adducts XXII and XXV follow from spectral and analytical data [25]. In the case of compound XXII, the data did not suffice to decide between the *exo* or *endo* configuration. By analogy to the formation of the corresponding Diels-Alder adducts of N-substituted *iso*-indoles [32], the *endo* configuration would be expected.

These findings demonstrate that the properties of the 2,3-benzofulvene system VIII differ substantially from those of the pentafulvenes; it behaves primarily as an *o*-quinoid system. It was to be expected that nonafulvenes (VII), formally obtained by opening of the bridging σ bond, would show very different properties of the 10-π-electron system. In order to investigate this suggestion experimentally, a synthesis of a simple nonafulvene was required. The preparation of 10,10-*bis*(dimethylamino)nonafulvene (XXVI) was particularly attractive, since, if nonafulvenes behave as do pentafulvenes, then the electron-donating groups should have a stabilizing effect [33]. The corresponding lower homologue — 6,6-*bis*(dimethylamino) pentafulvene (XXVII) — shows high chemical stability. Its dipole moment of 5.5 D indicates considerable participation of the dipolar form in resonance stabilization [30].

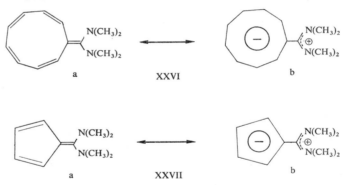

By analogy with the synthesis of XXVII, the corresponding nonafulvene XXVI should be obtainable by reaction of the cyclononatetraenyl anion XXVIII — first prepared by Katz [34–35], and also by Benson [36–37] — with chloro-*bis*(dimethylamino) iminium chloride (XXIX) [38]. It was possible, however, that the expected reaction intermediate, the cyclononatetraene derivative XXX, would, in a similar fashion to cyclononatetraene itself [16], undergo valence isomerization to the 3a,7a-dihydroindene derivative XXXI. None the less, the nonafulvene XXVI was obtained in 15% yield as light- and air-sensitive yellow crystals by reaction of XXVIII with XXIX in tetrahydrofuran at −70°C [39]. As a further product, the known *cis*-3a,7a-dihydroindene was isolated. This is formed by hydrolysis of the cyclononatetraenyl anion XXVIII and subsequent valence isomerization of the resulting cyclononatetraene. The structure of XXVI is in accord with elemental analysis, mass spectrum and quantitative catalytic hydrogenation, showing the presence of five double bonds.

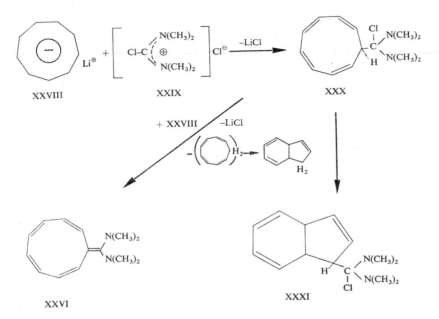

The question of the configuration and conformation of the new 10-π-electron cross-conjugated system is of special interest. Is the nonafulvene planar like the cyclononatetra-enyl anion? Does it show bond alternation and high polarizability like the pentafulvenes, or is it a truly conjugated system? Extended cyclic conjugation in the sense of the resonance structure XXVIb should correspond with resonance stabilization. On the other hand, an all-*cis* configuration in the 9-membered ring would be associated with high angle strain, which might, however, be at least partially compensated for by the potential delocalization energy. This is certainly the case for the cyclononatetraenyl anion XXVIII [40].

First indications, but not a final answer to these questions, were given by UV and NMR spectra of the nonafulvene derivative XXVI. These show interesting temperature and solvent dependence. The NMR spectrum of XXVI (Fig. 2) in CD_2Cl_2 at 35°C shows a complicated multiplet between $\tau = 3.63$ and $\tau = 4.67$ for the eight ring protons, and a sharp singlet at $\tau = 7.22$ for the 12 methyl protons. This suggests either non-planarity of the 9-membered ring or fast equilibrium between valence isomers. Otherwise, the signals for the ring protons would not show such large differences in chemical shift. At lower temperatures, the multiplet for the ring protons is both narrowed and shifted to lower field. There is concomitant broadening of the signal for the 12 methyl protons, and, with further decrease in temperature, the signal for these protons is resolved to two new singlets, which show a coalescence point at $-53°C$ and are sharpened at lower temperatures. These temperature-dependent effects may be due to an increase in diamagnetic ring current at low temperatures, that is, to an increase in cyclic conjugation in the sense of the resonance structure XXVIb. This will be associated with increasing restriction of rotation around the C–N bonds. The latter phenomenon would not be apparent were the two N atoms contained in a ring. Such a compound is currently being investigated [K. Hafner & H. Tappe, unpublished results].

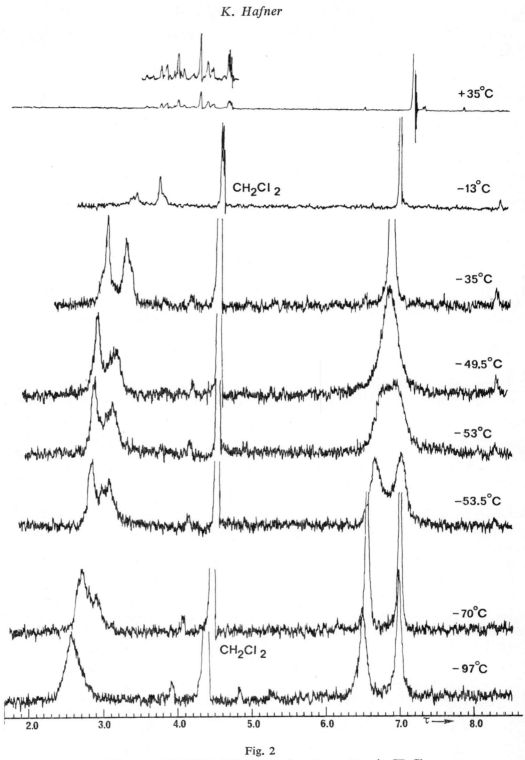

Fig. 2
NMR spectra (60 MHz) of XXVI at various temperatures in CD₂Cl₂

The solvent and temperature dependence of UV spectra of XXVI support the conclusion drawn from the NMR spectra. At room temperature, in *n*-hexane or in tetrahydrofuran, one absorption maximum is observed at about 330 nm; in methylene chloride, however, this is accompanied by a further absorption at 403 nm. With decrease in temperature, this further absorption also appears in the spectrum in tetrahydrofuran, but not in the hexane spectrum. In methylene chloride, a decrease in temperature causes the disappearance of the band at about 330 nm and an enhancement of the absorption at 403 nm. The results of preliminary SCF-CI calculations [H.J. Lindner, unpublished results] indicate that the long-wavelength absorption might tentatively be assigned to the planar nonafulvene derivative XXVI, and the shorter wavelength absorption to the valence isomer XXXII, formed from XXVI by a conrotatory electrocyclic reaction. XXVI and XXXII may be in fast equilibrium with each other. If this were so, then the bicyclic system XXXII is favoured in non-polar solvents, whereas the nonafulvene derivative XXVI is of increasing importance with increasing solvent polarity and decreasing temperature. It is hoped that further investigations, including an X-ray analysis [H.J. Lindner, unpublished results], will clarify the position.

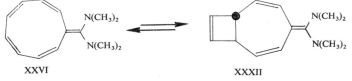

In benzene solution at room temperature, a slow and irreversible disrotatory electrocyclic reaction of the nonafulvene derivative XXVI takes place. This reaction is quite rapid at 60°C; the product is the *cis*-3a, 7a-dihydroindene derivative XXXIII, the structure of which was elucidated with the aid of analytical and spectroscopic data. Dehydrogenation of XXXIII with activated lead dioxide yields 6,6-*bis*(dimethylamino)-1,2-benzofulvene (XXXIV), which is identical to material prepared by reaction of sodium indenide (XXXV) with *bis* (dimethylamino) ethoxycarbonium fluoborate (XXXVI) [41].

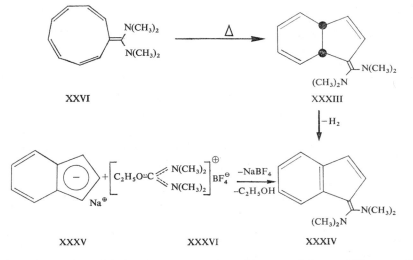

In conclusion, it can be said that the assumption that nonafulvenes, like pentafulvenes, should be stabilized by electron-donating groups at the exocyclic C atom is indeed justified.

265

Attempted preparations of 10,10-dimethyl- or diphenylnonafulvene have not been successful [K. Hafner & H. Tappe, unpublished results; 42]. It is to be hoped that, following the successful synthesis of XXVI, the investigation of other π-electron systems containing the 9-membered ring may be possible.

ACKNOWLEDGEMENT

It is a pleasure to acknowledge gratefully the co-operation of my able young associates W. Bauer and H. Tappe, who participated in the experimental efforts which I have discussed.

REFERENCES

1 K. Hafner (1963) *Angew. Chem.*, 75 : 1041; (1964) *Angew. Chem.* (*Int. Edn*), 3 : 165.

2 K. Hafner, R. Fleischer & K. Fritz (1965) *Angew. Chem.*, 77 : 42; *Angew. Chem.* (*Int. Edn*), 4 : 69.

3 K. Hafner, G. Hafner-Schneider & F. Bauer (1968) *Angew. Chem.*, 80 : 801; *Angew. Chem.* (*Int. Edn*), 7 : 808.

4 K. Hafner (1968) *Z. Chem.*, 8 : 74.

5 K. Hafner & H. Weldes (1957) *Liebigs Ann. Chem.*, 606 : 90.

6 K. Hafner, C. Bernard & R. Müller (1961) *ibid.*, 650 : 35.

7 K. Hafner, A. Stephan & C. Bernard, *ibid.*, p. 42.

8 K. Hafner & H. Schaum (1963) *Angew. Chem.*, 75 : 90; *Angew. Chem.* (*Int. Edn*), 2 : 95.

9 H. Prinzbach & L. Knothe (1967) *Angew. Chem.*, 79 : 620; *Angew. Chem.* (*Int. Edn*), 6 : 632.

10 Idem (1968) *Angew. Chem.*, 80 : 698; *Angew. Chem.* (*Int. Edn*), 7 : 729.

11 E. D. Bergmann (1968) *Chem. Rev.*, 68 : 41.

12 N. Norman & B. Post (1961) *Acta Crystallogr.*, 14 : 503.

13 H. Burzlaff, K. Hartke & R. Salamon (1970) *Chem. Ber.*, 103 : 156.

14 J. F. Chiang & S. H. Bauer (1970) *J. Am. Chem. Soc.*, 92 : 261.

15 P. A. Straub, D. Meuche & E. Heilbronner (1966) *Helv. Chim. Acta*, 49 : 517.

16 G. Boche (1969) *Angew. Chem.*, 81 : 565; *Angew. Chem.* (*Int. Edn*), 8 : 595.

17 K. Hafner, K. H. Häfner, C. König, M. Kreuder, G. Ploss, G. Schulz, E. Sturm & K. H. Völpel (1963) *Angew. Chem.*, 75 : 35; *Angew. Chem.* (*Int. Edn*), 2 : 123.

18 E. Heilbronner (1959) 'Azulenes', in : *Non-Benzenoid Aromatic Compounds* (ed. D. Ginsburg), Interscience, New York, p. 171.

19 A. J. Sadlej (1965) *Acta Phys. Polon.*, 27 (6): 859.

20 P. Yates (1968) 'Fulvenes', in : *Advances in Alicyclic Chemistry*, II, Academic Press, New York, p. 168.

21 K. Alder & F. Pascher (1942) *Ber. Deutsch. Chem. Ges.*, 75 : 1501.

22 K. Alder & M. Fremery (1961) *Tetrahedron*, 14 : 190.

23 K. Hafner & W. Bauer (1968) *Angew. Chem.*, 80 : 312; *Angew. Chem.* (*Int. Edn*), 7 : 297.

24 Z. Arnold (1965) *Coll. Czech. Chem. Comm.*, 30 : 2783.

25 W. Bauer (1968) Ph. D. Dissertation, Technische Hochschule Darmstadt.

26 L. A. Errede (1961) *J. Am. Chem. Soc.*, 83 : 949.

27 E. Sturm (1964) Ph. D. Dissertation, University of Munich.

28 M. H. Palmer (1967) *Structure and Reactions of Heterocyclic Compounds*, Arnold, London, p. 333.

29 J. D. White & M. E. Mann (1969) in : *Advances in Heterocyclic Chemistry*, X (eds. A. R. Katritzky & A. J. Boulton), Academic Press, New York, p. 113.

30 K. Hafner, G. Schulz & K. Wagner (1964) *Liebigs Ann. Chem.*, 678 : 39.

31 R. B. Woodward & R. Hoffmann (1969) *Angew. Chem.*, 81 : 797; *Angew. Chem.* (*Int. Edn*), 8 : 781.

32 R. Kreher & J. Seubert (1966) *Angew. Chem.*, 78 : 984; *Angew. Chem.* (*Int. Edn*), 5 : 967.

33 J. A. Berson, E. M. Evleth Jr & S. L. Manatt (1965) *J. Am. Chem. Soc.*, 87 : 2901.

34 T. J. Katz & P. J. Garrat (1963) *ibid.*, 85 : 2852.

35 *Ibid.* (1964) 86 : 5194.

36 E. A. LaLancette & R. E. Benson (1963) *ibid.*, 85 : 2853.

37 *Ibid.* (1965) 87 : 1941.

38 H. Eilingsfeld, G. Neubauer, M. Seefelder & H. Weidinger (1964) *Chem. Ber.*, 97 : 1232.

39 K. Hafner & H. Tappe (1969) *Angew. Chem.*, 81 : 564; *Angew. Chem.* (*Int. Edn*), 8 : 593.

40 H. E. Simmons, D. B. Chesnut & E. A. LaLancette (1965) *J. Am. Chem. Soc.*, 87 : 982.

41 H. Meerwein, W. Florian, N. Schön & G. Stopp (1961) *Liebigs Ann. Chem.*, 641 : 1.

42 K. F. Bangert & V. Boekelheide (1964) *J. Am. Chem. Soc.*, 86 : 1159.

Discussion

E. D. Bergmann:

I would like to report on some work carried out in our laboratory in the nonafulvene field, although we have not progressed as far as you did.

The anion of cyclononatetraene was condensed with *p*-dimethylaminobenzaldehyde and gives, undoubtedly, the desired product, as the mass spectrum indicates, but the work-up, even under all precautions, permitted only the isolation of the known 1-(4-dimethylaminobenzylidene)-indene, through spontaneous loss of two hydrogen atoms:

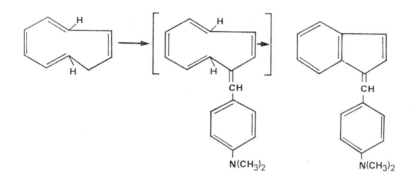

This reaction is very similar to your findings and also reminds us of the isomerization of *cis*-oxonin to *cis*-8, 9-dihydrobenzofuran; see S. Musamane, S. Takada & R. T. Seidner (1969), *J. Amer. Chem. Soc.*, 91:7769; S. Cohen & I. Agranat [unpublished results].

In order to stabilize the nonafulvene system and suppress such evasive reactions, we annelated benzene rings, in analogy to the recent work of Bindra and co-workers, in the oxonin and thionin series; see A. P. Bindra, J. A. Elix, P. J. Garratt & R. H. Mitchell (1968) *J. Amer. Chem. Soc.*, 90:7372. The following reaction of diphendialdehyde gave us 3, 4, 5, 6-dibenzocyclonona-1, 3, 5, 7-tetraene:

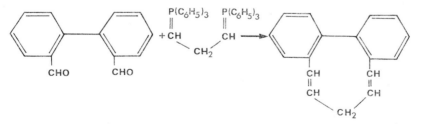

The olefinic double bonds in this compound have *cis* configuration, as IR and NMR spectra have shown: $8H_{ar}$ 420–440 cps, m; $4H_{olef}$ 320–390 cps, m; $2H_{CH_2}$ 160 cps; m(triplet of triplets). Hydrogenation of the compound gave dibenzocyclononane. In the presence of strong bases, the compound is isomerized to 3, 4, 5, 6-dibenzocyclonona-1, 3, 5, 8-tetraene:

267

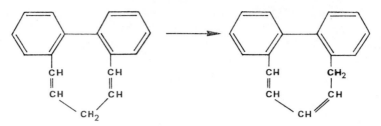

probably because the latter is more 'conjugated' than the isomer. This transformation indicates that in the anion the charge is delocalized (M. Rabinovitz & A. Garith, unpublished results). We believe that by your methods we can obtain nonafulvenes from this more stable isomer.

K. Hafner:

It might also interest you that we tried to prepare a dibenzocyclononatetraene anion as an analogue of the fluorene anion, but without success.

H. Prinzbach:

Just before coming here, I read that cyclononatetraene gives cycloaddition reactions. Does your nonafulvene react analogously?

K. Hafner:

We have not studied cycloaddition reactions with the nonafulvene until now. We will do so, but at the moment we are more interested in studying the monocyclic system and in clarifying the valence isomerization at different temperatures.

268

Sesquifulvalene and Calicene
Di-*t*-Butyl Derivatives; Dipole Moments

by H. PRINZBACH, H. KNÖFEL *and* E. WOISCHNIK

Chemisches Laboratorium der Universität Freiburg i. Br., Germany

INTRODUCTION

FOR A NUMBER OF YEARS already we have been dealing with cyclic cross-conjugated π systems of the general type I; these include calicene (II) [1–2], sesquifulvalene (III) [3], fidecene (IV) [4], pentaphenafulvalene (V) [5] and the 6π–14π member VI.

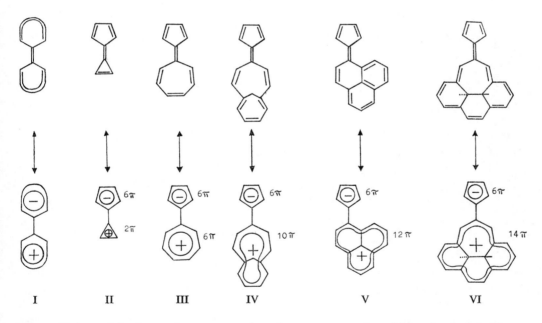

Some of the original questions concerning these compounds, which stimulated this research — such as charge separation, bond length, electron configuration and 'aromaticity' — have been answered in the meantime. Theory and experiment agree that these compounds are typical polyenes. The preparative-synthetic applications offered by α, ω cycloaddition reactions [6] with these systems have again rendered them highly attractive. The latter method, however, will be touched upon only briefly; this contribution is limited mainly to the following two points:

269

1. Synthesis of the so far most simple 'stable' derivatives of II and III [7];
2. Experimental evaluation of the dipole moments of II and III [8].

1. *7,9-Di-*t*-butylsesquifulvalene and 1,3-Di-*t*-butyl-5,6-dimethylcalicene*

The parent molecule III can exist only in highly dilute solutions [H. Prinzbach, W. Ross-wog & U. Fischer, unpublished work]. NMR data and dipole moments are therefore not available. Though a number of annelated and phenyl-substituted sesquifulvalenes have provided a large body of information, less highly substituted derivatives were nevertheless desirable. The situation is similar in the case of the calicene II. If we neglect the examples with hetero-substituents [9–11], the known pure hydrocarbon calicenes are stabilized either by annelation or by four-fold phenyl-substitution.

Because of the low tendency of di-*t*-butylcyclopentadiene to dimerize [12], and because of the increasing stability of the cyclopentadienones on going from IX to VII [13], protection of II and III by two *t*-butyl groups appeared sufficiently effective to render the compounds isolable.

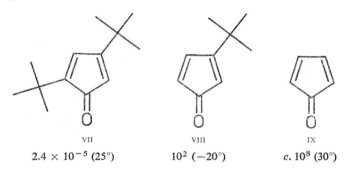

VII	VIII	IX	
2.4×10^{-5} (25°)	10^2 (−20°)	c. 10^8 (30°)	$k_{dimerization}$ (m^{-1} sec^{-1}) Garbisch & Sprecher [13]

Starting from the Li salt of di-*t*-butylcyclopentadiene and tropylium bromide, we isolated the dihydrosesquifulvalene X as the main tautomer. X is not amenable to hydride abstraction from C_1 by trityl salts. The attack of the Lewis acid on the five-membered ring causing skeletal transformation and polymerization is much faster. The trick is — and we had taken advantage of it upon earlier occasions [e.g., 14] — to favour H$^-$ abstraction by prior thermal or photochemical sigmatropic rearrangement. Both XI and XII (with minor amounts of other isomers) yield the cation XIV, if only in moderate yields.

Treatment of XIV or XIII — the latter is formed by slow tautomerization of XIV — with base at 0°, and chromatography under careful exclusion of oxygen and light, result in deeply red-coloured solutions of the sesquifulvalene XV. In the absence of impurities, such solutions can be concentrated up to 0.5 to 0.8 M. Total removal of the solvent seems impossible without partial decomposition.* Steric compression between the *t*-butyl group and H_6, probably the cause for the rearrangement XIV → XIII, is one possible explanation for the fact that the longest wavelength UV maximum is shorter by 20 nm than that of the parent molecule, III (λ_{max} [*iso*-octane] 391 nm).

---

* Recently, we have observed that, following the same approach, even the mono-*t*-butylsesquifulvalene can be made in useful quantity [7].

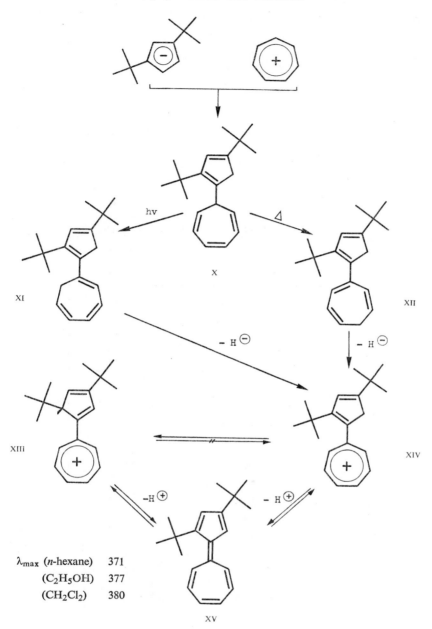

The NMR spectra, demonstrating the purity of XV as well as of XIII and XIV, are given in Fig. 1. On using benzene as solvent in place of CCl₄, the signals of the five- and seven-membered ring protons of XV are displaced, as may be expected, in the opposite direction, so that the AB quartet becomes well separated. Without going into details, our renewed interest in sesquifulvalenes and calicenes was prompted by the reactions of XV with tetracyanoethylene and dimethyl acetylenedicarboxylate. TCNE is added rapidly in a [12+2]-fashion to give XVI, the stereochemistry of which is postulated on the basis of the Woodward-Hoffmann rules. For reasons which are not yet fully understood, ADME

271

does not behave similarly; the attack occurs at positions 10/12, to give the spiro product XIX.

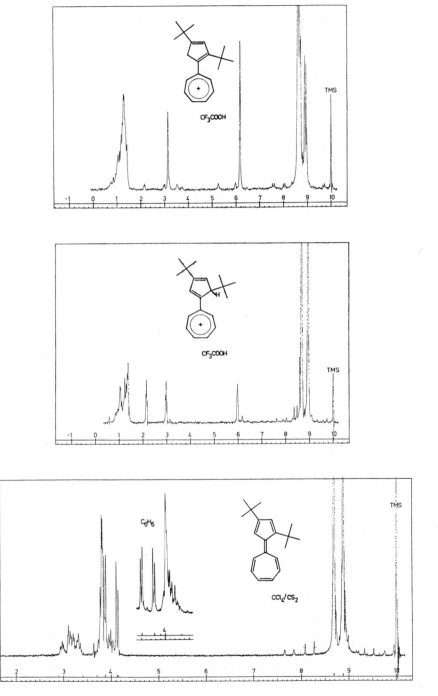

Fig. 1

NMR spectra of the sesquifulvalene XV and the conjugate acids XIII and XIV

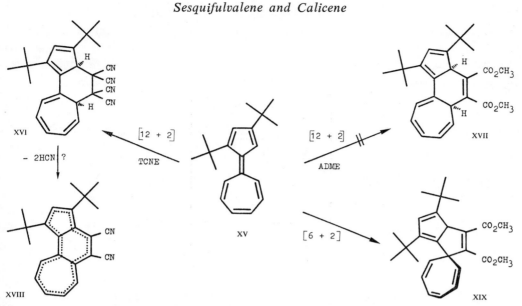

We have good indications that the adduct XVI might serve as starting material for the synthesis of the [14]annulene (XVIII), the structure of which is conformationally fixed by two zero bridges.

It was a surprise to find that the calicene XXIII — synthesized by following the sequence XX → XXI → XXII → XXIII — is moderately stable in crystalline form. So far, the limitation of the procedure is the very poor yield for the step XXI → XXII.

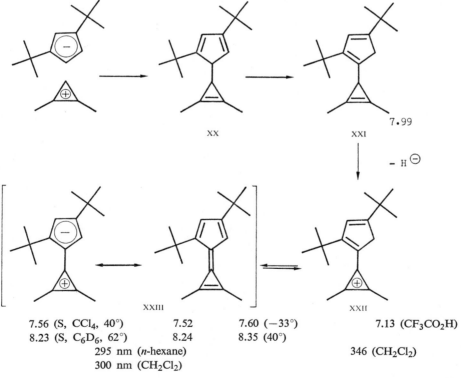

7.56 (S, CCl$_4$, 40°)	7.52	7.60 (−33°)	7.13 (CF$_3$CO$_2$H)
8.23 (S, C$_6$D$_6$, 62°)	8.24	8.35 (40°)	
	295 nm (*n*-hexane)		346 (CH$_2$Cl$_2$)
	300 nm (CH$_2$Cl$_2$)		

As we have noted earlier for the dibenzo- and tetraphenyldimethylcalicenes **XXXVII** and **XXXVIII**, the colourless crystals of compound **XXIII** are extremely sensitive to oxygen; the oxabicyclo[1.1.0]butane structure **XXIV** is a reasonable, though not as yet isolated, intermediate in the synthesis of the allene **XXV**.

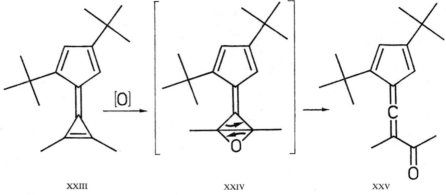

        XXIII   XXIV   XXV

In CCl_4 (40°), the six CH_3 protons of **XXIII** are isochronous; the two olefinic protons give rise to one sharp and one broadened singlet (Fig. 2). As we will see later, the spectra are clearly temperature-dependent.

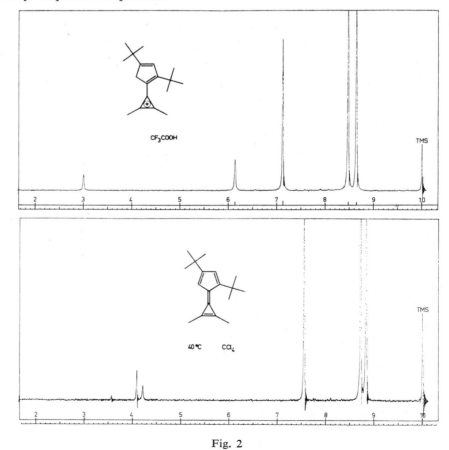

Fig. 2

NMR spectra of the calicene **XXIII** and the conjugate acid **XXII**

We may conclude this section by stating that, with the sesquifulvalene XV and the calicene XXIII, we now have derivatives at our disposal which allow a more profound study of the behaviour of the cross-conjugated systems II and III towards dienophiles.

2. *Dipole Moments of Sesquifulvalene and Calicene*

Extensive theoretical work with III (see table) has clearly shown that the drift of charge density is from the 7- to the 5-membered ring. Rather conflicting results are given, however, as far as dipole moments are concerned. Nakajima [15] reported a value of 5.1 D in 1963; Dewar [private communication] reported a value of 1.58 D in 1970.

Theoretical Data for Sesquifulvalene

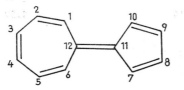

Dipole moment μ = 5.1 D (Nakajima [15])
1.58 D (Dewar)

Atom	Charge/Electron Density		Naka-jima [15]	Free Valence Naka-jima [15]	Bond	Bond Order				Bond Length Dewar [22]	
	Kuhn*					Pull-man [16]	Kuhn*		Naka-jima [15]	SPO	PPP
12	−0.18	−0.06	0.887	0.262	12–1	0.568	0.81	0.70	0.347	1.449	1.459
1	+0.19	+0.12	0.985	0.504	1–2	0.683	0.84	1.03	0.881	1.359	1.352
2	+0.15	+0.07	0.954	0.498	2–3	0.612	0.85	0.81	0.353	1.442	1.453
3	+0.17	+0.08	0.975	0.498	3–4	0.670	0.82	1.04	0.881	1.360	1.353
11	−0.42	−0.24	1.092	0.248	11–7	0.569	1.02	0.82	0.349	1.449	1.459
7	−0.10	−0.06	1.041	0.502	7–8	0.690	1.15	1.29	0.881	1.359	1.353
8	−0.11	−0.06	1.055	0.496	8–9	0.609	1.06	0.83	0.355	1.439	1.449
					11–12	0.446	0.80	0.83	0.776	1.372	1.362

* Private communication

Extrapolating from μ=5.20 D, measured for XXVI, a dipole moment of 3.7 to 4.1 D has been postulated for the annelated sesquifulvalene XXVII by Kitahara et al. [17].

Using the Guggenheim-Smith approximation, we have determined the dipole moments for the five sesquifulvalene derivatives, XV and XXVIII–XXXI, which we had synthesized

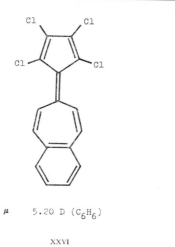

μ 5.20 D (C_6H_6)

XXVI

3.7 ~ 4.1 D

XXVII

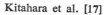
Kitahara et al. [17]

275

during the last years. The values range from 1.20 D for XXVIII to 2.60 D for XXXI, with the dipole moments of XV and XXX lying close together.

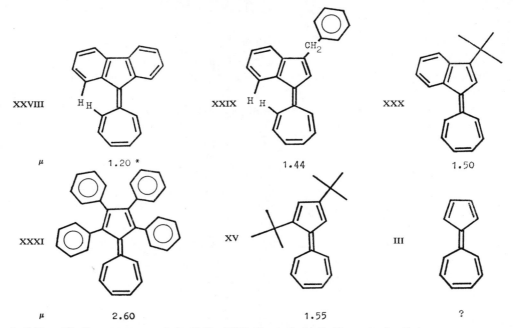

XXVIII	XXIX	XXX
μ	1.20 *	1.44	1.50

XXXI	XV	III
μ	2.60	1.55	?

* C_6H_{12}; all other measurements in C_6H_6, 20°C, Guggenheim-Smith approximation

In order to evaluate qualitatively the influence of the annelation, of the four phenyl groups and of one and two *t*-butyl groups, we examined the influence of the same substituents upon the dipole moments of 6,6-dimethyl-, 6-phenyl-, 6-*p*-methoxyphenyl- and 6-dimethyl-aminophenylfulvenes (Fig. 3). It may be concluded from the data collected in Fig. 3 that annelation results in a decrease in the dipole moment by about 0.25 D. The four phenyl groups increase the moment by about 0.45 D, one *t*-butyl group decreases it by about 0.5 D, while two such groups decrease the moment by about 0.75 D. Returning with these 'substituent constants' to the three sesquifulvalenes XXX, XXXI and XV (here one hopes

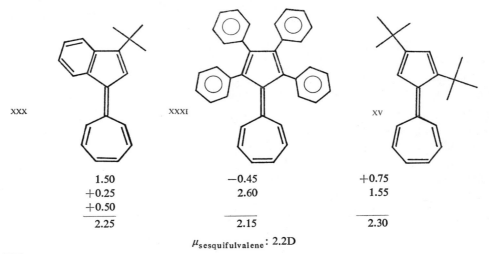

XXX	XXXI	XV
1.50	−0.45	+0.75
+0.25	2.60	1.55
+0.50		
2.25	2.15	2.30

$\mu_{sesquifulvalene}$: 2.2 D

a. *Annelation*

b. *4 Phenyl Groups*

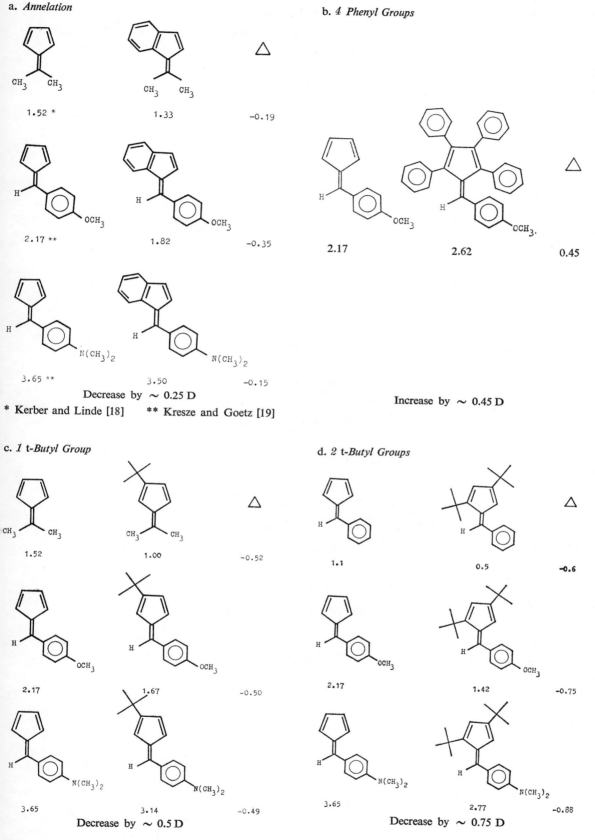

1.52 *

1·33

−0.19

2.17 **

1.82

−0.35

3.65 **

3.50

−0.15

Decrease by ∼ 0.25 D

* Kerber and Linde [18]   ** Kresze and Goetz [19]

2.17

2.62

0.45

Increase by ∼ 0.45 D

c. *1 t-Butyl Group*

1.52

1.00

−0.52

2.17

1·67

−0.50

3.65

3·14

−0.49

Decrease by ∼ 0.5 D

d. *2 t-Butyl Groups*

1·1

0·5

−0·6

2.17

1.42

−0.75

3.65

2.77

−0.88

Decrease by ∼ 0.75 D

Fig. 3

Dipole moments (D) of 6, 6-dimethyl-, 6-phenyl-, 6-*p*-methoxyphenyl- and 6-*p*-dimethylaminophenyl-
fulvene and of some annelated, phenyl- and *t*-butyl-substituted analogues

that the steric consequences of the specific substitution parallel the ones in the corresponding fulvenes), we calculate a moment of 2.2D for the parent sesquifulvalene III.

We expect a further test on the reliability of this value in the near future, when we will have the 8-*t*-butyl derivative at our disposal.

What is the situation in the case of the calicene II? Fig. 4 shows the results of the more recent theoretical work done in this area. Dipole moments between 4.4 and 6.13D have been proposed for the unsubstituted molecule. Single annelation decreases the moment by 0.4–0.6D, double annelation by 0.7–1.28D.

$\mu = 5.70$ D $\mu = 5.63$ D

Nakajima [20]   Dewar [private communication]

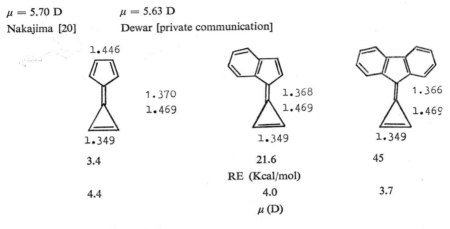

	RE (Kcal/mol)	
3.4	21.6	45
4.4	4.0	3.7

μ (D)

Dewar and Gleicher [21–22]

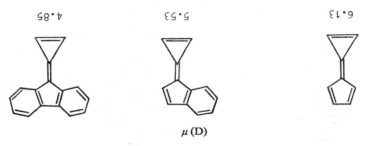

μ (D)

Zahradnik [private communication]

Fig. 4
Theoretical data for II and annelated derivatives

Before discussing the pure hydrocarbons, the effect of stabilizing hetero-substituents in the five-membered ring is documented by the examples XXXII–XXXVI.

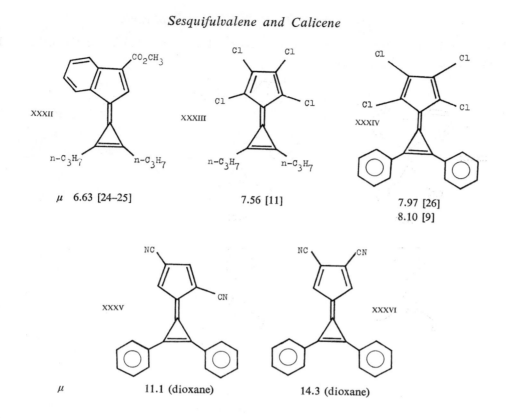

XXXII μ 6.63 [24–25]

7.56 [11]

7.97 [26]
8.10 [9]

μ 11.1 (dioxane)   14.3 (dioxane)

With the help of **XXXIII**, Kitahara et al. [17] estimated a dipole moment of 5.63 D for **II**.

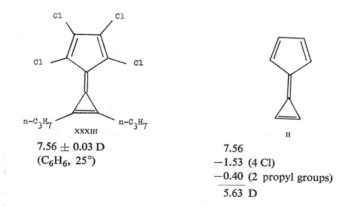

XXXIII
7.56 ± 0.03 D
(C_6H_6, 25°)

7.56
−1.53 (4 Cl)
−0.40 (2 propyl groups)
5.63 D

This value is obviously in very good accord with the ones proposed by Dewar [private communication] (5.63) and Nakajima [15] (5.70). When we apply the same 'substituent constants', we have used above, to the calicenes **XXXVII**, **XXXVIII** and **XXXIX** — bearing no hetero-groups — we obtain values ranging from 5.40 to 5.85 D for the structures stripped of the substituents on the five-membered ring. At present, we know of no experimental data which take into account the substituents fixed to the cyclopropene moiety. Unfortunately, we have no reliable data yet for the di-*t*-butyl-dimethylcalicene **XXIII**.

279

There can be no doubt, however, that a moment of about 5.6 D for the calicene **II** should turn out to be correct.

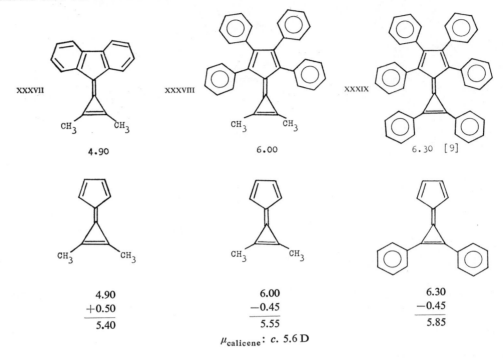

μ_{calicene}: *c.* 5.6 D

A point of special interest with **XXIII**, of some additional importance in connection with accepting **XXIII** as the so-far simplest model of **II**, is the temperature dependence of its NMR spectra (Fig. 5). Two separate doublets of equal intensity and two broad singlets, again of equal intensity, for the two olefinic protons and the two methyl groups, respectively, are shown at low temperatures. The doublets and singlets coalesce at higher temperature, duplicating qualitatively the spectra pictured in Fig. 2. With some reservation — other explanations, e.g., a barrier arising from steric interaction between *t*-butyl and methyl substituents, cannot be ruled out rigorously at the moment — we ascribe this behaviour

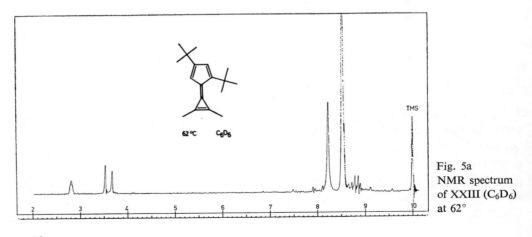

Fig. 5a
NMR spectrum
of XXIII (C₆D₆)
at 62°

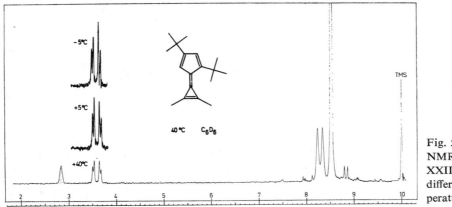

Fig. 5b
NMR spectra of
XXIII (C_6D_6) at
different tem-
peratures

to rotation around the bond between the rings. If this interpretation is correct, the low rotational barrier, as compared with that in ethylenes, certainly reflects the stability of the $C_5H_5^-$ and $C_3H_3^+$ ions in the transition state, and perhaps a deviation from co-planarity in the ground state.

A $\Delta F^{\neq}$ value of 18 to 19 Kcal/mole was reported by Kende et al. [25] for the rotation around the central bond of a calicene molecule, in which a formyl substituent in position 1 effectively withdraws electrons, thereby decreasing considerably the double-bond character of this central bond. Since the measurements presented in Fig. 5 must be considered as being preliminary, a more quantitative discussion must be postponed.

References

1 H. Prinzbach & E. Woischnik (1969) *Helv. Chim. Acta*, 52 : 2472.

2 H. Prinzbach & U. Fischer (1967) *ibid.*, 50 : 1692.

3 H. Prinzbach & H. Knöfel (1969) *Angew. Chem.*, 81 : 900.

4 H. Prinzbach & L. Knothe (1969) *Tetrahedron Lett.*, p. 2093.

5 H. Prinzbach & E. Woischnik (1969) *Angew. Chem.*, 81 : 901.

6 W. v. E. Doering & G. Schröder (1967) cited in: R. B. Woodward, 'The Conservation of Orbital Symmetry', *Aromaticity* (*Special Publication, No. 21*), Chemical Society, London.

7 H. Knöfel (1970) Ph. D. Thesis, University of Freiburg.

8 E. Woischnik (1970) Ph. D. Thesis, University of Freiburg.

9 E. D. Bergmann & I. Agranat (1966) *Tetrahedron*, 22 : 1275.

10 A. S. Kende, P. T. Izzo & P. T. McGregor (1966) *J. Am. Chem. Soc.*, 88 : 3359.

11 Y. Kitahara, I. Murata, M. Ueno, K. Sato & H. Watanabe (1966) *Chem. Comm.*, p. 180.

12 R. Riemschneider (1963) *Z. Naturf.*, 18b : 641, 645.

13 E. W. Garbisch Jr & R. F. Sprecher (1969) *J. Am. Chem. Soc.*, 91 : 6785.

14 R. W. Murray & M. L. Kaplan (1966) *ibid.*, 88 : 3527.

15 T. Nakajima & S. Katagiri (1963) *Molec. Phys.*, 7 : 149.

16 B. Pullman et al. (1952) *Bull. Soc. Chim. Fr.*, 19 : 73.

17 Y. Kitahara, I. Murata & S. Katagiri (1967) *Tetrahedron*, 23 : 3613.

18 R. C. Kerber & H. G. Linde (1966) *J. Org. Chem.*, 31 : 4321.

19 G. Kresze & H. Goetz (1957) *Chem. Ber.*, 90 : 2161.

20 T. Nakajima, quoted in [11], *loc. cit.*

21 M. J. S. Dewar (1967) *Aromaticity* (*Special Publication, No. 21*), Chemical Society, London.

22 M. J. S Dewar & G. J. Gleicher (1965) *J. Am. Chem. Soc.*, 87 : 685.

23 Idem (1965) *Tetrahedron*, 21 : 3423.

24 A. S. Kende & P. T. Izzo (1965) *J. Am. Chem. Soc.*, 87 : 1609, 4162.

25 A. S. Kende, P. T. Izzo & W. Fulmor (1966) *Tetrahedron Lett.*, p. 3697.

26 I. Murata, M. Ueno, Y. Kitahara & H. Watanabe (1966) *Tetrahedron Lett.*, p. 1831.

Discussion

S. Sarel:

I would like to ask about the oxidation reaction of the triapentafulvalene. I assume that the reaction is an oxidation — I mean, that it was carried out by a peroxy acid or something of this sort. It seems that the behaviour of the product is exactly the same as in the case of cyclopropene. We know that the peracid oxidation is an electrophilic reaction, and from charge distribution calculations it appears that the three-membered ring is a positive one. On the basis of that, I wonder how you correlate the theory with experiment.

M. Cais:

I would just like to put one formula on the board, because of a certain provocative statement that Mr Prinzbach has just made. The simplest unsubstituted sesquifulvalene derivative is a very stable compound, and there is no question at all that in this particular derivative the writing of the circles to which Mr Prinzbach objects is very much justified, as demonstrated by the NMR spectrum of this compound. By the way, the NMR spectrum consists of just three apparent singlets, one due to the tropylium ring, one due to the substituted cyclopentadiene ring and the third one due to the unsubstituted cyclopentadiene.

Exactly by the same approach we made what seems to be the simplest calicene derivative, unsubstituted in the five-membered ring.

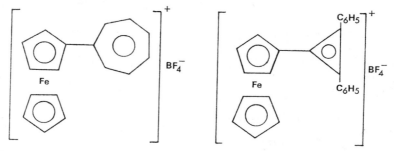

G. V. Boyd:

Unlike sesquifulvalene, its *iso*-π-electronic nitrogen analogues are very stable. They are formed directly by adding cyclopentadiene to a pyridinium salt in the presence of aqueous alkali:

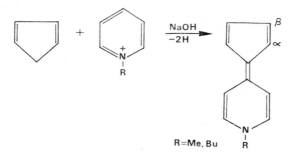

Discussion

The bases are protonated in the α and β positions in the ratio $1:4$. The NMR spectra of the indene analogues are temperature-dependent. Methyl substitution in the pyridine portion increases the ease of rotation round the central linkage.

Mrs A. Pullman:

What is the departure from co-planarity in sesquifulvalene?

H. Prinzbach:

Using models, based on the geometrical data we have for the fulvene molecule, we conclude that there is some deviation from planarity.

E. D. Bergmann:

The central double bond in sesquifulvalenes is obviously not a 'real' double bond, as otherwise there could be no free rotation about it. Furthermore, the theory predicts that the bond order of that central bond is exactly that of a benzene bond. You then have—as the dipole moment shows— a tendency for the five-membered ring to assume to some extent the form of a cyclopentadienide anion and for the three-membered ring to assume the form of the cyclopropenium cation. Do you not think that this contributes to 'aromaticity'?

The Reactions of Heptafulvenes

by Y. KITAHARA *and* M. ODA

Department of Chemistry, Tohoku University, Sendai, Japan

I. INTRODUCTION

IT IS A PRIVILEGE and great honour to be invited to speak about our research on heptafulvenes at this Symposium.

Although the parent hydrocarbon heptafulvene (I) and some of its derivatives, e.g. 8-vinyl-heptafulvene (II) and 8,8-dicyanoheptafulvene (III), have been synthesized, their chemical properties have scarcely been investigated, due to the instability of I and II and to the low reactivity of III under usual conditions [1–4]. These circumstances prompted us to inspect the aromaticity and chemical reactivity of heptafulvenes by using organic reactions. The heptafulvenes used for these investigations were primarily 8,8-dicyanoheptafulvene (III) and 8-cyanoheptafulvene (IV); the latter was first synthesized.

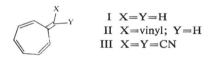

I X=Y=H
II X=vinyl; Y=H
III X=Y=CN

The processes for the syntheses of heptafulvenes here used are as follows [5–6]:

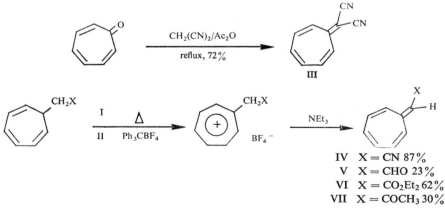

IV X = CN 87%
V X = CHO 23%
VI X = CO$_2$Et$_2$ 62%
VII X = COCH$_3$ 30%

Fig. 1

284

II. ELECTROPHILIC SUBSTITUTIONS [7]*

We have succeeded in three electrophilic substitution reactions with IV, giving VIII, IX and X, viz., nitration with tetranitromethane, trifluoroacetylation with trifluoroacetic anhydride and formylation under Vilsmaier conditions.

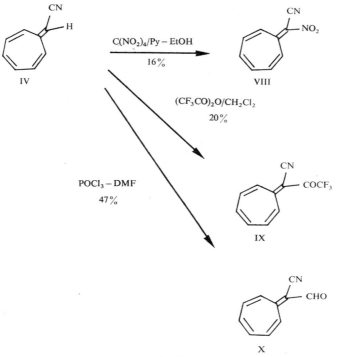

Fig. 2

The results indicate that IV has aromatic character even in the classical sense, which is due to a fairly large contribution by the $\sigma\pi$ polar aromatic sextet structure.

III. ADDITION REACTIONS

1. *Bromination* [8]

8,8-Dicyanoheptafulvene (III) shows resistance to bromination under usual conditions. This may indicate a substantially large aromatic character. However, heating III under reflux with an excess of bromine in carbon tetrachloride under photo-irradiation gave a hexabromo addition product (XI) in more than 90% yield. The compound XI easily underwent debromination and dehydrobromination with triethylamine or dimethylformamide to give a 2,4-dibromo compound (XII) or a 3-bromo compound (XIII) in good yields. These easy regenerations of the 8,8-dicyanoheptafulvene structure indicate its high stability.

---

* We have further found three substitutions: bromination [7], tropylation [*Chem. Comm.*, in press] and diphenylcyclopropenylation [to be reported].

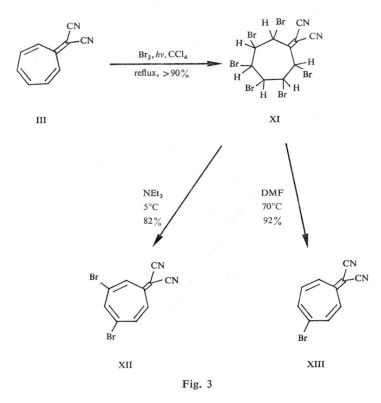

Fig. 3

2. Cycloaddition Reactions with Dienophiles

As was found with the parent hydrocarbon I, the 8-monosubstituted heptafulvenes underwent $8 + 2$-cycloadditions with dimethyl acetylenedicarboxylate at 140° in xylene, leading to azulene derivatives XIV, XV, XVI and XVII [5–6]. Other dienophiles, such as tetracyanoethylene*, benzyne [9] and *p*-benzoquinone* reacted in the same manner, giving XVIII, XIX and XX (see Fig. 4).

It has been reported that III does not give addition product with maleic anhydride. However, we found that it gives an *endo* 1, 4-cycloadduct (XXI) at higher temperatures (180–190°) in 74% yield.* Interestingly, III reacted with ethylene and styrene more easily, at 150° C and 80° C, respectively.* The former gave only a 1, 4-cycloadduct (XXII), while the latter gave a 1, 8-cycloadduct (XXIII) and a 1, 4-cycloadduct (XXIV) in a ratio of about 1 : 1. The structures of XXIII and XXIV were elucidated by 100 MHz NMR and NMDR spectra, and by comparison with the corresponding adducts from 1, 6-dideuterio-8, 8-dicyanoheptafulvene and styrene, or III and *cis*-1, 2-dideuteriostyrene. The reaction is stereo-selective (see Fig. 5). The easiness of reaction with electron-rich olefins may be rationalized as a Diels-Alder reaction with inverse electron demand [10].

---

* Kitahara and Oda, unpublished results.

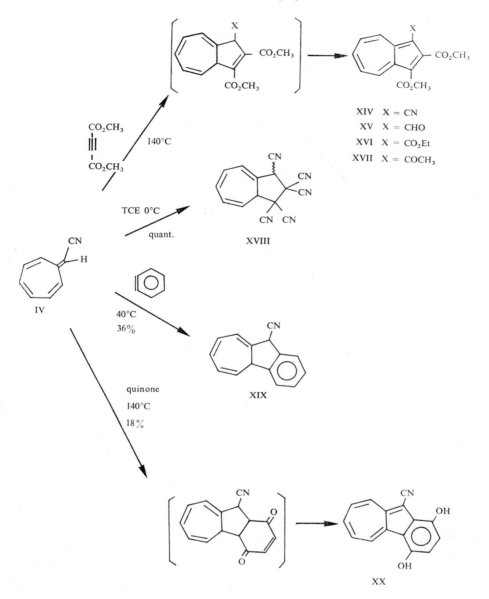

XIV X = CN
XV   X = CHO
XVI X = CO₂Et
XVII X = COCH₃

Fig. 4

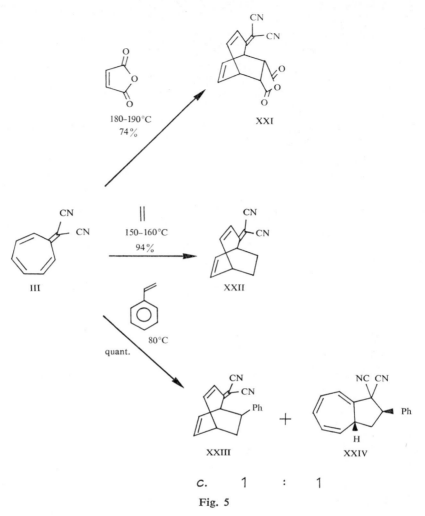

Fig. 5

3. Cycloaddition Reactions with Cyclic Dienes*

Recently, several thermal 6 + 4-cycloaddition reactions have been reported as being represented by the reaction of tropone and cyclopentadiene [11–12]. Compound III reacted more easily than tropone with cyclopentadiene and gave three types of products, according to the reaction temperature. Reaction at 20° C gave a 6 + 4-cycloadduct (XXV) in 90% yield. The structure of XXV was elucidated by the comparison of its NMR data with those of similar adducts from bromo-substituted 8, 8-dicyanoheptafulvenes.

On heating at 70° for 40 hr, the adduct XXV rearranged to give the second 1 : 1 adduct, XXVI and III, in 80% and 20% yield, respectively. From the data of 100 MHz NMR and NMDR spectra and the observation of 8% of NOE between H_a and H_b, we derive the structure, as shown in Fig. 5. Compound XXVI further rearranged at 140°C for 12 hr to give the third 1 : 1 adduct, XXVII and III, in 78% and 22% yield, respectively. The adduct XXVII was proved to be an about 1 : 1 mixture of double-bond isomers of an *endo* 4+2-type

* Kitahara and Oda, unpublished results.

cycloadduct, XXVIIa and XXVIIb, by its 100 MHz NMR and NMDR spectra and some chemical evidence.

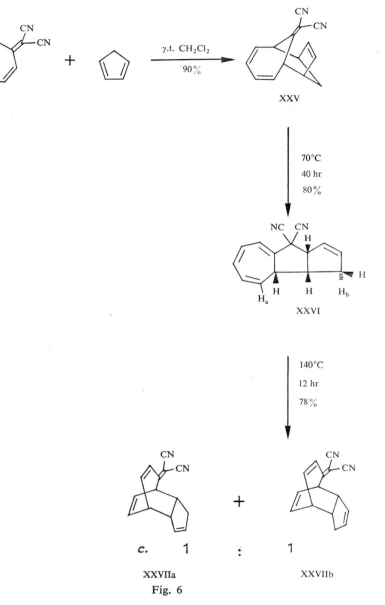

Fig. 6

It was inferred that the former mechanism of the rearrangement is intramolecular and the latter intermolecular from the following evidence:

1. The rearrangement of XXV at 70° in the presence of one equivalent of maleic anhydride gave 51% of XXVI, 41% of III and 16% of the adduct of cyclopentadiene and maleic anhydride (XXVIII). On the other hand, XXVI at 140° gave no trace of XXVII, but 94% of III and 77% of XXVIII.

2. The exchange reaction with one equivalent of 1,6-dideuterio-8,8-dicyanoheptafulvene,

289

Y. Kitahara and M. Oda

a mixture of $D_2 = 67.5\%$, $D_1 = 24.2\%$ and $D_0 = 8.5\%$, under the same conditions (the distribution of deuterium in the products was measured by mass spectrometry), gave the results summarized in the table. The intramolecular process may be explained by a Cope-type mechanism, and the intermolecular process by a dissociation and 4+2-recombination reaction.

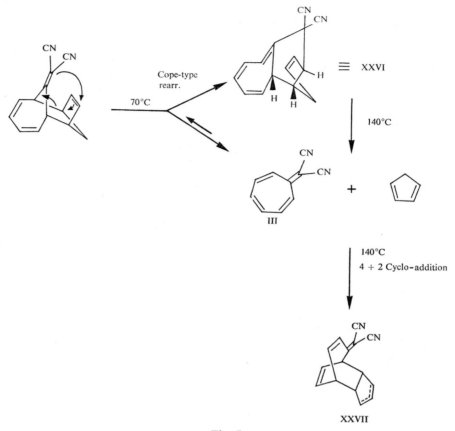

Fig. 7

The Results of the Exchange Reactions

Substrates	Temp. (°C)	Distribution of the Products (D;% ± 1)					
		XXVI			III		
XXV	70	D_2	D_1	D_0	D_2	D_1	D_0
		14.2	6.1	79.7	56.2	20.2	23.6
		XXVII			III		
XXVI	140	D_2	D_1	D_0	D_2	D_1	D_0
		37.2	13.3	49.5	40.1	14.4	45.5

With cyclohexadiene, a similar result was obtained, with minor differences.

290

4. *Reactions with Enamines* *

Because III and 8-cyano-8-ethoxycarbonyl heptafulvene (XXIX) are highly polar, they easily suffered nucleophilic attack by 1-morpholinocyclohexene at room temperature, to give reversible 1:1 cycloadducts, XXX and XXXI, in high yields. The less polar heptafulvene, IV, also reacted, but at higher temperature (80° to 140°), with various enamines to yield directly dihydroazulene derivatives, which are easily converted into the corresponding azulene derivatives by dehydrogenation with *p*-chloranil. One example, which led to XXXIII and XXXIV, is shown in Fig. 8. This procedure provides a new synthetic method for azulenoids.

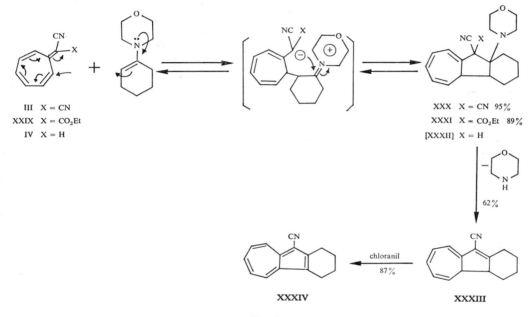

III   X = CN
XXIX   X = CO₂Et
IV   X = H

XXX   X = CN   95%
XXXI   X = CO₂Et   89%
[XXXII]   X = H

Fig. 8

5. *Reactions with Ketenes* *

8-Cyanoheptafulvene (IV) reacts with diphenylketene at 70° to give the compounds XXXV, XXXVI and XXXVII. With dichloroketene it gave only one isolable product (XXXVIII). The reaction may be rationalized as a two-step mechanism. The results are also consistent with the polar character of IV (see Fig. 9).

6. *Thermal Dimerization* *

8-Cyanoheptafulvene (IV) dimerized at 160° to 170° to give a dicyano-tetrahydro-*s*-heptindacene (XXXIX). Bromination of XXXIX, followed by dehydrobromination with lithium chloride-dimethylformamide, gave a dihydro derivative (XL), which might be a potential intermediate for a new cyclic conjugated system (XLII). The dimer XXXIX should be derived from the initial 8 + 8-type dimer (XLI), which is probably formed *via* an ionic mechanism (see Fig. 10).

---

* Kitahara and Oda, unpublished results; Oda, Tani and Kitahara [13].

291

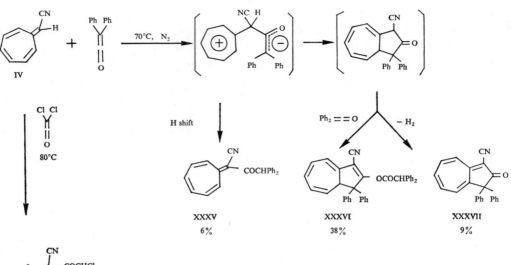

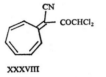

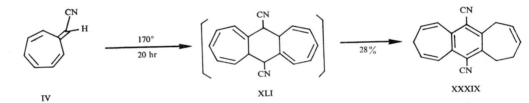

Fig. 9

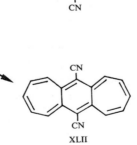

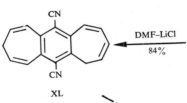

Fig. 10

292

7. *Photo-Oxygenations* [14]

8-Cyanoheptafulvene (IV) absorbs molecular oxygen under irradiation in the presence of a sensitizer to give a mixture of isomeric epidioxides, XLIVa and XLIVb, in high yield. On treatment with triethylamine, the epidioxides afforded 6-hydroxy-1-oxazulane-2-imine (XLV) as a sole product in 90% yield. 8,8-Dimethoxycarbonylheptafulvene (XLVI) also behaved in the same way, giving an epidioxide (XLVII) which was easily converted into 6-hydroxy-3-methoxycarbonyl-1-oxazulanone (XLVIII).

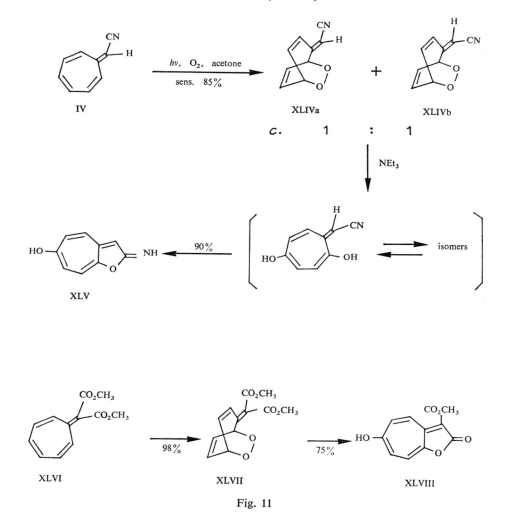

Fig. 11

In summary, heptafulvenes can be stabilized by the contribution of a polar 6π-electronic structure, and III and IV have an aromatic, as well as an appreciable olefinic, character.

REFERENCES

1 W. von E. Doering & D. W. Wiley (1960) *Tetrahedron*, 11 : 183.
2 D. J. Bertelli, C. Golino & D. L. Dreyer (1964) *J. Am. Chem. Soc.*, 86 : 3329.
3 T. Mukai, T. Nozoe, K. Osaka & N. Shishido (1961) *Bull. Chem. Soc. Japan*, 34 : 1384.
4 K. Hafner, H. W. Riedel & M. Danielisz (1963) *Angew. Chem.*, 75 : 344.
5 M. Oda & Y. Kitahara (1969) *Chem. Comm.*, p. 352.
6 Idem (1969) *Chemy Ind.*, p. 920.
7 Idem, *Bull. Chem. Soc. Japan* [in press].
8 M. Oda, M. Funamizu & Y. Kitahara (1969) *Chemy Ind.*, p. 75.
9 M. Oda & Y. Kitahara (1970) *Bull. Chem. Soc. Japan*, 43 : 1920.
10 J. Sauer & H. West (1962) *Angew. Chem.*, 74 : 353.
11 R. C. Cookson, B. V. Drake, J. Hudec & A. Morrison (1966) *Chem. Comm.*, p. 15.
12 S. Ito, Y. Fujise, T. Okuda & Y. Inoue (1966) *Bull. Chem. Soc. Japan*, 39 : 1351.
13 M. Oda, H. Tani & Y. Kitahara (1969) *Chem. Comm.*, p. 739.
14 M. Oda & Y. Kitahara (1969) *Angew. Chem.*, 81 : 702.

Discussion

E. D. Bergmann:

Diphenylcyclopropenone was condensed with cyanoacetic acid in the hope to obtain 4-carboxy-4-cyanotriafulvene (A), which was expected to decarboxylate to 4-cyanotriafulvene (B). This reaction appears to be spontaneous, but the product was not B. It had the structure C, and we assume that it is formed by the attack of the cyanoacetylium cation D on the 4-position of the triafulvene B.

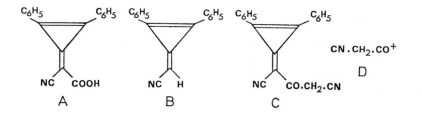

H. G. Marks (Hebrew University):

I was interested to see the dimerization product which you showed, and I was wondering if you had any other examples of such self-condensations of these cross-conjugated systems? You realize, of course, that these 1, 3-self-condensations are very unusual?

Y. Kitahara:

This is the only example I know.

The Study of Conformational Changes Related to Aromaticity and Anti-Aromaticity by Nuclear Magnetic Resonance Spectroscopy

by A. P. DOWNING, W. D. OLLIS, M. NÓGRÁDI *and* I. O. SUTHERLAND

Department of Chemistry, University of Sheffield

FULVENE (I), heptafulvene (II), tropone (III) and the related seven-membered ring systems (IV; X = S, NR, O) have been of interest to organic chemists for a considerable time [1–8] in view of the possibility that their properties might differ from those expected for the classical unsaturated systems. These differences may be associated with ground-state or excited-state properties. Although the latter are of considerable interest in that they may be related to differences in chemical reactivity, the ground-state properties have generally been scrutinized [9] to provide criteria for classifying compounds as aromatic or anti-aromatic [10–13]. These ground-state properties, which include structural [12–13], magnetic [14–16] and thermodynamic [12–13] properties, have been examined in some detail experimentally and also by theoretical methods. In spite of the rather misleading results of early calculations using the simple Hückel MO method [12–13, 17–20], the more recent use of various SCF–MO approaches [12–13, 21–27] has shown that I and II should be regarded as classical polyene systems in that their calculated heats of formation are close to those of the corresponding classical polyenes, and that their structures are predicted to show the alternation of bond lengths consistent with polyene character. The situation has been less clear with respect to the properties of III and IV.

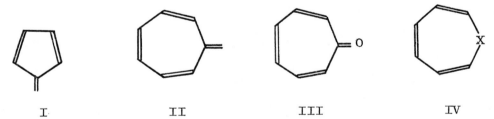

I   II   III   IV

In spite of the considerable amount of information available, it appeared that an examination of the conformational mobility of suitable derivatives of the systems I–IV might provide additional information concerning the ground-state properties of these interesting compounds. Thus, the fulvenes have often been considered in terms of the resonance representation V. In spite of the shortcomings of this representation, it does suggest that the 6-dialkylaminofulvenes (VI) would show enhanced dipolar character, as evidenced by partial double-bond character in both the C_6–N and C_1–C_6 bonds.

296

In order to examine this situation, we have investigated the energetics of rotational isomerism about these bonds [28–30]. Structural examinations by X-ray crystallography and other methods have shown that although 2-chlorotropone [31] and 8,8-dicyanoheptaful-vene [32] are planar, other heptafulvenes [33], substituted azepins [34] and thiepin-1,1-dioxide [35] are non-planar, and the alternation of bond lengths in compounds of this type has recently been fully discussed [36–37]. The ground-state properties of the seven-membered ring systems may therefore be considerably modified by their non-planarity, or by the angle strain in the seven-membered ring when a planar conformation is adopted. These considerations led us to examine the kinetics of ring-inversion in suitable derivatives of the ring systems II–IV [38].

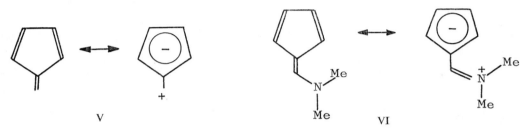

THE INVESTIGATION OF CONFORMATIONAL CHANGES IN 6-DIALKYLAMINOFULVENES

Rotational isomerism about the exocyclic C_1–C_6 bond and the C_6–N bond in VI may conveniently be examined by NMR line-shape methods [39–44], provided that the rotational barriers are of the correct magnitude (*c.* 8–25 Kcal/mole). These rotational changes are shown in Fig. 1. Rotation about the C_6–N bond (VII $\rightleftarrows$ VIIa) results in the exchange of the two N-methyl groups (Me and Me′) between two different environments (A and B) by way of a transition state (VIII) that may be considered as having the π-electron energy of isolated fulvene and dimethylamino systems. In VIII the C–N–C of the dimethylamino group is in a plane perpendicular to the plane containing the fulvene system; this description applies whether the nitrogen atom in the transition state is associated with a planar or pyramidal valence situation. Rotation about the C_1–C_6 bond (IX $\rightleftarrows$ IXa) results in the exchange of ring protons H_2 and H_5 between the environments A and D, and of H_3 and H_4 between the environments B and C; the transition state for this process (X) approximates in π-electron energy to that of a dipolar system that is formally (but not geometrically) related to the canonical form VIb. In the transition state X, the planar ene-ammonium grouping is perpendicular to the plane containing the five-membered ring. In the case of fulvenes with no ring substituents, or with a symmetrical ring substitution pattern, both processes are degenerate, that is, VII ≡ VIIa and IX ≡ IXa. This is an ideal situation for NMR study, since the magnetic sites involved in the exchange process are equally populated.

The series of fulvene derivatives, XI–XXI, were prepared according to literature methods [45], or by reaction of the appropriate dimethylaminoethoxycarbonium fluoroborate with sodium cyclopentadienide [28–30] for XVI–XVIII. The low-temperature NMR spectra of compounds XI–XIX showed two separate singlets for the N-methyl groups, which coalesced to a single singlet at higher temperatures. The rates of exchange of the methyl groups of XI–XV between the two environments A and B (Fig. 1) were determined using formulae appropriate to slow and fast exchange rates, or by line-shape comparison [39–44] over a temperature range of 70° to 80°. The examination of the compounds XVI–XIX was re-

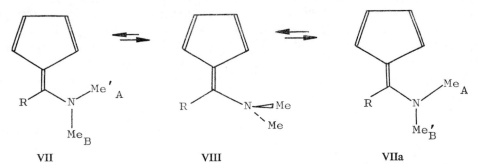

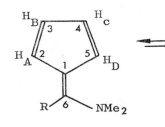

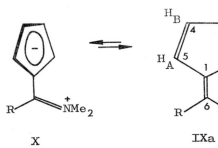

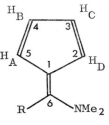

IX   X   IXa

Fig. 1

Rotational isomerism in 6-dimethylaminofulvenes

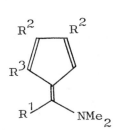

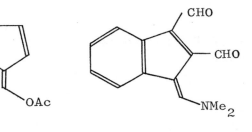

	R^1	R^2	R^3
XI	H	H	H
XII	H	H	CHO
XIII	H	CHO	H
XIV	H	CHO	Me
XVI	Me	H	H
XVII	Ph	H	H
XVIII	p-OMe-C_6H_4	H	H
XX	OEt	H	H

XXI   XV

XIX

298

stricted to the determination of the exchange rate at the temperature at which the two low-temperature N-methyl signals just coalesced [45]. The more complete kinetic study gave the Arrhenius parameters associated with the rotation process, but the coalescence method gave only the free energy of activation at the coalescence temperature (Table 1).

Table 1

Activation Parameters for Rotation about
C_6–N and C_1–C_6 Bonds of 6-Dimethylaminofulvenes XI–XIX

Com-pound	C_6–N Bond				C_1–C_6 Bond	
	$\Delta G^{\neq}$ (°C)* (Kcal/mole)	E_a (Kcal/mole)	$Log_{10}A$	Solvent	$\Delta G^{\neq}$ (°C)** (Kcal/mole $\pm$ 0.5)	Solvent
XI	13.5 ± 0.1 (0)	15.1 ± 1.0	14.1 ± 0.6	$CDCl_3$–C_6F_6 (1:3)	22.1 (148)	Me_2SO
XII	17.9 ± 0.3 (0)	20.9 ± 1.6	15.3 ± 1.0	C_2HCl_5		
XIII	21.5 ± 0.3 (0)	26.5 ± 2.0	16.8 ± 1.1	C_2HCl_5		
XIV	20.3 ± 0.2 (0)	27.3 ± 1.9	18.5 ± 1.0	C_2HCl_5		
XV	12.6 ± 0.2 (0)	17.3 ± 1.0	16.5 ± 1.0	$CDCl_3$		
XVI	10.7 ± 0.2 (−65)			$CDCl_3$	15.5 (20)	Me_2SO–D_2O (2:1)
					16.4 (38)	Me_2SO
					17.5 (57)	$(CD_3)_2CO$
					17.5 (57)	$CDCl_3$
XVII	11.8 ± 0.2 (−34)			$CDCl_3$	19.2 (100)	Me_2SO
XVIII	12.1 ± 0.2 (−28)			$CDCl_3$	18.8 (93)	Me_2SO
XIX	19.6 ± 0.4 (96)			C_2HCl_5		

* Based on coalescence data for XVI–XIX; in all other cases, $\Delta G^{\neq}$ at 0°C was estimated from the Arrhenius parameters

** Based on coalescence data only

The ring protons of the fulvenes XI and XVI–XVIII gave complex ABCD systems at low temperatures, which at high temperatures coalesced to AA′BB′ systems. The associated line-shape changes are complex, but the low-temperature spectra of the ring protons of XVII and XVIII were first order, and the approximation relating the exchange rate at the coalescence temperature to the site separation [45] could be used. The spectra for XI and XVI were, however, second order, and exchange rates could not be obtained from coalescence temperatures. The problem was simplified by partial exchange of the ring protons for deuterium by heating a solution of the fulvene in dimethyl sulphoxide-deuterium oxide at 100° for 12 hrs. The NMR spectra of the resulting deuteriated compounds, having approximately 70% deuteration at each ring position, showed broad singlets for each of the ring protons, and coalescence temperatures and exchange rates could readily be obtained.

The relevant details of the NMR spectra of compounds XI–XIX have been described elsewhere [28–29]; the calculated activation parameters for rotation about the C_6–N and C_1–C_6 bonds are listed in Table 1. Compounds XII, XIV and XV appear to exist in a single preferred conformation with respect to rotation about the C_1–C_6 bond. The fulvenes XX and XXI were also examined. The former showed an AA′BB′ spectrum for the ring protons and a single N-methyl singlet for the dimethylamino group down to −60°C, suggesting that both rotational barriers are low in this case (less than *c.* 11 Kcal/mole). The latter showed non-equivalence of the ring protons *cis* and *trans* to the acetoxyl substituent up to 200°C, indicating a C_1–C_6 rotational barrier greater than 27 Kcal/mole.

STERIC AND ELECTRONIC EFFECTS ON FULVENE ROTATIONAL BARRIERS

The free-energy barriers (Table 1) show considerable variation with structural variation in the fulvene system. The discussion that follows is based upon the observed free energies of activation. This is due to the lack of complete kinetic data in many cases, and also because free-energy data for rotational barriers obtained by the NMR method are less liable to systematic error than the entropies and enthalpies of activation. The effect of formyl substituents in increasing the C_6–N rotational barrier is predictable from a consideration of the resonance situation (compounds XI–XIII in Table 1). The slight lowering of this barrier by a ring methyl substituent (compound XIV) could be associated with the reputed inductive effect of a methyl group, or possibly be due to steric destabilization of the planar ground state. The effects of substituents at C_6 are exemplified by the series of compounds XI, XVI–XVIII (Table 1), and are both steric and electronic in origin. From a comparison of the effects of the same substituent on the fulvene C_6–N rotational barrier and the analogous rotational barrier for the N–CO bond of amides [46–49] it has been concluded that, in the former case, steric effects are predominant [28–30], although, in the case of the 6-aryl substituents, the effective size of the substituent depends upon its rotational orientation. The conjugation effect of a 6-alkoxy substituent lowers the C_6–N rotational barrier below the magnitude that could be studied by NMR methods (Table 1, compound XX). The effect of benzene annelation also lowers the C_6–N rotational barrier (Table 1, compound XV), presumably a consequence both of increased bond localization in the five-membered ring and also of steric destabilization of the planar ground state. The tropone derivative XIX, on the other hand, has a high C_6–N rotational barrier. This is of interest in that it stresses the importance of azulene character in this compound, as emphasized by the dipolar canonical form (see XIX).

The steric and electronic effects of 6-alkyl and aryl substituents upon the C_1–C_6 rotational barrier are illustrated by the series of compounds XI, XVI–XVIII. An analysis of these effects, which has appeared elsewhere [28–30], is consistent with the view that a dipolar transition state (X) is associated with the rotational process IX ⇌ IXa. This conclusion is in accord with the effect of solvent changes upon this rotational barrier, which decreases with increasing solvent polarity (Table 1, compound XVI). The estimated stabilizing effect of the 6-methyl substituent in the transition state X (*c.* 2.9 Kcal/mole) is very much smaller than the similar stabilizing effect of a methyl substituent (*c.* 7.4 Kcal/mole) in the transition state XXII for rotation about the olefinic double bond in the enamine XXIII [50]. In the latter case, this stabilization approaches the limiting value of 8.5 Kcal/mole observed in solvolytic reactions [51]. This difference in behaviour of the two systems is intriguing and could reflect a much larger degree of ground-state charge separation in the fulvene system; however, the dipole moments of 6-dialkylaminofulvenes are only about 25% of the value expected for complete charge separation [52–53]. The rather smaller lowering of the free energy of the transition state X (1.2–1.9 Kcal/mole) by the conjugating effect of the 5-aryl substituents [28–30] (Table 1, compounds XVII–XVIII) appears to be a consequence of the transition-state geometry, which does not permit a rotational orientation of the aryl substituent suitable for full benzylic stabilization of the C_6 carbonium centre of X.

300

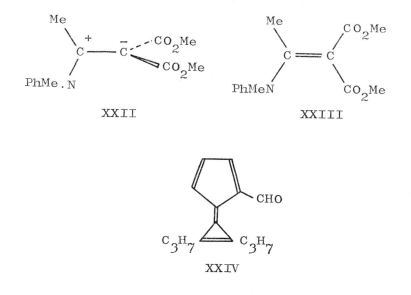

XXII   XXIII

XXIV

Measurements of rotational barriers for a number of the fulvenes discussed in this paper have been reported by two other groups [54–55]. In all cases, the results agree with those reported here. The NMR line-shape method has also been used to measure rotational barriers about formal olefinic double bonds in an extended quinone [56], enamines and enol ethers [50, 57–58], ketene mecaptals and imines [59] and a cumulene [60]. The NMR study of the formylcalicene, XXIV [61], is of particular interest, as in this case the rotational barrier about the interannular double bond was of the same order (18.0–19.4 Kcal/mole at 66° to 114° C) as that observed for 6-dimethylaminofulvene. The interannular double bond in calicenes, which have been examined by crystallography [62–63], is short (1.37 Å) and of the order predicted by SCF-MO calculations [64], which give results in accord with a polyene structure for calicenes, as for other fulvenes and fulvalenes. The calicene XXIV has, however, a formyl substituent, and, as far as we are aware, the possible relationship between the lengths of olefinic double bonds and the magnitude of the rotational barrier has not been directly investigated. The results quoted here suggest that rotational barriers may be more closely related to polarizability than to ground-state geometry or polarization.

COUPLING CONSTANTS AND BOND ORDERS IN FULVENES

The NMR spectra of fulvenes, including 6-dialkylaminofulvenes, have been the subject of a number of investigations [54–55, 64–68]. The relationship originally proposed for the *ortho*-coupling constants in polycyclic benzenoid compounds [69] has been extended to five-membered ring systems. For these systems, the simple equation $J_{ortho} = 7.12p - 1$ has been proposed [64], where p is the π bond order of the bond between the *ortho*-related protons. This relationship is exemplified by the coupling constants for cyclopentadiene XXV, in which the 2,3-bond (J_{23} 5.06 Hz) and the 3,4-bond (J_{34} 1.94 Hz) show the maximum observable difference in vicinal coupling constants [70], and for dimethoxycarbonylcyclopentadienide (XXVI), in which the vicinal coupling constant (3.5 Hz) is the mean of J_{23} and J_{34} of cyclopentadiene [71].

301

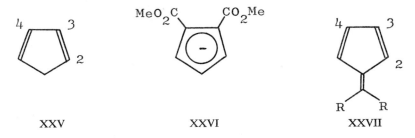

XXV	XXVI	XXVII

Fulvene itself and 6,6-dialkylfulvenes (XXVII) have similar values for J_{23} (5.1 Hz) and J_{34} (1.9 Hz) to cyclopentadiene [65–67]. The polyene structure of these fulvenes, consistent with these coupling constants, has recently been confirmed [72] by the electron diffraction examination of 6,6-dimethylfulvene. This molecule shows single and double bonds, as required by the representation XXVII, with bond lengths analogous to those of the non-terminal bonds in long-chain polyenes. The vicinal coupling constants of the 6-dialkylamino-fulvenes (J_{23}, J_{45} 4.40–4.75 Hz; J_{34} 2.45–2.50 Hz) [28–30] indicate a slightly lower degree of bond alternation than in the 6,6-dialkylfulvenes [see also 54–55, 68]. This lower degree of bond alternation in the five-membered ring of the 6-dialkylaminofulvenes is consistent with a lower barrier to rotation about the C_1–C_6 bond, as compared with the 6,6-dialkylfulvenes, but insufficient data are available to determine whether bond orders are directly related to rotational barriers.

From our results, the presence of a 6-dialkylamino substituent appears to be the minimum requirement for stabilization of the transition state X, so that the permitted rotation about the C_1–C_6 bond can be studied by NMR line-shape methods. It has recently been suggested [9] that the observed degree of bond alternation in cyclic conjugated systems may provide a rather better ground-state criterion for the presence or absence of aromatic character than the more frequently studied ring current effects [14–16]. The application of this criterion to the 6-dialkylaminofulvenes, in which the degree of bond alternation is clearly considerable, does not suggest that they should be regarded as aromatic compounds. A classification of this type, although easy to apply if structural data are available, has the disadvantage that it ignores the classical chemical significance of aromaticity.

DELOCALIZATION ENERGY OF 6-DIMETHYLAMINOFULVENE

The results described in the preceding sections of this paper show that the delocalization energies of 6-dimethylaminofulvene and its derivatives are considerably greater than would be expected for the classical structure VIa. Thus, the stabilization due to conjugation by the 6-dimethylamino substituent, which is of the order of 13.5 Kcal/mole for the parent compound XI (Table 1), follows from a comparison of the energy of the ground state VII and the transition state VIII. This estimate ignores steric effects, which tend to lower the C_6–N rotational barrier. Allowing for the non-bonded interaction between the C_6–N-methyl group and the adjacent C_5 hydrogen substituent, the total conjugation energy in 6-dimethyl-aminofulvene would be *c.* 20 Kcal/mole in excess of that in fulvene itself [28–30]. This strongly suggests that 6-dialkylaminofulvenes should be considered as having some aromatic character. Estimates of the resonance energy in 6,6-dimethylfulvene ($\sim$12 Kcal/mole) have been made [73] on the basis of thermal data. However, the significance of this figure is not easy to ascertain, and there seems to be no evidence, on the basis of physical and chemical properties of simple fulvenes, to suggest that these compounds should be regarded chemically as other than simple conjugated polyenes. Our results do show, however, that systems of the fulvene type may show thermodynamic properties consistent with aromatic character if suitable substituents are present to facilitate charge separation. This result is fully in accord with the chemical properties [8] of fulvenes with oxygen or nitrogen substituents at the 6-position, which are known to show a diminished tendency to undergo typical polyene reactions, such as the Diels-Alder reaction, and which may even show the classical property of aromatic systems and undergo substitution reactions.

The Investigation of Conformational Changes in 2,3:6,7-Dibenzo Derivatives of Cycloheptatriene, Tropone, Heptafulvene, Oxepin, Thiepin and Azepin

The chemical and physical properties of the seven-membered ring systems heptafulvene (II), tropone (III), oxepin (IV; X = O), thiepin (IV; X = S) and azepin (IV; X = NR) and their 2,3:6,7-dibenzo derivatives have been examined, in many cases in some detail. The results of the examinations show that the monocyclic systems II and IV behave as the corresponding classical unsaturated systems, although tropone (III) has often been quoted as having aromatic character. This lack of special properties associated with the cyclic conjugation in their structures has been demonstrated by their physical and chemical properties. Thus, heptafulvene (II) is very unstable [74] and polymerizes readily. Oxepin (IV; X = O) exists in tautomeric equilibrium with its highly strained benzene-epoxide isomer [75], an equilibrating system analogous to the much-studied cycloheptatriene-norcaradiene system [76]. Thiepin itself (IV; X = S) has not been prepared, but thiepin-1,1-dioxide (IV; X = SO$_2$) is known to be thermally labile [77]. Attempts to prepare azepin (IV; X = NH) have resulted only in the isolation of tautomers [78–79], and the N-substituted azepins that have been prepared are rather unstable compounds which undergo a variety of cycloaddition reactions [for a short review up to 1967, see 76]. Only the tropone system III has demonstrated chemical stability [80–81].

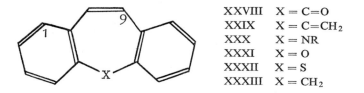

XXVIII	X = C=O
XXIX	X = C=CH$_2$
XXX	X = NR
XXXI	X = O
XXXII	X = S
XXXIII	X = CH$_2$

In general, benzene annelation is expected to increase the classical character of the seven-membered ring [12–13, 82–83]. Not surprisingly, the 2,3:6,7-dibenzo derivatives XXVIII–XXXII have properties consistent with limited delocalization of π electrons in the seven-membered ring. Thus, the dibenzotropone XXVIII adopts a non-planar conformation in the crystalline state [84]. The heptafulvene system in XXIX is also probably non-planar [82, 85]. Also, the non-planarity of the azepin system in XXX (R = H) has been suggested on the basis of its physical properties [86–87]. It seems likely that the dibenzoxepin XXXI [88–90] and dibenzothiepin XXXII [91] systems also have non-planar seven-membered rings. The observed physical properties of these ring systems may therefore be in part a consequence of their non-planarity.

In order to determine the extent to which this demonstrated non-planarity is a consequence of both the steric and electronic effects associated with the seven-membered ring system, we have investigated the kinetics of ring inversion in suitably substituted derivatives of the systems XXVIII–XXXII [38]. The results have been compared with the results of a similar examination of ring inversion in the dibenzocycloheptatriene system (XXXIII), in which the π-electron system of the seven-membered ring is interrupted by a methylene group.

Ring inversion of the seven-membered rings of the compounds XXVIII–XXXII is a degenerate process, which cannot be studied by NMR line-shape methods unless prochiral [92] substituents of the general type CR_2X are present. The use of a prochiral ring CR_2 group is excluded by obvious structural requirements, except for derivatives of the cycloheptatriene system XXXIII. The proposed investigation therefore required the synthesis of a series of derivatives of compounds XXVIII–XXXII with a suitably located prochiral substituent, and a 1-benzyl substituent was first investigated. 1-Benzyl-5H-dibenzo [a, d] cyclohepten-5-one (XXXIV) and the corresponding hydrocarbon XXXV were synthesized by methods which will be described elsewhere [38].

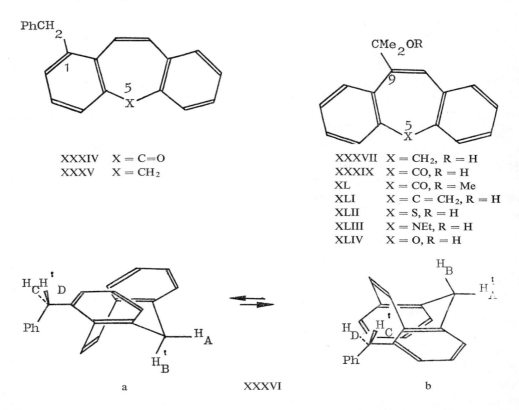

XXXIV	X = C=O
XXXV	X = CH₂

XXXVII	X = CH₂, R = H
XXXIX	X = CO, R = H
XL	X = CO, R = Me
XLI	X = C = CH₂, R = H
XLII	X = S, R = H
XLIII	X = NEt, R = H
XLIV	X = O, R = H

The NMR spectrum of the benzyl methylene group of XXXIV remained a singlet down to $-100°C$, either as a consequence of rapid ring inversion or the result of accidental chemical shift equivalence of the diastereotopic methylene protons H_C and H_D (cf. XXXVI) in a slowly inverting ring system. Both the 5-methylene group and the methylene group of the 1-benzyl substituent of the hydrocarbon XXXV gave singlets in the NMR spectrum at 35° C. At low temperatures ($< -51°$ C), the ring 5-methylene group gave an AB system consistent with a slow rate, on the NMR time scale, for the inversion process XXXVIa $\leftrightarrows$ XXXVIb. The 1-benzyl methylene group remained a singlet down to $-80°$ C. This apparent chemical shift equivalence of H_C and H_D (see XXXVI), together with the low inversion barrier for the hydrocarbon XXXV (11.0 Kcal/mole), indicated that the 1-benzyl substituent would probably be unsuitable for the more general study.

Accordingly, the use of a second prochiral substituent was investigated, and 10-(1-hydroxy-isopropyl)-5H-dibenzo $[a,d]$ cycloheptene (XXXVII) was synthesized. The temperature dependence of the NMR spectrum of this compound was more encouraging. Thus, at low temperatures, evidence for slow ring inversion (XXXVIIIa $\leftrightarrows$ XXXVIIIb) on the NMR time scale was obtained from the non-equivalence of the *endo*- and *exo*-protons (H_A and H_B) at C_5 and of the diastereotopic methyl groups (Me_A and Me_B) of the prochiral hydroxyiso-propyl substituent at C_{10}. As the temperature of the solution was increased, the NMR spectrum changed, indicating exchange of H and H' and Me and Me' between the sites differentiated by the subscripts A and B in XXXVIII. The barrier to ring inversion (17.5 Kcal/mole) also indicated that this was a suitable system for more general study, and compounds XXXIX–XLIV were synthesized. The temperature dependence of the NMR spectrum of the prochiral CMe_2OR group was investigated for each compound. The results of this study are summarized in Table 2.

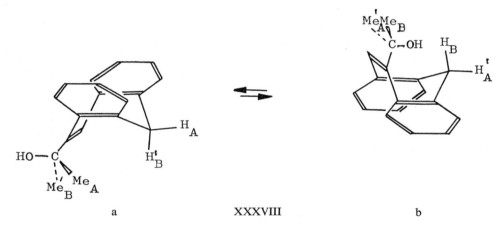

$$a \qquad\qquad \text{XXXVIII} \qquad\qquad b$$

In addition, at this stage of our study it was also necessary to investigate the temperature dependence of the NMR spectrum of the 5-methylene group of 5H-dibenzo $[a,d]$ cyclohep-tene (XXXIII), which was required as a reference compound (see below). During the analysis of our results, in terms of steric and electronic effects, it became evident that other compounds were required to complete our study. Accordingly, the 9H-tribenzo $[a,c,e]$ cycloheptene derivatives XLV–XLVII were synthesized, and their NMR spectra were investigated.

THE NMR SPECTRA OF COMPOUNDS XXXIII, XXXV, XXXVII AND XXXIX–XLVII

The relevant details of the temperature dependence of the NMR spectra of all of these compounds are summarized in Table 2. The behaviour of the ring methylene group signals and the CMe_2OR signals has been discussed. The ethyl groups of the tribenzocyclohepta-

305

A.P. Downing, W.D. Ollis, M. Nógrádi and I.O. Sutherland

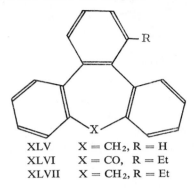

XLV	X = CH₂, R = H
XLVI	X = CO, R = Et
XLVII	X = CH₂, R = Et

XLV   X = CH_2, R = H
XLVI   X = CO, R = Et
XLVII   X = CH_2, R = Et

triene derivatives XLVI and XLVII gave ABX_3 systems in their NMR spectra at 35° C, corresponding to slow inversion of the non-planar conformation of the seven-membered ring. The spectrum of XLVII did not change up to 200° C, indicating that ring inversion was still slow at the highest accessible operating temperature. The spectrum of the tropone XLVI did show temperature dependence, and the AB portion of the ABX_3 spectrum coalesced to give an A_2X_3 spectrum above 125° C.

Table 2

NMR Spectral Parameters (60 and 100 MHz) and Free Energies of Activation for Conformational Inversion for Seven-Membered Ring Systems

Compound	Solvent*	Prochiral Group	$\nu_A - \nu_B$ (Hz)	J_{AB} (Hz)	T_C (°C)	$\Delta G^{\neq}$ at T_C** (Kcal/mole)
XXXIII	CS_2	CH_2	c.11	12.5	−90	9.0
XXXIII***	CS_2	CH_2	15	12.5	−85	9.2
XXXV	$CDCl_3$: CS_2 (1:3)	CH_2	14.3	12.5	−51	11.0
XXXVII	$CDCl_3$	CMe_2	2.6		44	17.5
XXXVII	$CDCl_3$	CH_2	11.1	12.2	67	17.1
XXXIX	CS_2	CMe_2			< −90	< 9
XL	CS_2	CMe_2			< −110	< 9
XLI	$PhNO_2$	CMe_2	2.7		109	21.1
XLII	C_6F_6	CMe_2	3.5		51	17.7
XLIII	C_2HCl_5	CMe_2	2.4		116	21.7
XLIV	$CDCl_3$: CS_2 (1:2)	CMe_2	16		−69	10.3
XLV	Ph.Ph	CH_2	13.1	12.5	196	23.8
XLV***	Ph.Ph	CH_2	21.8	12.5	202	24.0
XLVI***	Ph.Ph	CH_2 of Et	13.3	14.0	125	20.0
XLVII***	Ph.Ph	CH_2	25.1	12.5	>200	>27.7
XLVII***	Ph.Ph	CH_2 of Et	21.6	14.5	>200	>27.7

   * The range of solvents used was necessary to give mobile solutions of sufficient concentrations at the appropriate temperatures. The selection of solvents was also determined by the need to produce observable chemical shift non-equivalence between the diastereotopic protons and methyl groups
  ** Errors in $\Delta G^{\neq}$ are difficult to estimate, but should not exceed ±0.5 Kcal/mole
*** Data from 100 MHz spectra; other spectra determined at 60 MHz

STERIC AND ELECTRONIC FACTORS
AFFECTING THE INVERSION BARRIERS

The inversion of the non-planar seven-membered rings (see XXXVI and XXXVIII) is presumed to involve a planar transition state with respect to the geometry of the seven-membered ring in which three sources of π-electron delocalization can operate more efficiently than in the non-planar ground state. These are: (i) stilbenoid conjugation; (ii) π–p or π–π aryl-X-aryl conjugation; and (iii) conjugation associated with aromatic or anti-aromatic character in the seven-membered ring. The object of comparing the energies of activation listed (Table 2) is to assess the possible importance of conjugation of type (iii). To do this, it is first necessary to consider the selection of a reference compound and then, for all the compounds studied, to assess the non-bonded interactions in the transition state, relative to those of the transition state for inversion of the reference compound. It is also necessary to allow for the relative angle strain energies E_θ(rel.) resulting from the development of the planar transition state for each compound, relative to the angle strain energy of the planar transition state of the reference compound.

5H-Dibenzo [a,d] cycloheptene (XXXIII) was selected as the reference compound ($\Delta G^{\neq} = 9.1$ Kcal/mole). The types of non-bonded interactions encountered in the transition states for inversion of the other compounds studied are indicated in Fig. 2.

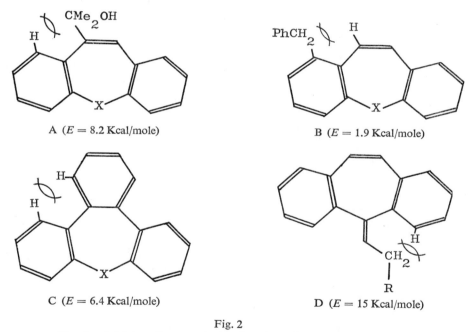

A ($E = 8.2$ Kcal/mole)

B ($E = 1.9$ Kcal/mole)

C ($E = 6.4$ Kcal/mole)

D ($E = 15$ Kcal/mole)

Fig. 2
Non-bonded interaction energies for situations of types A, B, C and D

The magnitudes of the non-bonded interactions of types A–D, indicated in Fig. 2, were assessed in the following way. The inversion barriers for XXXVII (17.3 Kcal/mole, average value) and XXXIII differ by a quantity (8.2 Kcal/mole), which must represent the effect of interaction A upon the transition state for inversion of XXXVII. We assume that interaction A has a constant value, regardless of the nature of the group X. This is obviously an approximation, but we believe that our conclusions, based upon this assumption, give a reasonably precise view of the electronic interactions in the transition states. Interaction B is similarly assessed as 1.9 Kcal/mole by comparing the inversion barriers for **XXXIII** (9.1 Kcal/mole) and **XXXV** (11.0 Kcal/mole). Interaction C is derived from the difference between the inversion barriers for **XXXIII** (9.1 Kcal/mole) and **XLV** (23.9 Kcal/mole, average value); this difference is clearly approximately 2 C. Before C can be assessed, allowance must be made for the effect of the third benzene ring upon the angle strain in the planar seven-membered ring of **XLV**, as compared with **XXXIII**. The effect is estimated as 2 Kcal/mole, giving a value for C of 6.4 Kcal/mole. The magnitude of interaction D cannot be estimated from the results in Table 2, but a comparison of the inversion barrier for **XLI**, obtained in the present study, with those for **XLVIII** and **XLIX**, obtained in a study of their racemization rates by Ebnöther, Jucker and Stoll (see Fig. 3) [93], provides the necessary information. The difference between the activation energies for racemization of the heptafulvenes **XLVIII** and **XLIX** (7.5 Kcal/mole) is approximately equal to the quantity D–C. Interaction C, in this case, may be approximated by the geometrically similar non-bonded interaction C, referred to in Fig. 2 (6.4 Kcal/mole), giving a value for D of 13.9 Kcal/mole).

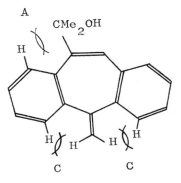

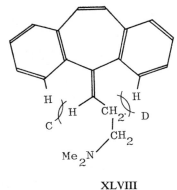

XLVIII

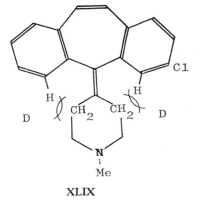

XLIX

Fig. 3
Non-bonded interactions in the transition states for ring inversion
of dibenzoheptafulvene derivatives

An alternative approach to this problem is a comparison of the inversion barriers for XLI and XLVIII. The former involves the non-bonded interactions A+2C in the transition state (21 Kcal/mole), and the latter the interactions C+D. Assuming that in both cases the inversion barriers for the unsubstituted fulvene systems would be identical, the value obtained for D is 16.0 Kcal/mole. The agreement between the values of D obtained by these two methods is encouraging in view of the very large non-bonded interactions that are involved, and we take as the average value for D, 15 Kcal/mole.

Having analysed the non-bonded interactions in the transition states in terms of these basic non-bonded interaction types, it is also necessary to consider the relative angle strain in the transition state. This may be done by using the general approach that has been discussed previously [94–95]. However, in view of the crudeness of this approach and the general lack of reliable information concerning force constants, appropriate functions to describe the variation of angle strain with changes in angles between bonds, and even the strain-free values for these angles, we felt that a rather simpler approach would be justified in this work. We have therefore assumed, in our assessment of transition-state angle strain E_θ, that the benzene rings remain undistorted and that there are no changes in bond lengths. Using literature values [94–95] for force constants, the geometries of the planar seven-membered rings of the transition states were adjusted so that the angle strain E_θ, summed over all the variable angles, was minimized. Absolute values of E_θ in the transition state are, of course, rather dependent upon the potential functions used. However, relative values are fairly insensitive to the choice of parameter values, and therefore only relative values of E_θ are used in the discussion that follows.

These calculations also indicated that the carbon framework of the seven-membered ring adopts a rather similar geometry for E_θ minimization in the planar system, regardless of the nature of X. This suggests that the use of standard values for the interactions A, B, C and D is a reasonable procedure. The various values obtained for E_θ (rel.), the transition state angle strain relative to that of the reference compound XXXIII, are given in Table 3.

Table 3

Comparison of the Calculated Inversion Barrier $E_{inv.}$(calc.) and the Observed Inversion Barrier ($\Delta G^{\neq}$) for Seven-Membered Ring Systems

Compound	E_θ (rel.) (Kcal/mole)	Additional Non-Bonded Interaction in Planar Transition States (Kcal/mole)	$E_{inv.}$(calc.) (Kcal/mole)	$\Delta G^{\neq}$ (Kcal/mole)	$E_{inv.}$(calc.) $- \Delta G^{\neq}$ (Kcal/mole)	Estimated DE for Ar-X-Ar (Kcal/mole) [96–98]
XXXIII	0	—	—	9.1	—	—
XXXV	0	B (1.9)	—	11.0	—	—
XXXVII	0	A (8.2)	—	17.3	—	—
XXXIX	−2	A (8.2)	15.3	< 9	> 6	11.2
XLI	−2	A (8.2) + 2C (12.8)	28.1	21.2	7	11.6
XLII	+4	A (8.2)	21.3	17.7	4	7.3
XLIII	+1	A (8.2) + 2B (3.8)	22.1	21.7	0	14.9
XLIV	+1	A (8.2)	18.3	10.3	8	11.3
XLV	+2	2C (12.8)	—	23.9	—	—
XLVI	0	C (6.4) + D (15)	30.5	20.0	11	11.2
XLVII	+2	C (6.4) + D (15)	32.5	>27.7	< 5	—
XLVIII	−2	C (6.4) + D (15)	28.4	22.6	6	11.6
XLIX	−2	2D (30)	37.1	30.1	7	11.6

In general, the value of E_θ (rel.) is quite small, and our general conclusions would not be changed if this term were neglected.

Given values for the non-bonded interactions A, B, C, D and E_θ(rel.), relevant to the transition state for inversion of each of compounds XXXIX and XLI–XLIX, it was now possible to calculate the activation energy for the inversion process $E_{inv.}$ (calc.) for each of these compounds (Table 3). This was done by summation of the activation energy for the reference compound XXXIII ($\Delta G^{\neq} = 9.1$ Kcal/mole), the estimated change in transition state angle strain E_θ (rel.) relative to that for the reference compound, and the relevant non-bonded interactions A, B, C and D, or their equivalent. This procedure allows for the change in stilbenoid conjugation energy (*a*), which is approximately identical for all of the compounds studied. The difference between $E_{inv.}$ (calc.) and $\Delta G^{\neq}$ (see last but one column of Table 3) provides a measure of the stabilization of the transition state by increased π–p or π–π aryl-X-aryl conjugation (*b*) and aromatic or anti-aromatic type conjugation (*c*), associated with the planar seven-membered ring. From Table 3, this difference usually falls within the range 4–8 Kcal/mole, which is compatible with the increase in conjugation energy to be expected for π–p and π–π aryl-X-aryl conjugation.

It is interesting to compare the various values of $[E_{inv.}$ (calc.) $- \Delta G^{\neq}]$ with the resonance energies that have been obtained experimentally [96–97] for the Ar-X-Ar system (Table 3, last column). In general, the relationship between the two values is consistent with some ground-state conjugation of type *b*, considerably enhanced in the transition state for ring inversion. The compounds examined, excluding XLIII, do not show a significant increase, or decrease, in delocalization energy, which could be a consequence of cyclic conjugation in the planar seven-membered ring.

The result $[E_{inv.}$(calc.) $- \Delta G^{\neq}] - c.$ 0 Kcal/mole for the azepin derivative XLIII — is surprising in view of the large p–π-conjugation energy estimated for N-methyl-diphenyl-amine [96]. This difference could possibly be associated with some anti-aromatic character in the planar azepin ring, or it may be a consequence of an unexpectedly large steric interaction, considerably greater than the estimated value of 3.8 Kcal/mole (2B) between the alkyl substituent on the nitrogen atom and the adjacent hydrogen atoms on C_4 and C_5. The extensive studies of Tochtermann's group [99–100], which are complementary to those described in this paper, do suggest that interactions of this type may be important.

The results and conclusions discussed in this paper indicate that, with the possible exception of XLIII, the dibenzo derivatives of the various cyclic conjugated seven-membered ring systems II–IV are best regarded in terms of their classical structures, and that, in general, the geometry they adopt is a consequence of steric effects and electronic effects associated with stilbenoid and Ar-X-Ar conjugation. Similar conclusions have been drawn for the non-annelated ring systems on the basis of NMR coupling constants [36–37]. The latter method, although providing a sensitive probe for ground-state structural characteristics, does not enable a decision to be made regarding the magnitude of conjugation effects in the planar ring systems.

REFERENCES

1 J. H. Day (1953) *Chem. Rev.*, 50 : 167.

2 E. D. Bergmann (1955) *Prog. Org. Chem.*, 3 : 81.

3 P. L. Pauson (1959) *Non-Benzenoid Aromatic Compounds* (ed. D. Ginsburg), Interscience, New York, p. 22.

4 D. Lloyd (1966) *Carbocyclic Non-Benzenoid Aromatic Compounds*, Elsevier, London.

5 J. A. Moore & E. Mitchell (1967) *Heterocyclic Compounds*, IX (ed. R. C. Elderfield), Wiley, New York, p. 224.

6 E. D. Bergmann (1968) *Chem. Rev.*, 68 : 41.

7 P. Yates (1968) *Adv. Alicyclic Chem.*, 2 : 60.

8 K. Hafner, K. H. Hafner, C. Konig, M. Kreuder, G. Ploss, G. Schulz, E. Stürm & K. H. Vöpel (1963) *Angew. Chem. (Int. Edn)*, 2 : 123.

9 A. J. Jones (1968) *Rev. Pure Appl. Chem.*, 18 : 253.

10 R. Breslow, J. Brown & J. J. Gajewski (1967) *J. Am. Chem. Soc.*, 89 : 4383.

11 R. Breslow & W. Washburn (1970) *ibid.*, 92 : 428.

12 A. Streitwieser Jr (1961) *Molecular Orbital Theory for Organic Chemists*, Wiley, New York.

13 M. J. S. Dewar (1969) *The Molecular Orbital Theory of Organic Chemistry*, McGraw-Hill, New York.

14 J. A. Elvidge & L. M. Jackman (1961) *J. Chem. Soc.*, p. 859.

15 J. A. Pople & K. G. Untch (1966) *J. Am. Chem. Soc.*, 88 : 4811.

16 H. J. Dauben, J. D. Wilson & J. Laity (1968) *ibid.*, 90 : 811.

17 G. W. Wheland (1934) *J. Chem. Phys.*, 2 : 474.

18 Idem (1941) *J. Am. Chem. Soc.*, 63 : 2025.

19 G. Berthier & B. Pullman (1949) *Trans. Faraday Soc.*, 45 : 484.

20 A. Pullman, G. Berthier & B. Pullman (1950) *Bull. Soc. Chim. France*, 17 : 1097.

21 T. Nakajima & S. Katagiri (1963) *Molec. Phys.*, 6 : 149.

22 M. J. S. Dewar & G. J. Gleicher (1965) *J. Am. Chem. Soc.*, 87 : 685, 692.

23 P. B. Empedocles & J. W. Linnett (1966) *Theor. Chim. Acta*, 4 : 377, 390.

24 G. Doggett (1966) *Molec. Phys.*, 10 : 225.

25 M. D. Newton, F. P. Boer & W. N. Lipscomb (1966) *J. Am. Chem. Soc.*, 88 : 2367.

26 H. Kuroda & T. Kunii (1967) *Theor. Chim. Acta*, 7 : 220.

27 M. J. S. Dewar (1967) *Aromaticity (Special Publication, No. 21)*, The Chemical Society, London, p. 177.

28 A. P. Downing, W. D. Ollis & I. O. Sutherland (1967) *Chem. Comm.*, p. 143.

29 *Ibid.* (1968) p. 329.

30 Idem (1969) *J. Chem. Soc.* (B), p. 111.

31 E. J. Forbes, M. J. Gregor, T. A. Hamor & D. J. Watkin (1966) *Chem. Comm.*, p. 114.

32 H. Shimanouchi, T. Ashida, Y. Sasada, M. Kakudo, I. Murata & Y. Kitahara (1966) *Bull. Chem. Soc. Japan*, 39 : 2322.

33 H. Shimanouchi, Y. Sasada, C. Kabuto & Y. Kitahara (1968) *Tetrahedron Lett.*, p. 5053.

34 I. C. Paul, S. M. Johnson, L. A. Paquette, J. H. Barrett & R. J. Haluska (1968) *J. Am. Chem. Soc.*, 90 : 5023.

35 F. A. L. Anet, C. H. Bradley, M. M. Brown & W. L. Mock (1969) *ibid.*, 91 : 7782.

36 D. J. Bertelli & T. G. Andrews Jr (1968) *ibid.*, 91 : 5280.

37 D. J. Bertelli, T. G. Andrews Jr & P. O. Crews, *ibid.*, 91 : 5286.

38 M. Nógrádi, W. D. Ollis & I. O. Sutherland (1970) *Chem. Comm.*, p. 158.

39 J. A. Pople, W. G. Schneider & H. J. Bernstein (1959) *High-Resolution Nuclear Magnetic Resonance*, McGraw-Hill, New York, Chap. 10.

40 L. W. Reeves (1965) *Adv. Phys. Org. Chem.*, 3 : 187.

41 J. E. Anderson (1965) *Q. Rev.*, 19 : 426.

42 A. Allerhand, H. S. Gutowsky, J. Jonas & R. A. Meinzer (1966) *J. Am. Chem. Soc.*, 88 : 3185.

43 J. D. Roberts (1966) *Chem. in Britain*, 2 : 529.

44 G. Binsch (1968) *Topics Stereochem.*, 3 : 97.

45 H. S. Gutowsky & C. H. Holm (1956) *J. Chem. Phys.*, 25 : 1228.

46 C. W. Fryer, F. Conti & C. Franconi (1965) *Ricerca Scient.*, 8 : 788.

47 M. T. Rogers & J. C. Woodbrey (1962) *J. Phys. Chem.*, 66 : 540.

48 R. C. Neuman Jr & V. Jonas (1968) *J. Am. Chem. Soc.*, 90 : 1970.

49 M. Rabinovitz & A. Pines (1969) *ibid.*, 91 : 1585.

50 Y. Shvo & H. Shanon-Atidi, *ibid.*, pp. 6683, 6689.

51 A. Streitwieser Jr (1962) *Solvolytic Displacement Reactions*, McGraw-Hill, New York, p. 72.

52 H. Meerwein, W. Florian, N. Schön & G. Stopp (1961) *Annalen*, 641 : 1.

53 K. Hafner, K. H. Vöpel, G. Ploss & C. Konig (1963) *ibid.*, 661 : 52.

54 A. Mannschreck & U. Koelle (1967) *Tetrahedron Lett.*, p. 863.

55 J. H. Crabtree & D. J. Bertelli (1967) *J. Am. Chem. Soc.*, 89 : 5384.

56 H. Kessler & A. Rieker (1966) *Tetrahedron Lett.*, p. 5257.

57 Y. Shvo, E. C. Taylor & J. Bartulin (1967) *ibid.*, p. 3259.

58 Y. Shvo (1968) *Tetrahedron Lett.*, p. 5923.

59 G. Isaksson, J. Sandström & I. Wennerbeck (1967) *ibid.*, p. 2233.

60 R. Kuhn, B. Schulz & J. C. Jochims (1966) *Angew. Chem. (Int. Edn)*, 5 : 420.

61 A. S. Kende, P. T. Izzo & W. Fulmor (1966) *Tetrahedron Lett.*, p. 3697.

62 H. Shimanouchi, T. Ashida, Y. Sasada, M. Kakudo, I. Murata & Y. Kitahara (1967) *Tetrahedron Lett.*, p. 61.

63 O. Kennard, D. G. Watson, J. K. Fawcett, K. A. Kerr & C. Romers (1967) *Tetrahedron Lett.*, p. 3885.

64 W. B. Smith, W. H. Watson & S. Chiranjeevi (1967) *J. Am. Chem. Soc.*, 90 : 811.

65 M. Neuenschwander, D. Meuche & H. Schaltegger (1964) *Helv. Chim. Acta*, 47 : 1023.

66 D. Meuche, M. Neuenschwander, H. Schaltegger & H. U. Schlunegger, *ibid.*, p. 1211.

67 W. B. Smith & B. A. Shoulders (1964) *J. Am. Chem. Soc.*, 86 : 3118.

68 M. L. Heffernan & A. J. Jones (1966) *Aust. J. Chem.*, 19 : 1813.

69 N. Jonathan, S. Gordon & B. P. Dailey (1962) *J. Chem. Phys.*, 36 : 2443.

70 C. Ganter & J. D. Roberts (1966) *J. Am. Chem. Soc.*, 88 : 741.

71 A. S. Kende, P. T. Izzo & P. T. MacGregor, *ibid.*, p. 3359.

72 J. F. Chiang & S. H. Bauer (1970) *ibid.*, 92 : 261.

73 J. H. Day & C. Oestreich (1957) *J. Org. Chem.*, 22 : 214.

74 W. von E. Doering & D. W. Wiley (1960) *Tetrahedron*, 11 : 183.

75 E. Vogel & H. Gunther (1967) *Angew. Chem. (Int. Edn)*, 6 : 385.

76 G. Maier (1967) *Angew. Chem. (Int. Edn)*, 6 : 402.

77 W. L. Mock (1967) *J. Am. Chem. Soc.*, 89 : 1281.

78 R. Huisgen, D. Vossius & M. Appl (1958) *Chem. Ber.*, 91 : 1.

79 R. Huisgen & M. Appl, *ibid.*, p. 12.

80 W. von E. Doering & F. L. Detert (1951) *J. Am. Chem. Soc.*, 73 : 876.

81 H. J. Dauben Jr & H. J. Ringold *ibid.*, p. 876.

82 C. Jutz (1964) *Chem. Ber.*, 97 : 2050.

83 M. Rabinovitz, I. Agranat & E. D. Bergmann (1966) *Tetrahedron*, 22 : 225.

84 H. Shimanouchi, T. Hata & Y. Sasada (1968) *Tetrahedron Lett.*, p. 3573.

85 E. D. Bergmann, M. Rabinovitz & I. Agranat (1968) *Chem. Comm.*, p. 334.

86 E. D. Bergmann & M. Rabinovitz (1960) *J. Org. Chem.*, 25 : 827.

87 R. Huisgen, E. Laschtuvka & F. Bayerlein (1960) *Chem. Ber.*, 93 : 392.

88 F. A. L. Anet & P. M. G. Bavin (1956) *Can. J. Chem.*, 34 : 991.

89 *Ibid.* (1957) 35 : 1084.

90 R. F. H. Manske & A. E. Ledingham (1950) *J. Am. Chem. Soc.*, 72 : 4797.

91 E. D. Bergmann & M. Rabinovitz (1960) *J. Org. Chem.*, 25 : 828.

92 K. R. Hanson (1966) *J. Am. Chem. Soc.*, 88 : 2731.

93 A. Ebnöther, E. Jucker & A. Stoll (1965) *Helv. Chim. Acta*, 48 : 1237.

94 F. H. Westheimer (1956) *Steric Effects in Organic Chemistry* (ed. M. S. Newman), Wiley, New York, p. 523.

95 E. L. Eliel, N. L. Allinger, S. J. Angyal & G. A. Morrison (1965) *Conformational Analysis*, Interscience, New York, p. 433.

96 I. P. Romm, E. N. Guryanova & K. A. Kocheshkov (1969) *Tetrahedron*, 25 : 2455.

97 R. W. Taft & M. M. Kreevoy (1957) *J. Am. Chem. Soc.*, 79 : 4016.

98 M. M. Kreevoy (1959) *Tetrahedron*, 5 : 233.

99 W. Tochtermann, H. Küppers & C. Franke (1968) *Chem. Ber.*, 101 : 3808.

100 W. Tochtermann & C. Franke (1969) *Angew. Chem. (Int. Edn)*, 8 : 68.

Discussion

E. D. Bergmann:

We have investigated the dibenzazepin molecule and we found that it is not planar. One of the difficulties with nitrogen-containing systems of this kind is that you have two possibilities: it can either be inside the cavity of the molecule, or outside it. In the compound we have tested the nitrogen seems to be inside, and there seems to be some interaction between the carbon–carbon bridge and the nitrogen. One can, by changing the size of the substituent group, e.g. by introducing the sulphonamide group, change the barrier for the inversion of the molecule. This is only a remark which might complement a little bit what you have said in your very beautiful exposition.

I. O. Sutherland:

We do not know whether the substituent on nitrogen is in the *exo* or *endo* positions in our azepin derivative, but this would not have much effect on the inversion barrier. The change in conjugation energy during the inversion process does show a discrepancy as compared with the N–Ar conjugation energy of diphenylamine which we are unable to explain, unless we say that this is a special property of the cyclic conjugated system.

Structure and Chemistry of Heterofulvalenium Ions

by M. A. BATTISTE *and* J. H. M. HILL*

Department of Chemistry, University of Florida, Gainesville

I. INTRODUCTION

THE FULVALENES, as cross-conjugated examples of C_nH_{n-2} cyclic polyenes, have attracted the attention of organic chemists for over a half century [1]. Whether the attraction is totally aesthetic or derives in large part from the promise of unusual chemistry for such polyenes is not clear. The recent [2] and lamented [3] charge that this attention is undeserved on theoretical grounds and the plea [2] for chemists to resist the siren call of these pseudo-aromatic systems has, not unexpectedly, had a reverse effect on the research activity in these areas.

Much of the recent activity in the fulvalene series has been focused on the mixed ring systems pentatriafulvalene (I) and heptapentafulvalene (II) [1]. The hydrocarbon II is known to be extremely unstable [4], while similar instability for I can be inferred from the behaviour of its lightly substituted derivatives [5–6]. The feasibility of isoperimetric (as opposed to isosteric) hetero-atom analogues of I and II, having increased stability over the carbocyclic systems, takes precedence from the many examples [7–8] of stable heterocyclic derivatives of the unstable hydrocarbon pentafulvalene (III). In view of this, and in view of theoretical interest in the effect of hetero-atoms on charge distribution, the synthesis of mixed heterofulvalenes was deemed a worthy objective.

In the course of our synthetic efforts toward the above goal, it became increasingly apparent that the intermediate heterofulvalenium ions IV (X = NH) are of significantly greater interest than as simply precursors to the corresponding fulvalenes. As charged analogues of neutral fulvalene systems, the heterofulvalenium ions may be considered as iso-electronic models for the less stable carbocyclic systems. The extent to which such a comparison is valid depends on the degree of charge delocalization and the relative magnitude of interannular π bonding in the heterofulvalenium ion structure.

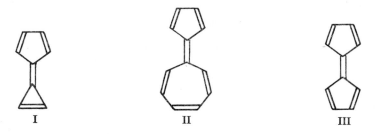

I   II   III

* On leave of absence from Hobart and William Smith Colleges, Geneva, N.Y., 1967–1968.

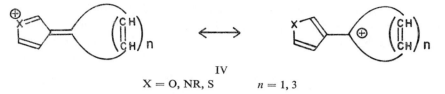

IV

$$X = O, NR, S \qquad n = 1, 3$$

We have now prepared a series of azapentatriafulvalenium ions, IV (X = NR; $n = 1$), and obtained the necessary spectral and thermodynamic data to allow an assessment of the fulvalenium character of these salts, as discussed below.

II. The Azapentatriafulvalenium Ion

Simple molecular orbital calculations by the crude, but nevertheless instructive, Hückel method suggest a clear π-electronic relationship between the neutral hydrocarbon, pentatriafulvalene (I), and the 2-azapentatriafulvalenium cation (V). Both systems show a low intercyclic π-bond order (P_{78}), with cation V having only slightly less double-bond character. The calculated charge distribution patterns within the three-membered rings are very similar for the two systems.

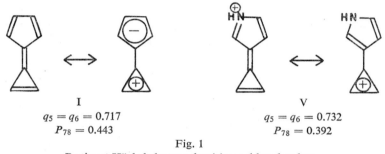

$$\text{I}$$
$$q_5 = q_6 = 0.717$$
$$P_{78} = 0.443$$

$$\text{V}$$
$$q_5 = q_6 = 0.732$$
$$P_{78} = 0.392$$

Fig. 1
Pertinent Hückel electron densities and bond orders
for pentatriafulvalene (I) and azapentatriafulvalenium cation (V)

1. Synthesis

The first synthetic entry into the 2-azapentatriafulvalene system was achieved by Kende, Izzo and MacGregor [9] by reaction of indolylmagnesium iodide with the ethoxycyclopropenium salt VI. The resulting product, obtained as the iodide salt in 21% yield, was formulated as the 3-indolylcyclopropenium salt VII.

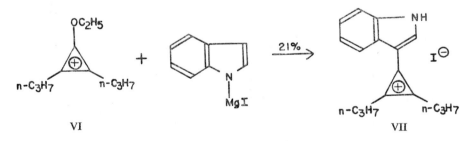

VI

VII

In view of the low yields obtained by Kende's route, we sought a more general procedure, which would involve generation of the azapentatriafulvalenium cation under non-basic conditions. Now, it was known from the work of Föhlisch and Bürgle [10] and West and

315

co-workers [11] that activated aromatic compounds, such as phenols and phenolic ethers, are alkylated by 1-chlorocyclopropenium cations (VIII; Z = Cl) under relatively mild conditions to give the corresponding arylcyclopropenium cations IX. On the other hand, the ethoxycyclopropenium cation VIII (R = Ph; Z = OC_2H_5) appeared to be sufficiently electrophilic to alkylate only the most activated aromatic nucleus, such as N,N-dimethyl-aniline [10]. This is not a serious problem to our needs, however, since pyrrole and its derivatives are easily as reactive as the anilines toward electrophilic substitution. The direct alkylation of substituted indoles and pyrroles by cyclopropenium cations VIII therefore appeared to be an attractive route to the desired azapentatriafulvalenes.

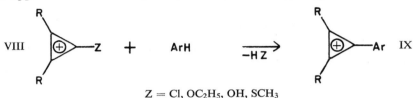

$$Z = Cl, OC_2H_5, OH, SCH_3$$

Before attempting the alkylation reactions with the chloro- and alkoxylcyclopropenium salts, which must be generated in a separate step, we were encouraged to examine the reaction of cyclopropenones with aromatic heterocycles under acidic conditions. The reasoning for this was quite simple: (i) cyclopropenones are readily protonated under acidic conditions [12]; (ii) hydroxycyclopropenium salts should be at least as reactive in the alkylation of aromatic substrates as the corresponding alkoxy salts; (iii) indoles and pyrroles are weakly basic and, in protic solvents, should be largely unprotonated even at fairly high acid concentrations. In fact, as recently reported [13], we found that in the presence of acid diphenylcyclopropenone (X; R^3 = Ph) undergoes ready condensation with indole (XI; $R^1 = R^2$ = H) and various methyl- and phenyl-substituted indoles (XI) to yield the corresponding indolium salts, XII, in high isolated yields (see Fig. 2). The

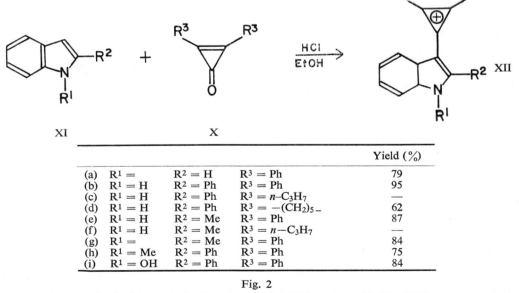

				Yield (%)
(a)	R^1 =	R^2 = H	R^3 = Ph	79
(b)	R^1 = H	R^2 = Ph	R^3 = Ph	95
(c)	R^1 = H	R^2 = Ph	R^3 = n-C_3H_7	—
(d)	R^1 = H	R^2 = Ph	R^3 = $-(CH_2)_5-$	62
(e)	R^1 = H	R^2 = Me	R^3 = Ph	87
(f)	R^1 = H	R^2 = Me	R^3 = n-C_3H_7	—
(g)	R^1 =	R^2 = Me	R^3 = Ph	84
(h)	R^1 = Me	R^2 = Ph	R^3 = Ph	75
(i)	R^1 = OH	R^2 = Ph	R^3 = Ph	84

Fig. 2

Synthetic route to 3,4-benz-2-azapentatriafulvalenium chlorides (XII) dialkylcyclopropenones $X(R^3 = n\text{-}C_3H_7)$ and $X(R^3 = -(CH_2)_5-)$ can be similarly condensed with substituted indoles, although the yields are somewhat lower than in the diphenylcyclopropenone cases. Typically, a dry ethanolic solution of both reactants is cooled to 0°, saturated with dry HCl, and then allowed to stand at room temperature for several hours. The crystalline yellow salts, isolated as the chlorides, are obtained on dilution of the reaction mixture with ether.

The general scope of this alkylation method with reactive aromatic heterocycles may be illustrated by using representatives of two quite different ring systems: 2,4,5-triphenyl-pyrrole (XIII) and 2-phenyl-7-methylindolizine (XIV). Reaction of diphenylcyclopropenone with XIII and XIV under the above conditions led to the novel azapentatriafulvalenium salts XV and XVI in 88% and 91% yield, respectively. In a similar fashion, 3-methylindole surprisingly underwent electrophilic substitution at the 2-position to yield XVII (64%). It should be noted that XVI and XVII are examples of the previously unknown 1-aza-pentatriafulvalenium system.

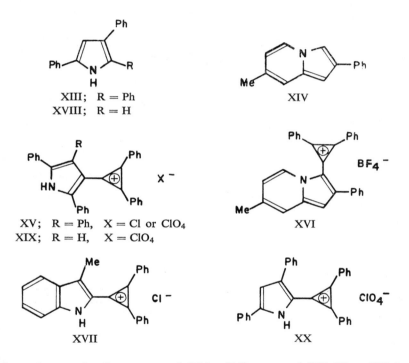

XIII; R = Ph
XVIII; R = H

XIV

XV; R = Ph, X = Cl or ClO₄
XIX; R = H, X = ClO₄

XVI

XVII

XX

In an independent study Gompper and Weiss [14] prepared XV (X = ClO₄) as well as XIX and XX by reaction of XIII and 2,4-diphenylpyrrole (XVIII) with methylthiodiphenyl-cyclopropenium perchlorate (VIII; R = Ph; Z = SMe) in acetic acid, or with 3,3-dichloro-1,2-diphenylcyclopropene in acetonitrile. These authors formulated their ionic compounds in the maximally double-bonded form (pentatriafulvalenoid), but did not comment on the structures.

2. *Structure*

As stated earlier, in order to correctly assess the π-electronic structure of the azapenta-triafulvalenium ions, we will require information relating to the degree of charge delocalization between the two rings and the relative amount of double-bond character of the interannular bond. Evidence bearing on the latter question will be considered in the next section, while here we treat the question of charge distribution, as revealed by spectral (NMR) and thermodynamic (pK_{R+}) measurements.

On the basis of the rather large chemical shift difference (c. 1.3 ppm) between the α and β methylenes of the propyl chain, Kende had assigned full cyclopropenium character to his dipropylazapentatriafulvalene cation VII [9]. The chemical shift difference between the α and β methylenes is assumed to be 'a measure of the positive charge on the propyl-bearing cyclopropene carbon atoms' [9]. However, no allowance is made for possible incremental changes in ring-current deshielding effects within a series of cyclopropenium compounds. In addition, we are surely in the position of trying to decide between 50% v. 80% to 90% cyclopropenium cation character, and this spectral index may well be experimentally unreliable in this region. In any event, consideration of the spectral data in Table 1 suggests a different interpretation.

To probe the question of charge delocalization from the three- into the five-membered ring of an azapentatriafulvalenium cation, it would seem more instructive to examine the chemical shifts of substituents on the heterocyclic ring. In this regard, the series of indole derivatives listed in Table 1 are particularly enlightening [15].

Table 1

Chemical Shift and pK_{R+} Values
for some Azapentatriafulvalenium Ions and Model Systems

Compound		Chemical Shift (τ values)*		
		R¹	R²	pK_{R+}**
XII	a, R¹ = R² = H	−4.96	0.29	5.21
	e, R¹ = R² = Me	5.92***	6.91***	4.87
	g, R¹ = H; R² = Me	−4.94	6.91	4.91
	h, R¹ = Me; R² = Ph	5.98***	—	—
XI	R¹ = R² = H	2.40	3.31	—
	R¹ = H; R₂ = Me	2.50	7.81	—
	R¹ = R² = Me	6.70	7.80	—
XVII	R¹ = Me; R² = Ph	6.47	—	—
	R¹ = H; R² = Me	−1.35	6.94	2.24
XXI	a, R¹ = R² = Me	5.90***	7.70***	—
	b, R¹ = Me; R² = Ph	6.00***	—	—
Ph₃C⁺ 1.62				
An₃C⁺ 5.24				

* CDCl₃ as solvent, unless otherwise noted   ** 50% aqueous ethanol   *** CF₃COOH as solvent

Comparison of each of the four azapentatriafulvalenium salts XIIa, XIIe, XIIg and XIIh with the correspondingly substituted free base XI reveals a marked down-field chemical shift for all protonic groups attached to the heterocyclic ring. Thus, while the N–H protons of indole and 2-methylindole are found at 2.40–2.50 τ, those for salts XIIa and XIIg are shifted down-field by over 7 ppm. Similarly, the N–CH$_3$ signals of XIIe and XIIh are *c.* 1 ppm down-field from those of 2-methyl- and 1,2-dimethylindole. It is critical to our argument to note that the magnitude of this down-field shift of the N–CH$_3$ is identical to that observed for the 3-benzhydrylidene-3H-indolium salts (XXIa and XXIb), which must certainly have the positive charge localized largely on nitrogen. Equally interesting is the surprisingly low field position of the other heterocyclic ring proton (R^2 = H) of XIIa. This proton experiences a down-field shift of 3.0 ppm relative to the C$_2$ proton of indole, which seems too large an effect to explain by any single deshielding mechanism not invoking appreciable positive charge density in the heterocyclic ring. The protons of the C$_1$ methyl groups (R^2 = Me) of ions XIIe and XIIg are likewise shifted to lower fields by *c.* 0.9 ppm.

From the above NMR considerations it seems reasonable to suggest extensive charge delocalization into the heterocyclic ring of azapentatriafulvalenium ions. This contention is amply supported by the results of pK_{R+} measurements on ions XII, XV and XVII in 50% aqueous ethanol. Some of these values, obtained spectrophotometrically, are listed in Table 1, along with comparison values for the triphenyl- and tri-*p*-anisylcyclopropenium cations. It is seen that substitution of one of the phenyl groups in triphenylcyclopropenium cation by an indole (or pyrrole) ring results in an apparent increase in stability of from 3 to 4 *p*K units. This gain in stability is equivalent to the extra effect of three *p*-methoxy group substitutions on Ph$_3$C$_3^+$.

When the cyclopropenium ring is located in the 2-position of the indole nucleus, as in XVII, direct charge delocalization onto nitrogen should be markedly diminished, as this would involve contributions from the *ortho*-quinonoidal indolium ion XXII. Thus, as expected, a 2-indolyl group adds only slightly greater stability relative to phenyl. This result parallels the observation that the down-field shift for the N–H proton of XVII is only about half of that found for ions XIIa and XIIg. Ion XVII may be more properly considered as a 2-substituted styryldiphenylcyclopropenium cation.

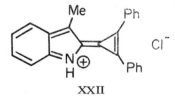

XXII

Evidence that the azapentatriafulvalenium ions considered above also have significant cyclopropenium ion structure is readily provided by inspection of the chemical shift pattern of the group, phenyl or propyl, bound to cyclopropenyl carbons. The diphenylcyclopropenium protons are, for example, observed at lower overall field than the corresponding protons of diphenylcyclopropenone, but not as low as those for the diphenylcyclopropenium cation [16]. An almost identical aromatic pattern, both in shape and position, is exhibited by hydroxydiphenylcyclopropenium cation in FSO$_3$H [17]. Similar considerations hold for the *n*-propyl-triafulvalenium ions presented in Table 2.

Table 2

Limiting Chemical Shift Data (τ) for RCH_2 $\triangle$ CH_2R / Ar_N
(25% TFA–SO$_2$)

R	Ar_N	T, °C*	CH$_2$**	CH$_2$***	CH$_3$**
C$_2$H$_5$	2-Ph Indolyl	−40°	6.51, 7.19	7.90, 8.45	8.80, 9.13
		+80°	6.91	8.22	8.97
C$_2$H$_5$	2-Me Indolyl	−40°	6.58, 6.65	7.86	8.81
		+40°	6.60	7.86	8.81
C$_2$H$_5$	2, 4, 5-Ph$_3$ Pyrrolyl	−40°	7.31, 7.72	8.62△△	9.16△△
		+40°	7.57	8.62	9.16
−(CH$_2$)$_3$⁻	2-Ph Indolyl	−30°	6.68, 7.10△	7.86△	
		+33°	6.89△	7.89△	

* Limiting spectra for the inner –CH$_2$– were obtained at these temperatures   ** Triplet   *** Sextet
△ Broad singlet   △△ Signal broadened, J_{ab} and $J_{a'b}$ are 7Hz

3. Interannular Rotational Studies

The above NMR spectral data and pK_{R+} measurements for the indolyl series of tria-fulvalenium ions indicate that both cyclopropenium and 3H-indolium ion structures are major π-electronic contributors. Consequently, the interannular (7–8) bonds of these systems should have some double-bond character and would be expected to demonstrate restricted rotation about the intercyclic bond axis. By means of variable-temperature NMR studies [M.A. Battiste & J.H.M. Hill, unpublished results], we have determined the free energy activation barriers ($\Delta G^{\neq}$) for four representative alkyl-substituted aza-pentatriafulvalenium ions, i.e., XIIc, XIId, XIIf and XXIII.

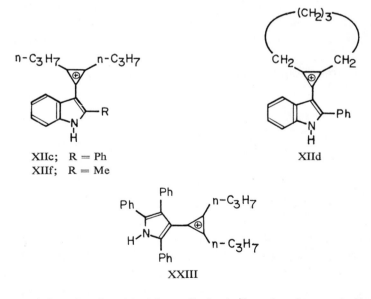

XIIc; R = Ph
XIIf; R = Me

XIId

XXIII

The NMR spectral data for the above four alkyl-substituted cations at the limiting temper-ature ranges for fast and slow exchange are summarized in Table 2. The fairly large magnetic non-equivalence of the two *n*-propyl groups in the low-temperature spectrum of XIIc stems primarily from diamagnetic deshielding of the propyl group on the same side and in the plane of the fused benzene ring of indole, while the other propyl group experiences a strong shielding effect from the twisted phenyl substituent at the 2-position of the indole ring. At higher temperatures, rotation about the interannular bond increases until, at some point above coalescence, a time-averaged signal is obtained for both propyl groups. These changes with temperatures are illustrated graphically in Fig. 3.

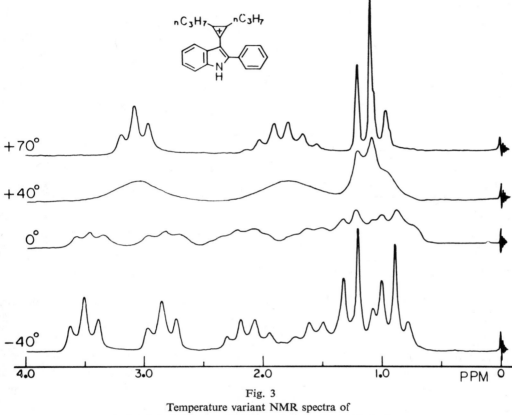

Fig. 3
Temperature variant NMR spectra of
1-phenyl-5, 6-di-*n*-propyl-3, 4-benz-2-azapentatriafulvalenium
chloride in liquid SO_2 containing *c*. 25% trifluoroacetic acid

Comparison of the low-temperature spectra of XIIc and XIIf shows that replacement of 2-phenyl by 2-methyl considerably reduces the magnetic non-equivalence of the propyl groups. The low-temperature configuration now shows non-equivalence for only the inner $-CH_2-$, due to deshielding by the benzo grouping. The outer $-CH_2-$ and $-CH_3$ groups are sufficiently distant from the benzo group to show their magnetic equivalence by temperature invariant signals.

In XIId the $-CH_2-$ groups attached to the cyclopropene ring are similarly shielded and deshielded as those of XIIc, but the remainder of the methylene bridge is sufficiently remote from the magnetic environment of either the benzo group or 2-phenyl ring to show only temperature-independent signals.

At low temperatures both propyl groups of XXIII are shielded, as evidenced by their up-field positions relative to those of XIIc, XIId and XIIf. Here non-equivalence of the α-methylene groups is a consequence of their being flanked by one and two phenyl groups, respectively. Models show that the phenyl rings in XXIII cannot be co-planar with the pyrrole ring, and that the *ortho* phenyls (C_3, C_4) are more severely twisted than the isolated phenyl at C_1. This would then suggest that the more shielded α-methylene is the one flanking the more heavily phenylated side of the pyrrole ring. Interestingly, the difference in shielding of the outer methylene and methyl is insufficient to show signal broadening at a temperature ($-40°$) that produces the maximum separation of the inner methylene signals.

From the Gutowsky-Holm peak separation equation [18], $k_t = \pi V_m^2 - V_t^2/2$, where V_m and V_t are the maximum and intermediate temperature peak maxima separations, respectively, the exchange rates, k_t, for the α-methylene protons were calculated over a temperature range of 15° to 20°, including coalescence. Table 3, however, contains only the values of k_t at coalescence ($V_t = 0$), $^\nabla$ which were then used to calculate the free energies of activation at coalescence, $\Delta G_c^{\neq}$, using the Eyring equation and a transmission coefficient of unity.

Table 3

Kinetic Parameters for Interannular Rotation *

Compound	$V_m - V_t$, Hz	T_c, °K**	K_t (sec^{-1})	ΔG_c(Kcal/mole)***
XIIc	41.0	297	91.1	14.8
XIIf	3.9	280	8.55	15.3
XIId	28.1	304	62.1	15.3
XXIII	29.6	266	65.7	13.2

* Solvent 25% TFA–SO$_2$ ** ± 1°   *** ± 0.6 Kcal/mole

With exception of the slightly lower value for the pyrrole derivative XXIII, the activation barrier for rotation about the interannular bond is *c.* 15 Kcal/mole. Barriers of comparable size have been observed in other compounds, which presumably have rotation constrained by partial double-bond character [20–21]. Of greater significance to the problem at hand, Kende has measured the activation barrier, $\Delta G_c^{\neq}$, for rotation about the interannular bond of pentatriafulvalene XXIV [22]. Although somewhat solvent dependent, a figure of 18–19 Kcal/mole was obtained, which is abnormally low for a fullfledged double bond [23–24]. Since the 2-azapentatriafulvalenium ions in the present study have rotational barriers only 3–4 Kcal/mole lower in energy than XXIV, one is drawn to the conclusion that the former ions have appreciable fulvalenium character and are more accurately represented by the composite structure XXV.

$^\nabla$ The Arrhenius activation energies for these interannular rotations have been determined from plots of ln k_t v. $1/T$. Although good linear relationships were obtained over the available temperature range, the approximations commonly assumed for evaluation of exchange rates from high resolution spectra by the peak separation method have been shown to be subject to systematic error [19]. Since the magnitude of errors in the determination of rates by NMR is most evident in the energies of activation and frequency factors, we choose not to report these data until a more complete analysis of these rate processes has been obtained by computer-aided complete line-shape fitting methods.

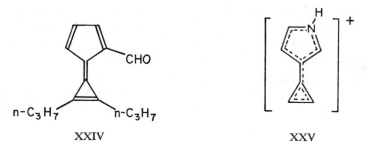

XXIV   XXV

4. Chemistry

The chemistry of the azapentatriafulvalenium salts prepared in this study has as yet been little explored. The primary efforts have revolved around attempts to convert these salts to neutral azapentatriafulvalene derivatives. In our hands treatment of salts XIIb and XIIe with a variety of bases (triethylamine, pyridine, lithium carbonate, potassium *t*-butoxide) in aprotic solvents gave only transitory colours, which might be ascribable to fulvalene XXVI. On work-up, no monomeric products could be detected. Rather, the only product isolated was a yellow polymeric solid, which still retained the disubstituted cyclopropene chromophore in the infra-red at 1,810 cm^{-1}. On the basis of the fact that this polymer was efficiently, though not quantitatively, converted back to the original monomeric salt XII by ethanolic HCl, the structure XXVII has been proposed for the polymer [13]. In strongly acidic solutions, protonation on a nitrogen terminus could initiate depolymerization, with reformation of XII.

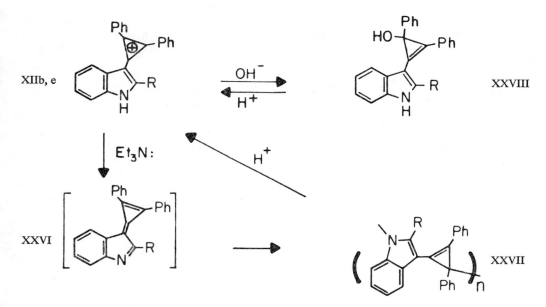

In contrast to our deprotonation results, Gompper and Weiss [14] report that treatment of the pyrrole systems XV, XIX and XX with Hunig's base or, in the case of XV, potassium *t*-butoxide, yields the corresponding azapentatriafulvalenes as unstable red oils or solutions. The results from both laboratories nevertheless confirm the highly reactive nature of the azapentatriafulvalene system.

323

Dissolution of ions XIIb and XIIe, as well as others in this series, in weakly basic or acidic aqueous buffer solutions results in the reversible formation of triarylcyclopropenols, as evidenced by well-defined isosbestic points for the electronic spectra of these solutions. Since the long-wavelength maxima for these solutions are bathochromically shifted from that of the 1,2-diphenylcyclopropene chromophore, the major product alcohols must have structure XXVIII. Other nucleophiles, such as cyanide and borohydride, react similarly with various azapentatriafulvalenium cations to give stable, colourless products, whose structures are at present under investigation.

ACKNOWLEDGEMENT

The authors are sincerely grateful to the donors of the Petroleum Research Fund, administered by the American Chemical Society, and the Alfred P. Sloan Foundation for financial support of this research.

REFERENCES

1 E. D. Bergmann (1968) *Chem. Rev.*, 68 : 41.

2 M. J. S. Dewar & C. J. Gleicher (1965) *Tetrahedron*, 21 : 3423.

3 A. S. Kende (1966) *Trans. N. Y. Acad. Sci.*, 28 (8) : 981.

4 H. Prinzbach & W. Rosswog (1961) *Angew. Chem.*, 73 : 543.

5 H. Prinzbach, D. Seip & V. Fischer (1965) *ibid.*, 77 : 258, 621.

6 W. M. Jones & R. S. Pyron (1965) *J. Am. Chem. Soc.*, 87 : 1608.

7 J. H. M. Hill (1963) *J. Org. Chem.*, 28 : 1931.

8 *Ibid.* (1967) 32 : 3214.

9 A. S. Kende, P. T. Izzo & P. T. MacGregor (1966) *J. Am. Chem. Soc.*, 88 : 3359.

10 B. Föhlisch & P. Bürgle (1967) *Ann. Chem.*, 701 : 58, 67.

11 R. West, D. C. Zecher & W. Goyert (1970) *J. Am. Chem. Soc.*, 92 : 149, 155.

12 R. Breslow et al. (1965) *ibid.*, 87 : 1326.

13 J. H. M. Hill & M. A. Battiste (1968) *Tetrahedron Lett.*, p. 5537.

14 R. Gompper & R. Weiss (1968) *Angew. Chem.*, 80 : 277.

15 M. A. Battiste & J. H. M. Hill (1968) *Tetrahedron Lett.*, p. 5541.

16 D. G. Farnum & C. F. Wilcox (1967) *J. Am. Chem. Soc.*, 89 : 5379.

17 P. O'Brien (1967) Ph. D. Dissertation, University of Florida.

18 H. S. Gutowsky & C. H. Holm (1956) *J. Chem. Phys.*, 25 : 1228.

19 A. Allerhand, H. S. Gutowsky, J. Jonas & R. A. Meinzer (1966) *J. Am. Chem. Soc.*, 88 : 3185.

20 H. S. Gutowsky, J. Jonas & T. H. Siddall III (1967) *ibid.*, 89 : 4300.

21 T. H. Siddall III & W. E. Stewart (1969) *J. Org. Chem.*, 34 : 2927.

22 A. S. Kende, P. T. Izzo & W. Fulmor (1966) *Tetrahedron Lett.*, p. 3697.

23 J. F. Douglas, B. S. Rabinowitch & S. F. Looney (1955) *J. Chem. Phys.*, 23 : 315.

24 G. B. Kistiakowsky & W. R. Smith (1934) *J. Am. Chem. Soc.*, 56 : 638.

Discussion

E. D. Bergmann:

I would like to make a remark which may have some bearing on what you said. Mr Agranat in our laboratory has studied the 4-azatriafulvene system — for instance, the DNP derivative of diphenyl-cyclopropenone — and we find that protonation does not attack the three-membered ring, but only the nitrogen atom. The triafulvene system does not change into a cyclopropenium cation.

B. Binsch:

Just a very small observation: I would like to point out that the Gutowsky-Holm equation does not apply at all to your spin system because of spin–spin coupling.

M. A. Battiste:

Not between the methylenes. We are aware of the problem. You can overcome this very simply by using the appropriate computer programme or by mapping the shapes exactly. In our case, we have three non-equivalent lines; we assume that these are three uncoupled spectra, which are simply superimposed, going to three single lines. We have three groups of two lines, uncoupled, each giving one single line under limiting conditions.

W. S. Tobey:

In compounds of Type I, in which you show that R′ (in particular) shows strong NMR deshielding due to charge delocalization *via* resonance form I′, it seems to me the methylene chemical shifts on the 'cyclopropenium' propyl groups should move upfield into the cyclopropenone region out of the cyclopropenium ion region. Did you observe this?

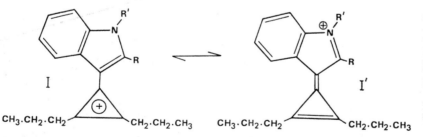

M. A. Battiste:

The chemical shift is, of course, affected by the shielding and deshielding factors within the heterocyclic system, so it is pretty hard to pull out the exact chemical shift of the methylenes. But, again, the methylene difference is between 1 and 1.2 ppm, and the trouble is here that there are different ring current contributions to the deshielding of these methylene groups, and these have not been isolated. There could well be changes in this ring current effect in going from one cyclopropenium system to another. And so, I do not know if this is an important factor. It is also in the range where we are trying to decide between 50% to 60% *versus* 90% cyclopropenium character, and it may well be that these chemical shifts are insensitive to this difference.

The Conformational Analysis of Heptalene and 8,8-Diphenylheptafulvene

by D. J. BERTELLI

Department of Chemistry, Memphis State University, Tennessee

I. INTRODUCTION

ALTHOUGH heptafulvene and heptalene are two fundamental examples of pseudo-aromatic compounds, there are no conformational data available on them. The reasonably good correlation between vicinal NMR coupling constants and bond lengths (Fig. 1) [1–2] indicates that conformational data on heptafulvenes and heptalene would be derivable from their NMR spectra.

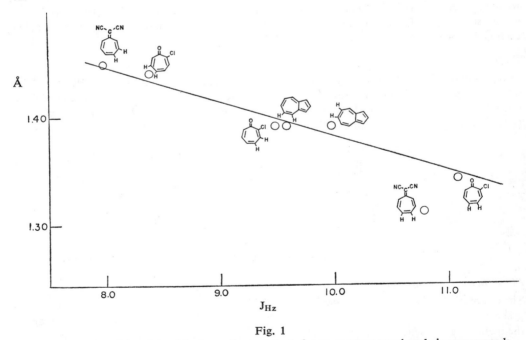

Fig. 1

Plot of bond length *v.* vicinal coupling constants for some seven-membered ring compounds

II. Results and Discussion

1. *Analysis of the NMR Spectrum of 8,8-Diphenylheptafulvene and Conformational Assignment*

A computer-generated analysis of the NMR spectrum of 8,8-diphenylheptafulvene (Fig. 2) yields the coupling constants contained in Table 1.* The value of J_{34} found for 8,8-diphenyl-heptafulvene, when compared with suitable reference compounds, indicates that the seven-membered ring of this molecule is not planar [1–2]. However, these NMR data do indicate that 8,8-diphenylheptafulvene is considerably closer to planarity than cycloheptatriene [1–2, 4]. Molecular models indicate that there are severe steric interactions between the

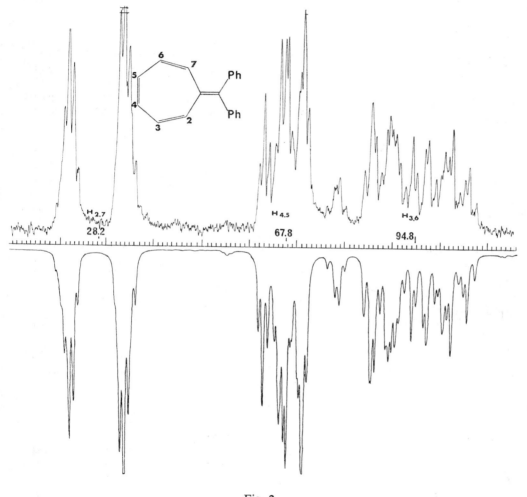

Fig. 2
Experimental and computer-generated NMR spectra for the seven-membered ring of 8,8-diphenylheptafulvene

* The NMR spectra were analysed by the method of Swalen and Riley [3] and are estimated to be accurate to ± 0.2 cps.

Table 1

Compound	Relative	Chemical Shift	Coupling	Constant
8,8-Diphenylheptafulvene	1, 6	28.2	12 = 56	12.18
	2, 5	94.77	13 = 46	0.99
	3, 4	67.78	14 = 36	0.65
			15 = 26	−0.28
			16	2.45
			23 = 45	7.3
			24 = 35	0.95
			25	0.12
			34	11.33
Dihydroheptalene	1, 2	273.12	12 ∼	−12.0
	3	86.96	13 = 23	7.15
	4	50.30	14 = 24	0.13
	5	28.09	15 = 25	0.09
	6	9.45	16 = 26	0.09
			34	9.19
			35	0.69
			36	0.82
			45	5.47
			46	0.31
			56	10.99
1-Heptalenium ion	1, 2	290.08	12 ∼	−12.0
	3	101.77	13 = 23	6.77
	4	78.09	14 = 24	0.32
	5	31.16	15 = 25	0.12
	6	11.24	16 = 26	0.08
			34	9.38
			35	0.74
			36	0.59
			45	5.68
			46	0.85
			56	11.34
Heptalene	1, 5	61.22	12 = 45	9.23
	2, 4	14.33	13 = 35	0.61
	3	21.84	14 = 25	0.45
	6, 10	80.0	15 = 610	−0.30
			16 = 510	1.02
			110 = 56	0.310
			23 = 34	8.8
			24	0.88
			26 = 410	0.010
			210 = 46	0.01
			36 = 310	0.14

two phenyl rings and between the phenyl rings and the 1,6-hydrogens of the seven-membered ring. These steric repulsions can be partially alleviated by distortion of the seven-membered ring, consistent with the NMR data. This evidence leads to the conclusion that unsubstituted heptafulvene is planar. The UV spectra of 8,8-diphenylheptafulvene [5] and of heptafulvene [6] are consistent with the same conclusion. The former molecule exhibits its longest-wavelength absorption at 333 mμ [5], which is shifted 93 mμ — to shorter wavelength than the latter compound (426 mμ) [6].

2. *Analysis of the NMR Spectrum of Heptalene and a Conformational Assignment*

A rigorous computer analysis of the NMR spectrum of heptalene is complicated by the large inter-ring coupling constants, which would necessitate treating a ten-spin system. For this reason, we have performed an approximate solution to the NMR spectrum of heptalene, treating it as a seven-spin system and assigning the 6,10-hydrogens to a remote chemical shift (Fig. 4). This approximation will not yield exact vicinal coupling constants, but the error introduced by this oversimplification is estimated to be no greater than ± 0.2 cps. The coupling constants resulting from this approximation are listed in Table 1. For additional model compounds, we have also carried out analyses of the NMR coupling constants for dihydroheptalene and the 1–5 hydrogens of the 1-heptalenium ion.

It is possible to predict the vicinal NMR coupling constants of heptalene, based on existing data from various model compounds [1–2]. The solution of the NMR spectrum required treating heptalene as a symmetrical ABB'CC' system, rather than an ABCDE system. Thus, the bond shift process $2a \rightleftarrows 2b$ is rapid on the NMR time scale. For this reason, the coupling constant J_{12} of heptalene will be the average of the formal single bond–formal double bond coupling constants shown in Fig. 3. Using 8,8-dicyanoheptafulvene, 8,8-diphenylheptafulvene, 1,2-benzazulene and dihydroheptalene (i.e., using 12.0 rather than the observed 9.2 for the 2–3 bond) as models, the estimated values for J_{12} and J_{23} of heptalene, shown in Fig. 3, were calculated.

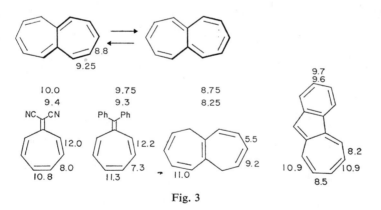

Fig. 3

These data indicate that heptalene is non-planar, and exhibits a conformation intermediate between that of dihydroheptalene and a planar system.

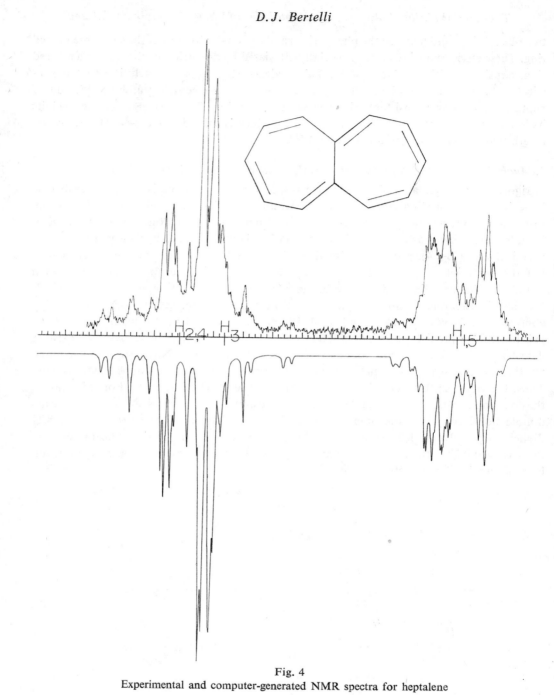

Fig. 4
Experimental and computer-generated NMR spectra for heptalene

The conformational inversion $1a \rightleftarrows 1b$ of heptalene does not serve to change the relative magnetic environment of the hydrogens. Therefore, the NMR spectrum of heptalene will

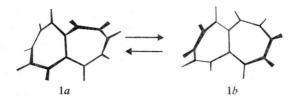

1a   1b

be independent of the rate of this process (except as noted below). However, the bond shift process $2a \rightleftarrows 2b$ will invert the relative magnetic environment of the hydrogens about the molecular centre of symmetry, and the NMR spectrum will be dependent upon the rate of this conversion. Assuming that the transition states for both the inversion

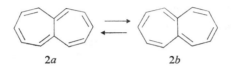

2a   2b

$1a \rightleftarrows 1b$ and bond shift $2a \rightleftarrows 2b$ are planar, the activation energy for the latter process must be equal to, or greater than, that of the former. Only in the limiting cases, where both interconversions occurred through the same transition state, could the NMR spectrum be dependent on the process $1a \rightleftarrows 1b$. Alternatively, the activation energy for the bond shift $2a \rightleftarrows 2b$ may be larger than that for the conformational inversion $1a \rightleftarrows 1b$.*

III. CNDO/2 CALCULATIONS OF HEPTAFULVENE AND HEPTALENE

CNDO/2 calculations were performed on heptafulvene and heptalene using both planar and non-planar models and assuming C=C, C–C and C–H bond lengths of 1.34, 1.46 and 1.08, respectively (Table 2). A CNDO/2 calculation was also performed on cyclo-heptatriene for a planar carbon skeleton and boat conformation, using the electron diffraction bond lengths and angles (in the latter model) [7–9].**
These calculations predict that boat cycloheptatriene should be more stable than the planar form by 3 Kcal. However, this theory predicts planar heptafulvene to be more stable than boat heptafulvene by 3.7 Kcal. Finally, the calculations predict planar heptalene to be more stable by 25 Kcal than the non-planar form.
An analysis of these calculated energies yields an interesting prediction, based on CNDO/2 theory. Considering the total π energies calculated for these systems, the non-planar models are always the more favoured. However, when the Coulombic repulsion terms are added to the total energy, the planar models become energetically favoured in heptafulvene and heptalene. Furthermore, in the two boat conformations used for heptafulvene, it can

---

* The alternative that the bond shift process could proceed through a non-planar transition state of lower energy appears unlikely.
** The bond lengths were those listed in Pople and Gordon [9], unless otherwise noted. The program written by G. A. Segal was obtained through the Quantum Chemical Exchange at Indiana University.

Table 2

Compound	Conformation	Electronic Energy	Total Energy	ΔE (Kcal)
Cycloheptatriene	Boat	−195.2914	−55.7115	
	Planar	−193.2226	−55.7066	3
Heptafulvene	Boat, 40°	−229.0622	−62.6681	
	Boat, 35°	−228.2330	−62.6740	−3.8
	Planar	−225.9078	−62.6799	−3.8
Heptalene	Boat	−401.9578	−92.4732	
	Planar	−395.5330	−92.5153	−25
Tropone	Boat	−238.2986	−72.4478	
	Planar	−235.8343	−72.4570	−5.5
Butadiene	Planar-*cis*	−90.7294	−32.7234	
	90° twist	−89.6554	−32.7282	3
	Planar-*trans*	−89.3099	−32.7283	
Benzene	Bond alternant	−150.0725	−47.0691	
	Regular hexagon	−150.0315	−47.0940	−15.5

be seen that increasing the C_2–C_3 dihedral angle from 35° to 40° lowers the total electronic energy, but this factor is overcompensated for by a proportionally larger increase in the repulsion terms. Thus, CNDO/2 theory predicts that the important driving force towards planarity in these systems is not due to π-energy consideration, but rather to the alleviation of Coulombic repulsion interactions in the planar models. Additional calculations on tropone, based on planar and non-planar models, yield the same predictions.

To further test these ideas, calculations were performed on *cis*- and *trans*-butadiene, butadiene twisted 90° about the C_2–C_3 bond, and on bond alternant and normal models for benzene. These results are somewhat disappointing in that they predict *trans*-butadiene and the 90° twisted model of butadiene to be of equal energy. The calculations for benzene indicate a tendency of CNDO/2 calculations to favour bond alternant models in the electronic energy portion of the calculations, which in this case is compensated for by the Coulombic repulsion terms. Whether this is a general trend in the theory must await further use.

The apparent failure of the CNDO/2 calculations to correctly predict the conformation of heptalene may be partially due to the inadequacy of the method to account for angle strain effects. Dauben estimated that the strain energy of a planar heptalene would be 21 Kcal [10]. If we correct the calculated energy for this angle strain, then the two estimated energies are much closer, although still favouring the planar model by ∼ 4 Kcal.

Molecular models indicate that the conformational inversion 1a ⇌ 1b must be a synchronous process, involving a completely planar transition state. Making the assumption that the activation energy for this process is approximately two times that of the inversion of cycloheptatriene (5.7 or 6.3 Kcal) [11–12], one might reasonably estimate the inversion energy barrier for heptalene at ∼ 10 Kcal. This would place the activation energy of the bond shift process 2a ⇌ 2b at some value greater than this, and should be experimentally available from isotopically labelled variable temperature NMR data. Determination of the activation energy of this process would be of interest, since the transition state would be symmetrical heptalene, and the difference in energy between the ground state and the transition state should provide an answer to the possible anti-aromaticity of planar symmetrical heptalene.

ACKNOWLEDGEMENT

The author would like to express his indebtedness to the National Science Foundation for Research Grant GP–14692 which supported this work, and for Travel Grant GP–21048. He would also like to express his appreciation to the Israel Academy of Sciences and Humanities and the Institut de Biologie Physico-Chimique, Fondation Edmond de Rothschild, Paris, for the invitation to attend this Symposium.

REFERENCES

1 D. J. Bertelli, T. G. Andrews Jr & P. O. Crews (1969) *J. Am. Chem. Soc.*, 91 : 5286.

2 M. A. Cooper & S. L. Manatt, *ibid.*, p. 6325.

3 J. D. Swalen & C. A. Riley (1960) *J. Chem. Phys.*, 37 : 21.

4 M. Tratteberg (1964) *J. Am. Chem. Soc.*, 86 : 4765.

5 R. B. Medz (1964) Ph. D. Thesis, University of Washington, p. 296.

6 W. von E. Doering & D. W. Wiley (1960) *Tetrahedron*, 11 : 183.

7 J. A. Pople, D. P. Santry & G. A. Segal (1965) *J. Chem. Phys.*, 43 : 5129.

8 J. A. Pople & G. A. Segal (1966) *ibid.*, 44 : 3289.

9 J. A. Pople & M. Gordon (1967) *J. Am. Chem. Soc.*, 89 : 4253.

10 D. J. Bertelli (1961) Ph. D. Thesis, University of Washington, p. 75.

11 F. L. Anet (1964) *J. Am. Chem. Soc.*, 86 : 458.

12 F. R. Jensen & L. A. Smith, *ibid.*, p. 956.

Discussion

J. J. C. Mulder:

In my opinion, one thing is very clear: if you go from azulene to benzazulene, you go from a situation of two completely equivalent structures to one of two completely non-equivalent structures.

D. J. Bertelli:

I think you have a sort of a battle here. The energy that you gain by symmetrizing the azulene ring would cost you energy in terms of asymmetrizing the benzene ring.

E. D. Bergmann:

If 8, 8-dicyanoheptafulvene is planar and 8, 8-diphenylheptafulvene almost planar, why should heptafulvalene be non-planar? Would one not at least expect one of the seven-membered rings in this fulvalene to be planar?

D. J. Bertelli:

Perhaps this is exaggerated a little, but the interaction here is probably reasonably large if the molecule is planar. I do not really know what the structure is in the crystal, but there is certainly a steric effect.

E. D. Bergmann:

The molecule could also avoid this effect just by having one ring not planar.

D. J. Bertelli:

That is true. I could, I guess, always go back to the old 'crystal effect' argument, namely, that there may be specific effects in the crystal.

Mrs A. Skancke:

Heptafulvalene would probably have different structures in the crystalline and in the vapour phase. The Van der Waals forces around the central bond should be approximately equal to that for biphenyl.

Mrs A. Pullman:

Regarding the CNDO results which you reported, I would like to make two remarks:
1. In the case of butadiene, the electronic energy would lead to the result that the *cis* isomer is more stable. This would be an exception to your suggested rule that the electronic energy alone is a better measure of stability.
2. As a rule, it seems to me that keeping the same geometry for different isomers in a molecule might be a source of errors in the CNDO results, because the energy differences are small and may very well be revised when allowance is made for changes in geometry. We have a number of unpublished calculations which support this possibility.

Discussion

D. J. Bertelli:

I believe you are quite right. I think that an even more serious problem is the bond angle, because there are so many bond angles in these molecules. First of all, if you look at heptafulvene—and I have found this very difficult myself—no matter what structure you picture, you can never have 120° bond angles. And you have to make an arbitrary assumption as to how much your angles are opened out, and this has a large effect. I think it would be a very difficult process to try and do this, because there are so many variables one can change, and I am sure they will really alter the energy. So, I don't really claim any significance for the CNDO calculation. I am not a theoretician, I just like to square up with the troubles.

H. Prinzbach:

May be I did not get the point, but on one of your slides I saw formulae of tropilidene and heptafulvene, and you gave the coupling constants of the single bond for various angles. Now, is your conclusion relative to planarity? Are you adopting the same bond order?

D. J. Bertelli:

No, all of them had the same bond length within two hundredths of an Ångstrom. But the coupling constants went from 5.5 to 8.0, and that was very nicely correlated with the dihedral angle.

H. Prinzbach:

But you have to know the length before?

R. H. Martin:

In heptalene, you mentioned a large inter-ring coupling constant. What is the order of this coupling?

D. J. Bertelli:

One cycle to 1.2 cycles.

R. H. Martin:

If this coupling involves H_1 and H_6, could you not include it in a six- or seven-spin system?

D. J. Bertelli:

I actually considered a seven-spin system, but the only way I could do that was to assign two hypothetical hydrogen atoms. I solved it as a seven-spin system by showing these 6, 10-hydrogen atoms as remote chemical shifts, which is not 'legal', but it is the only thing I could do. I did include couplings between these two sites, but there will be a lot of virtual coupling which in no way could reproduce all the fine structure.

R. H. Martin:

In view of the sensitivity of methyls to bond orders, have you done any work on methyl-substituted heptalenes?

D. J. Bertelli:

No, we have not. I know that things like that have been published. But, for example, if you look at a recent piece of work where the NMR spectrum of toluene has been measured, you get a different picture. It is said that if the methyl group sits on something that looks like a double bond, the methyl-vinyl coupling constant goes up. People have been saying for a long time that the methyl ring coupling constants in toluene are very small, but recent work has shown that they are of the order of 0.8 cps.

335

Electron Spin Resonance and Aromaticity of Hydrocarbon Free Radicals

by G. VINCOW

Department of Chemistry, University of Washington, Seattle

To the memory of
Hyp Joseph Dauben Jr

I. INTRODUCTION

A LARGE NUMBER of experimental and theoretical criteria have been applied to determine the aromatic character of closed-shell (diamagnetic) cyclic conjugated molecules. Let us start by reviewing the applicability of these criteria to the open-shell case of doublet states, i.e., neutral free radicals and radical ions (see Table 1).

Table 1
Criteria of Aromaticity for Free Radicals

Criterion of Aromatic Character	Applicability
1 Empirical resonance energy	Yes
2 MO calculations: delocalization energy and other indices	Yes
3 Wavelength shifts in absorption spectra	Yes
4 Changes in bond length and charge distribution associated with π-electron delocalization	Yes
5 Reactivity: benzene-like stability and chemical behaviour	Sometimes correlates ($C_7H_7\cdot$) and sometimes does not (hydrocarbon radical anions)
6 Aromatic sextet	No. Odd number of electrons
7 Hückel rule	No. Odd number of electrons
8 NMR chemical shift	No. Spectrum dominated by contact hyperfine interaction
9 NMR diamagnetic susceptibility exaltation	No. Molecules are paramagnetic
10 Anisotropy of diamagnetic susceptibility	No. Molecules are paramagnetic
11 Craig symmetry criteria	No. Derived for closed-shell molecules

336

Empirical resonance energies of a few neutral hydrocarbon free radicals have been determined by Benson and co-workers, using kinetic and thermodynamic data for the reaction of iodine with hydrocarbons [1–4]. MO calculations are useful in the estimation of π-electron delocalization energies [5–6]. It seems reasonable that correlations can be made of the wavelength shifts of the absorption spectra relative to the spectra of olefinic compounds. The criterion of reactivity is of limited significance, as has been discussed previously [7], sometimes correlating with other properties, as in the case of $C_7H_7\cdot$, and other times being in disagreement, e.g., for hydrocarbon anion radicals.

A number of criteria are obviously not applicable to free radicals, since they possess an odd number of electrons, are consequently paramagnetic, and since their NMR spectra are very broad and are dominated by the contact hyperfine shifts. These are: (i) aromatic sextet; (ii) Hückel rule; (iii) NMR chemical shift; (iv) NMR diamagnetic susceptibility exaltation; (v) anisotropy of diamagnetic susceptibility; and (vi) the Craig symmetry criteria.

We conclude from the above listing that only a few of the traditional tests of aromaticity are available to us in our consideration of free radicals. It is therefore important that we develop these as far as possible. Our approach toward this end has been to apply the technique of electron spin resonance spectroscopy (ESR).

II. ESR

I would like to start with a very brief review of some of the basic ideas of the ESR spectroscopy of π-electron free radicals in solution. The measured spectral parameters are: (i) the g value (resonance field of the centre of the spectrum, usually very close to that for a free electron); (ii) line positions and relative intensities (from which we obtain Fermi contact nuclear hyperfine splitting constants and the number of equivalent magnetic nuclei giving rise to each splitting); (iii) line shapes; and (iv) line widths. The concentration of free radical can be obtained by measurement of the absolute intensity of the signal.

The interpretation of the nuclear hyperfine splittings has been one of the most interesting and fruitful pursuits in the field of ESR. For example, some years ago we investigated the thioxanthone S,S-dioxide anion radical in an effort to shed light on the problem of sulphur valence-shell expansion [8]. In Fig. 1 we show the structure of this radical and values of the proton hyperfine splitting constants that are extracted from its spectrum. These constants are a measure of the interaction between the delocalized electron spin magnetic moment and the localized moments of the various magnetic nuclei, in this case protons.

Thioxanthone S, S-Dioxide Mononegative Ion

Splitting (Gauss)	Spin Density	
$a_1 = 2.35 \pm 0.04$	0.102	
$a_2 = 0.52 \pm 0.01$	0.023	
$a_3 = 3.10 \pm 0.06$	0.135	
$a_4 = 0.84 \pm 0.02$	0.036	

Fig. 1

Nuclear framework of the thioxanthone S, S-dioxide anion radical, proton hyperfine splitting constants and values of ρ_{kk}^{π}

337

Specifically, we are dealing with the Fermi contact hyperfine interaction, which is proportional to the spin density at the magnetic nucleus, i.e., the difference between the number of electrons/cm^3 with spin α and the number of electrons/cm^3 with spin β at the position of the nucleus. The general expression for the hyperfine splitting arising from nucleus N is

$$a^N = \frac{8\pi}{3} g_N \beta_N [<\Psi| \sum_k (S_z)^{-1} S_{zk} \delta(r_k - r_N)| \Psi>], \tag{1}$$

where the quantities in front of the bracket are constants related to the magnetic nucleus, and the quantity within brackets is just the spin density at the magnetic nucleus [9].

To return to the specific radical under discussion, analysis of the measured spectrum reveals only that the radical contains four groups of two equivalent protons each, with splitting constants of magnitude 2.35, 0.52, 3.10 and 0.84 G. How, then, can we ascribe the spectrum to the molecule shown? The numbers of equivalent protons fit the structure and symmetry of the postulated radical. But what about the magnitudes of the splittings? To interpret these, we must invoke the following approximate equation (McConnell relationship), which is a brilliant simplification of Equation (1),

$$a_k^H \cong Q\rho_{kk}^\pi. \tag{2}$$

The proton hyperfine splitting is approximately proportional (semi-empirical constant of $c. -25\,G$) to ρ_{kk}^π, which is called the π spin density at the carbon atom adjacent to the proton. This quantity is actually not a spin density at all, but is rather a diagonal element of the matrix of coefficients for the expansion of the SCF π-electron spin density in an atomic orbital basis [9]. Simply put, in Hückel molecular orbital (HMO) theory terms, ρ_{kk}^π denotes the occupation of the unpaired electron in the $2p$ π atomic orbital centred at carbon atom k. Thus, from the splittings and a semi-empirical value of Q, we compute approximate experimental values of ρ_{kk}^π.

We can also easily calculate values of ρ_{kk}^π from HMO, VB, SCF-MO and various other semi-empirical π-electron theories. To the extent that these calculations agree with the experimental a_k^H/Q, we develop confidence that we have correctly identified the radical.

More often than not, the identity of the radical is not seriously in question. In such cases, the semi-quantitative agreement often obtained between theory and experiment has been interpreted as a verification of the accuracy of the ground-state wave functions predicted by approximate π-electron theory.

To complete our review of ESR, reference should be made to the considerable versatility and usefulness of this technique as a spectroscopic probe of the properties of organic molecules. Applications from the work of my group include chemical kinetics (rate of radical recombination [10] and of intramolecular thermal isomerization [11]), molecular vibrations (torsional oscillations of alkyl groups [12–13]), molecular geometry (angle of twist of a phenyl substituent [14]) and molecular electronic structure (molecular orbital degeneracy [15]).

In the following sections we discuss applications of ESR to the question of the aromaticity of free radicals.

III. CYCLOHEPTATRIENYL RADICAL

Few experimental determinations of the empirical resonance energy (RE) of free radicals have been reported to date [1–4]. We have evaluated the resonance energy of the cycloheptatrienyl (tropenyl) radical, $C_7H_7\cdot$ [16]. The method employed involves the measurement of the enthalpy of cleavage, ΔH^0, of bitropenyl using ESR spectroscopy.

Let us begin by discussing the enthalpy determination. We have detected the ESR spectrum of $C_7H_7\cdot$ produced by homolytic thermal cleavage of its neat liquid dimer (see Fig. 2) [17] and have measured the radical concentration as a function of temperature in the range $120°$ to $140°C$.

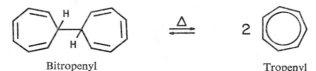

Bitropenyl   Tropenyl

Fig. 2

Homolytic thermal cleavage of biscycloheptatrienyl (bitropenyl)
to form the cycloheptatrienyl (tropenyl) radical

Using the Van't Hoff equation and the equilibrium constant expression for the dimer–monomer equilibrium, one can derive the following relationship between the radical concentration and ΔH^0:

$$2 \ln C_A\cdot (T) - 2 \ln \rho_{A_2}(T) = -\Delta H^0/RT + \text{constant,} \qquad (3)$$

where $C_A\cdot$ is the concentration (moles/litre) of the radical, $\rho_{A_2}(T)$ is the density of the dimer, and ΔH^0 is assumed to be constant over the $20°$ interval employed in this work. An important advantage of our system involving the neat dimer as solvent, in comparison with a dimer and monomer both in an inert solvent, is that uncertainties associated with the use of the concentration equilibrium constant are removed from the above-mentioned derivation. We use the equilibrium constant in terms of activities and take advantage of the fact that the activity of the dimer — essentially a pure liquid — is unity, that Henry's Law applies for the radical ($\sim 10^{-6}$ M at $133°C$ [18]), and that the density of the solution is equal to that of the dimer.

Since the ESR line width is constant in the interval $120° \leq t \leq 140°C$, we used the product of peak height and absolute temperature (to incorporate Curie's Law) as a measure of $C_A\cdot (T)$. An approximate measurement of the density of bitropenyl was performed and the result used in Equation (3). Ten runs for the determination of ΔH^0 were made. For each run numerous amplitude measurements were made at each of two temperatures in the $20°$ range cited above. The average value of ΔH^0 obtained from the ten determinations is 35.0 ± 1.0 Kcal/mole.

This result for ΔH^0, which measures the central C–C bond strength in bitropenyl, is strikingly small. By way of comparison, biisopropyl, which also has secondary carbons involved in its central bond, has an enthalpy of cleavage 40 Kcal/mole higher. We ascribe the weakened bond in bitropenyl to resonance stabilization of the tropenyl radicals formed in the dissociation. Using this assumption, we can estimate the empirical resonance energy of the radical.

We define the empirical RE by reference to the structures shown in Fig. 3. It is given by

$$RE \equiv \left| \Delta H_f^0(I) - \Delta H_f^0(III) \right|. \qquad (4)$$

Structure I is the planar, symmetrical tropenyl radical, and III is the hypothetical planar radical with the unpaired electron localized and 3 non-interacting double bonds.

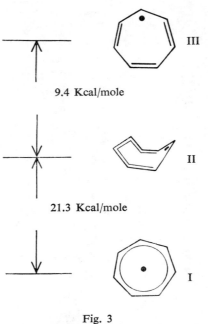

9.4 Kcal/mole

21.3 Kcal/mole

Fig. 3
Structures involved in the determination of the
empirical RE of $C_7H_7\cdot$ and estimated differences
in their standard enthalpies of formation (ΔH_f^0)

The RE is evaluated in two stages, the differences $[\Delta H_f^0(\mathrm{I}) - \Delta H_f^0(\mathrm{II})]$ and $[\Delta H_f^0(\mathrm{II}) - \Delta H_f^0(\mathrm{III})]$ being summed to give the final result. The intermediate II denotes a hypothetical structure with the σ-bond framework of cycloheptatriene, a delocalized triene π system and an unpaired electron localized on the carbon corresponding to the methylene carbon of cycloheptatriene. The magnitude of the quantity $[\Delta H_f^0(\mathrm{II}) - \Delta H_f^0(\mathrm{III})]$ is essentially the empirical resonance energy of cycloheptatriene, which is 9.4 Kcal/mole, as obtained from heats of hydrogenation data [19–20].
The quantity $[\Delta H_f^0(\mathrm{I}) - \Delta H_f^0(\mathrm{II})]$ is obtained from the following equations:

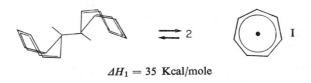

$\Delta H_1 = 35$ Kcal/mole

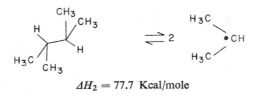

$\Delta H_2 = 77.7$ Kcal/mole

340

We approximate the enthalpy of the following cleavage reaction, involving the hypothetical structure II, by the enthalpy of cleavage of biisopropyl, ΔH_2.

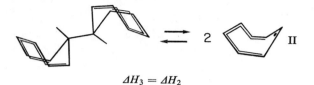

$$\Delta H_3 = \Delta H_2$$

Then, $[\Delta H_f^0(I) - \Delta_f^0 H(II)] = \frac{1}{2}(\Delta H_1 - \Delta H_2) = -21.3$ Kcal/mole. Adding to this -9.4 Kcal/mole, we obtain the total empirical RE of $C_7H_7\cdot$, 31 Kcal/mole (± 4 Kcal/mole estimated error).

The principal conclusion to be drawn is that $C_7H_7\cdot$ is an extensively resonance-stabilized molecule. Our perspective in making this statement can be illustrated by noting that the resonance energy of $C_7H_7\cdot$ is comparable with that of benzene, 36 Kcal/mole. Further, as shown in Table 2, the extra RE of $C_7H_7\cdot$ — $[\Delta H_f^0(I) - \Delta H_f^0(II)]$; RE obtained by delocalization of the unpaired electron as the dimer dissociates — is next to the highest among free radicals investigated to date.

Table 2

Extra Resonance Energy of Free Radicals

Radical	Extra RE (Kcal/mole)
Cyclohexadienyl*,**	24
Tropenyl	21
Triphenylmethyl**	16
Pentadienyl**	15.4
Benzyl**	12.5
Allyl**	10.2

* James and Stuart [21]
** Benson [1], Egger and Benson [2],
 Walsh et al. [3], Golden et al. [4]

The magnitude of the RE obtained for $C_7H_7\cdot$ is consistent with the lack of reactivity found for this radical [22], and with the HMO delocalization energy (DE) of 2.54β (42 Kcal/mole) and the VB resonance energy of 1.08α (32 Kcal/mole).

It is also of interest to compare our result with that of Hedaya [23] for the stability of the cyclopentadienyl radical, $C_5H_5\cdot$. This investigator has estimated the endothermicity of the dissociation of 9,10-dihydrofulvalene (bicyclopentadienyl) to cyclopentadienyl as 40–45 Kcal/mole. Thus, there is appreciable resonance stabilization for this radical, in agreement with the HMO DE of 1.85β (30 Kcal/mole).

Another application of the ΔH^0 measurement made in this work is the computation of a number of important thermochemical quantities, namely, $\Delta H_f^0(C_7H_7\cdot_{gas})$, $\Delta H_f^0(C_7H_7{}^+{}_{gas})$ and RE ($C_7H_7{}^+$). We continue to make the approximation that the enthalpy of dissociation of neat liquid bitropenyl is the same as that in the gas phase.

The heat of formation of the tropenyl radical is given by

$$\Delta H_f^0(C_7H_7 \cdot_{gas}) = \tfrac{1}{2}[\Delta H_f^0(C_7H_7 - C_7H_{7gas}) + \Delta H^0]. \tag{5}$$

$\Delta H_f^0(C_7H_7 - C_7H_{7\,gas})$ has been evaluated previously [24], using the Franklin group-equivalent method and $\Delta H_f^0(C_7H_8) = 43.47$ Kcal/mole; the value obtained is 94.6 Kcal/mole. From this and $\Delta H^0 = 35$ Kcal/mole, $\Delta H_f^0(C_7H_7 \cdot_{gas}) = 64.8$ Kcal/mole.

This heat of formation can be used to compute the heat of formation of the tropenium ion. A number of attempts have been made to evaluate $\Delta H_f^0(C_7H_7{}^+_{gas})$, because $C_7H_7{}^+$ is the prototype of the heptagonal aromatic system. The results obtained vary from 207 to 244 Kcal/mole [24–31]. From $\Delta H_f^0(C_7H_7 \cdot_{gas}) = 64.8$ Kcal/mole and a spectroscopic ionization potential for $C_7H_7 \cdot (I_p = 6.237 \pm 0.01\,eV)$ [32–33], we compute $\Delta H_f^0(C_7H_7{}^+_{gas}) = 209$ Kcal/mole. This value is in excellent agreement with an estimate made from the appearance potentials of the benzyl halides, i.e., 207–210 Kcal/mole [25–31].

Using this heat of formation, one can estimate the empirical RE of $C_7H_7{}^+$. By analogy with the treatment presented for $C_7H_7 \cdot$, the extra RE of $C_7H_7{}^+$ is given by

$$(RE)_{extra} = \left| [\Delta H_f^0(C_7H_7{}^+_{gas}) - \Delta H_f^0(C_7H_{8\,gas})] + \right.$$
$$\left. - [\Delta H_f^0(i - C_3H_7{}^+_{gas}) - \Delta H_f^0(C_3H_{8\,gas})] \right|. \tag{6}$$

Since $\Delta H_f^0(C_3H_8) = -24.83$ Kcal/mole [34], $\Delta H_f^0(C_7H_8) = 43.47$ Kcal/mole [35] and $\Delta H_f^0(i - C_3H_7{}^+_{gas}) = 190$ Kcal/mole [36], $(RE)_{extra} = 49$ Kcal/mole. Adding to this the resonance energy of cycloheptatriene (9 Kcal/mole), we obtain 58 Kcal/mole for the total empirical RE of $C_7H_7{}^+$. It is interesting to note that this value is considerably larger than the empirical RE of benzene (36 Kcal/mole).

The resonance energy of $C_7H_7{}^+$ is much larger than that of $C_7H_7 \cdot$, a result which is in accord with the fact that $C_7H_7{}^+$ satisfies the Hückel $4n+2$ rule. The HMO theory correctly predicts a larger DE for $C_7H_7{}^+$ than for $C_7H_7 \cdot$, i.e., 2.99β (49 Kcal/mole) v. 2.54β.

IV. Heptafulvalene Cation and Anion Radicals

The non-benzenoid hydrocarbon heptafulvalene (IV) has an empirical RE of only 28 Kcal/mole [37], which is considerably less than the DE predicted by HMO theory (4β; 64 Kcal/mole). This apparent discrepancy is explained by the fact that a uniform C–C bond length is assumed in HMO theory, whereas more refined calculations [38–40] have demonstrated that IV consists of alternate single and double bonds. The occurrence of such bond alternation is accompanied by a reduction in the stabilization energy.

IV

HMO theory also predicts large DE's for the cation and anion radicals of heptafulvalene, namely, 4.18β and 3.56β, respectively. Using ESR, we have found two pieces of evidence which indicate that the ion radicals may have considerable delocalization energy, as predicted, and may thus differ in their degree of aromaticity from heptafulvalene itself [41]. We proceed to discuss these findings below.

1. *Dissociation of Dimers of Heptafulvalene Cation and Anion Radicals*

One-electron oxidation of **IV** with silver ion generated the heptafulvalene cation radical **VI**, which dimerized to produce the dication **V**. The cation radical was regenerated by equilibrium thermal cleavage of **V** in nitroethane. The ESR signal was detected at temperatures as low as 30°C.

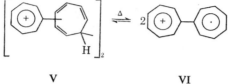

V	**VI**

Heptafulvalene anion radical was produced by reduction of **IV** with potassium metal and was detected in several solvents at $-80°$. As the temperature was raised, the ESR signal of the radical anion first disappeared ($c. -50°$) and then reappeared at higher temperature ($c. +30°$). This sequence is ascribed to radical dimerization, and then thermal cleavage of the dimer.

We measured the equilibrium constant for the cleavage of the dianion dimer in THF, relative to that for the homolytic dissociation of neat liquid bitropenyl to form C_7H_7·. The equilibrium constant is $\approx 10^{-7}$ at 66° and is approximately nine orders of magnitude greater than that for bitropenyl, extrapolated to this temperature.

The fact that the dimers of the heptafulvalene cation and anion radicals are both appreciably dissociated into radicals at room temperature is interpreted as a qualitative indication that these radicals are extensively stabilized by resonance.

2. *Proton Hyperfine Splittings*

The second line of evidence for resonance stabilization of the ion radicals comes from comparison of the measured proton hyperfine splitting constants with calculated values [for details, see 25]. The main result is that the application of approximate π-electron theory, in which a uniform nuclear framework is assumed, leads to the calculation of hyperfine splittings that are in good agreement with experiment. This concordance argues against the occurrence of alternate single and double bonds in the ion radicals, with concomitant reduction of delocalization energy.

There is one novel aspect of the ESR spectra worthy of further discussion. It should be recalled that the proton hyperfine splitting is approximately proportional to ρ_{kk}^{π}, the π spin density at the carbon atom adjacent to the proton. Our interpretation of the above-mentioned spectra is that, for the cation, the unpaired spin is delocalized throughout the molecule, whereas for the heptafulvalene anion the π spin density is essentially localized on a single ring.

Let us first discuss the experimental results that led to this proposal. If one were asked to predict the general pattern for the ESR spectra of these ion radicals, the obvious answer would be spectra consistent with three groups of four equivalent protons each. This is indeed what was observed for the heptafulvalene cation in nitroethane at 50°C, and for the anion in DMF at $-70°$C and in THF at $+90°$C. The splitting from one of the three groups of four protons was too small to be resolved in the case of the anion. The spectrum of the anion in DMF at $-70°$C is shown at the top of Fig. 4; it is quite ordinary in appearance. At $-90°$C in THF, the middle spectrum was detected; it shows 'alternating line width effects'. Our suspicion was that this was due to the K$^+$ counterion hopping back and forth between the two rings at a rate comparable to the hyperfine frequency. A concerted effort was made to freeze this motion out. We accomplished this by using a less polar solvent, MTHF, at −115°, and observed the spectrum shown at the bottom of Fig. 4. To our great surprise, this spectrum arises from two groups of only two protons each, with just double the splittings observed at the higher temperature.

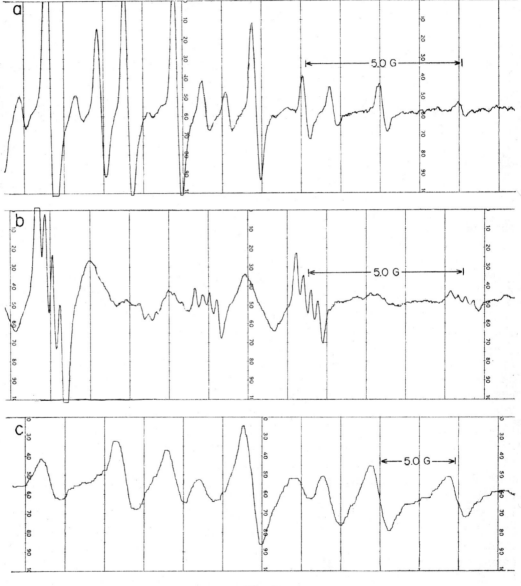

Fig. 4

Hyperfine spectrum of the heptafulvalene anion radical

a. DMF at −70° (half the spectrum shown); b. THF at −90° (half the spectrum shown);
c. MTHF at −115° (entire spectrum shown)

We associate this result with localization of the unpaired spin, and use HMO theory to rationalize such a novel π spin distribution. To understand the HMO results, consider the following model for the radical ions of heptafulvalene, namely, that they behave as if heptafulvalene were a dimer consisting of two almost independent tropenyl halves (see Fig. 5). In Column a of Fig. 5 the Hückel molecular-orbital energy levels of $C_7H_7\cdot$ are shown; in Column b those for a non-interacting dimer are portrayed. The bond between carbon atoms 7 and 8 constitutes a weak perturbation of the dimer, which lifts some, but not all, of the molecular-orbital degeneracy. The symmetry of this perturbation is such that the appropriate zero-order linear combinations of the fourfold-degenerate molecular orbitals are either symmetric (designated S_1 and S_2) or anti-symmetric (designated A_1 and A_2) with respect to reflection in a plane perpendicular to the aromatic plane and passing through the C_7–C_8 bond. At the 7- (or 8-) position, the symmetric C_7H_7 orbitals have non-zero coefficients, whereas the anti-symmetric orbitals have nodes. In the HMO approximation, a perturbation applied at that position will lift the degeneracy of the S orbitals and will mix them (S_1 and S_2 yield S_{12} and S'_{12}), but will neither perturb nor mix the anti-

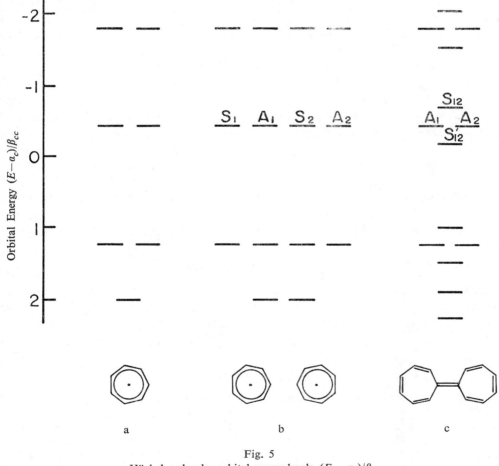

Fig. 5
Hückel molecular orbital energy levels, $(E - \alpha_c)/\beta_{cc}$
a. C_7H_7; b. non-interacting dimer of C_7H_7; c. heptafulvalene

symmetrical orbitals, nor will it lift their degeneracy. The resulting molecular-orbital pattern for the perturbed dimer is shown in Column c and is qualitatively identical with that predicted by HMO theory for heptafulvalene.

What are the consequences of the HMO theory? The heptafulvalene cation has thirteen π electrons, and so the unpaired electron finds itself in orbital S'_{12}, a delocalized orbital formed by mixing two perturbed symmetrical $C_7H_7\cdot$ orbitals.

The anion of heptafulvalene has fifteen π electrons, and so the unpaired electron is in either of two degenerate MO's, A_1 or A_2, which have not changed in the formation of the dimer, due to the node at positions 7 and 8. For example, HMO theory predicts that A_1 has coefficients identical with those of the A orbital of $C_7H_7\cdot$ on one ring, and zero coefficients (no spin density) on the other. That is the explanation for the two groups of only two equivalent protons with splittings of 8.2 and 5.0 G. These splittings are indeed in good agreement with the prediction of HMO theory for the A orbital of an isolated $C_7H_7\cdot$ ring, namely, 7.1 and 4.5 G, respectively. In addition, we have experimental evidence concerning the A orbital of C_7H_7 from our previous work on alkyl-substituted tropenyl radicals [15]. Extrapolation of our splittings for tri-*tert*-butyltropenyl to 0°K, at which temperature only the state with the unpaired electron in orbital A is populated, leads to the approximate values 7.8 and 5.0 G. These splittings are in good agreement with our results for the HF anion. Thus, we see how a simple molecular-orbital picture explains the localization of electron spin and how striking can be the influence of molecular orbital symmetry on spin-density distributions.

A few additional comments should be made regarding the various spectra observed for the anion. The K^+ counterion likely lifts the A_1, A_2 MO degeneracy. Calculations indicate that including the counterion does not modify the localization of electron spin. The location of the counterion dictates which ring will possess the unpaired spin. As the potassium ion hops back and forth from ring to ring, the unpaired spin also jumps to accommodate the change in position of the perturbation. The splittings are just halved in the limit of fast exchange, because we are averaging a splitting from one ring, with zero from the other. Since both rings are involved, there are twice as many protons interacting. This explains the spectrum at high temperature as the fast-exchange limit, and also the one showing alternating line width effects as a case of intermediate hopping rate.

ACKNOWLEDGEMENTS

I wish to thank the US Army Research Office, Durham, the donors of the Petroleum Research Fund, administered by the American Chemical Society, and the National Science Foundation for support of this research. I would also like to express my appreciation to my collaborators who are named in the references in connection with their various contributions.

REFERENCES

1 S. W. Benson (1965) *J. Chem. Educ.*, 42 : 502.

2 K. W. Egger & S. W. Benson (1966) *J. Am. Chem. Soc.*, 88 : 241.

3 R. Walsh, D. M. Golden & S. W Benson, *ibid.*, p. 650.

4 D. M. Golden, A. S. Rodgers & S. W. Benson, *ibid.*, p. 3196.

5 A. Streitwieser Jr (1961) *Molecular Orbital Theory for Organic Chemists*, Wiley, New York.

6 L. Salem (1966) *The Molecular Orbital Theory of Conjugated Systems*, Benjamin, New York.

7 A. J. Jones (1968) *Rev. Pure Appl. Chem.*, 18 : 253.

8 G. Vincow (1962) *J. Chem. Phys.*, 37 : 2484.

9 S. Y. Chang, E. R. Davidson & G. Vincow (1968) *ibid.*, 49 : 529.

10 M. L. Morrell & G. Vincow (1969) *J. Am. Chem. Soc.*, 91 : 6389.

11 W. V. Volland & G. Vincow (1968) *ibid.*, 90 : 4537.

12 M. K. Carter & G. Vincow (1967) *J. Chem. Phys.*, 47 : 302.

13 M. D. Sevilla & G. Vincow (1968) *J. Phys. Chem.*, 72 : 3647.

14 Idem, *ibid.*, p. 3641.

15 G. Vincow, M. L. Morrell, F. R. Hunter & H. J. Dauben Jr (1968) *J. Chem. Phys.*, 48 : 2876.

16 G. Vincow, H. J. Dauben Jr, F. R. Hunter & W. V. Volland (1969) *J. Am. Chem. Soc.*, 91 : 2823.

17 G. Vincow, M. L. Morrell, W. V. Volland, H. J. Dauben Jr & F. R. Hunter (1965) *ibid.*, 87 : 3527.

18 W. V. Volland (1967) Ph. D. Thesis, University of Washington.

19 J. B. Conn, G. B. Kistiakowsky & E. A. Smith (1939) *J. Am. Chem. Soc.*, 61: 1868.

20 R. B. Turner et al. (1957) *ibid.*, 79 : 4127.

21 D. G. L. James & R. D. Stuart (1966) *Chem. Comm.*, p. 484.

22 F. R. Hunter (1966) Ph. D. Thesis, University of Washington.

23 E. Hedaya (1969) *Acc. Chem. Res.*, 2 : 367.

24 A. G. Harrison, L. R. Honnen, H. J. Dauben Jr & F. P. Lossing (1960) *J. Am. Chem. Soc.*, 82 : 5593.

25 F. P. Lossing, K. U. Ingold & I. H. S. Henderson (1954) *J. Chem. Phys.*, 22 : 1489.

26 F. H. Field & J. L. Franklin, *ibid.*, p. 1895.

27 S. Meyerson & P. N. Rylander (1957) *ibid.*, 27 : 901.

28 P. N. Rylander, S. Meyerson & H. M. Grubb (1957) *J. Am. Chem. Soc.*, 79 : 842.

29 S. Meyerson, P. N. Rylander, E. L. Eliel & J. D. McCullum (1959) *ibid.*, 81 : 2606.

30 R. B. Turner, H. Prinzbach & W. von E. Doering (1960) *ibid.*, 82 : 3451.

31 K. R. Jennings & J. H. Futtrell (1966) *J. Chem. Phys.*, 44 : 4315.

32 B. A. Thrush & J. L. Zwolenik (1963) *Discuss. Faraday Soc.*, 35 : 196.

33 F. A. Elder & A. C. Parr (1969) *J. Chem. Phys.*, 50 : 1027.

34 *Selected Values of Physical and Thermodynamic Properties of Hydrocarbons and Related Compounds (American Petroleum Institute Research Project, No. 44)* (1953) Carnegie Press, Pittsburgh.

35 H. L. Fink, D. W. Scott, M. E. Cross, J. E. Messerly & G. Waddington (1953) *J. Am. Chem. Soc.*, 75 : 2819.

36 F. H. Field & J. L. Franklin (1957) *Electron Impact Phenomena*, Academic Press, New York, p. 260.

37 R. B. Turner (1959) *Theoretical Organic Chemistry (Kekulé Symposium)*, Butterworth, London, p. 67.

38 A. J. Silvestri, L. Goodman & J. A. Dixon (1962) *J. Chem. Phys.*, 36 : 148.

39 T. Nakajima & S. Katagiri (1963) *Molec. Phys.*, 7 : 149.

40 M. J. S. Dewar & G. J. Gleicher (1965) *J. Am. Chem. Soc.*, 87 : 685.

41 M. D. Sevilla, S. H. Flajser, G. Vincow & H. J. Dauben Jr (1969) *ibid.*, 91 : 4139.

Discussion

B. Binsch:

First, I have a question. Do you have a rough idea of what the hopping frequency is?

G. Vincow:

Yes, an approximate hopping frequency for K^+ was measured in THF at $-90°$ C. It is $\approx 7 \times 10^8$ sec^{-1}.

B. Binsch:

Secondly, I would like to comment on the degeneracy of the two anti-symmetric levels. I completely agree that the cation must play an important role, but in principle, even without this, you would expect a lifting of the degeneracy due to asymmetric distortion of the molecule.

G. Vincow:

I agree completely; I was merely trying to show how a simple model might explain the phenomena. In fact, Mr Nakajima has done these calculations, and he will present them tomorrow. I do not want to pre-empt what he is going to say, and it does not really materially affect the picture; there are other possibilities going beyond Hückel theory where you might expect other effects. It is, after all, not a real degeneracy, but rather an accidental one.

R. West:

Mr Vincow has reported some very interesting results, and his explanation seems convincing in view of the line-width alternation found at intermediate temperatures. I have two questions: First, if the spin is localized in one ring, as in your model, there should be three sets of two protons, instead of two, as you found.
Is one of the splitting constants then very small?

G. Vincow:

Yes, even in the high-temperature limit spectrum recorded ($-70°$ C in DMF), for which the line width is 0.25 G, this splitting is not resolved.

R. West:

And is the small splitting constant in agreement with predictions from HMO theory?

G. Vincow:

MO calculations do indicate that the third splitting should be small.

R. West:

Second, I wonder if you have tried to check your model involving spin localization by potassium ion by using a very large counterion, such as tetrabutylammonium, which would not bind so strongly to the anion, and so might not show the localization? The tetrabutylammonium derivative of the anion-radical could be produced by electrolytic reduction.

348

G. Vincow:

No, we have not yet tried generating the radical by electrolytic reduction.

V. Boekelheide:

I was wondering about the assumption that $C_7H_7\cdot$ is planar. Do you have some experimental knowledge on this?

G. Vincow:

Are you asking about a time-average behaviour or are you asking about a static equilibrium conformation?

V. Boekelheide:

How do you know that you are not just looking at some time-average or an isomerization or association effect so that, in fact, the molecule spends little or no time in that form?

G. Vincow:

We have also investigated the carbon-13 hyperfine splitting, which would be much more sensitive to bending. The carbon-13 splitting is not consistent with a very significant deviation from planarity.

V. Boekelheide:

You did not try out all the other possible models? I mean, if you say your model is consistent with your experience, it does not necessarily rule out other models, unless you have looked at them.

G. Vincow:

Yes, that is right. These have not really been ruled out. But the carbon-13 splitting, which is an average on a time scale long compared with nuclear motions, indicates that contributions from significantly non-planar structures are not quantitatively important. It seems likely that models with appreciable contributions from such structures can be ruled out.

S. M. Sprecher:

What resonance energy do you attribute to the corresponding open-chain radicals?

G. Vincow:

Do you mean experimentally or theoretically?

S. M. Sprecher:

Whichever way you want it. Is it consistent with the resonance energy attributed to the cyclic structures?

G. Vincow:

Well, let me just say, if I understood your question, that I really do not like to go into definitions of resonance energy. I try to use the simplest possible ones that are presented by the practitioners of the art, and my contribution is merely an experimental one.

S. M. Sprecher:

The reason I ask is that one of the practitioners of the art (M. J. S. Dewar) defines the resonance energy as the delocalization energy you get in a system like this upon cyclization of the corresponding open-chain radical, and which should be close to zero in an odd non-alternant cyclic radical like cycloheptatrienyl, by this definition.

349

H. Prinzbach:

Do you know the structure of your dication dimer?

G. Vincow:

Some features of the structure were confirmed by mass spectroscopy and NMR. We do not have any information regarding the isomer(s) involved.

M. Cais:

Just one short question. In your bitropenyl-tropenyl dimer-monomer equilibrium, you do not find any other product. Was this done in a completely deoxygenated medium?

G. Vincow:

Yes, but a most interesting feature of this system is that $C_7H_7 \cdot$, in equilibrium with neat dimer, is not oxygen sensitive. You can work with it in an open tube.

Origin of the Dipole Moment and
Basicity of Cyclopropenones

by S. W. TOBEY

Eastern Research Laboratory, the Dow Chemical Company, Wayland, Massachusetts

THE HIGH DIPOLE MOMENTS (4.7–5.1 D) and basicities ($pK \sim -2$) of cyclopropenones are generally ascribed to an unusually high contribution of cyclopropenium oxide structure (Ib) to cyclopropenone ground-state electronic structures [1].

Ia   Ib

This abnormally high ground-state polarization is, in turn, ascribed to strong stabilization of the positive end of the dipole in the cyclopropenium nucleus by 'aromatic' charge delocalization. This picture of cyclopropenones, as generally understood, also presumes that this cyclopropenium ion stabilization results in an abnormally high fractional degree of charge separation across the carbonyl group. That is, cyclopropenones are generally accepted to have an unusually high negative charge on the carbonyl oxygen and a correspondingly high partial positive charge in the ring [2–3].

The purpose of this paper is to show that this presumption is neither necessary to explain the observed high dipole moments and basicities of cyclopropenones nor consistent with NMR and spectral evidence.

That no abnormal degree of charge separation is required to explain the facts can be demonstrated as follows. The dipole moments of all simple aliphatic and aromatic ketones, including cyclopropanone [3], lie in the range 2.8 ± 0.2 D [4]. To a first approximation, then, the dipole moment is a property of the carbonyl group itself, with the positive pole fairly well localized on the carbonyl carbon. The carbonyl distance in ketones is 1.22±0.2 Å. Let us assume simply that in cyclopropenones the negative charge on oxygen is identical to that in ordinary ketones, but that the positive end of the carbonyl dipole is effectively delocalized over the cyclopropenium nucleus, with a radius of 0.8 Å [5], as pictured in Ib. The predicted dipole moment for cyclopropenones is then

$$2.8 \times \frac{(1.22+0.80)}{1.22} \text{ D, or } 4.65 \text{ D.}$$

This result is completely in agreement with the observed values, and suggests that in cyclopropenones the distance over which charge separation takes place is large and that the fractional degree of charge separation is entirely comparable to that in ordinary ketones. The special feature of cyclopropenones apparently lies in the ability of the cyclopropenium nucleus to delocalize the (modest) positive charge over all three carbon atoms of the ring. Bertelli has shown quite clearly that the analogous delocalization of positive charge over the 'tropylium' nucleus of tropone does *not* occur [6], as has been routinely proposed. Part of this difference must lie in the relative abilities of the Pi-stabilization energies to compensate for the electrostatic energies required to separate the centres of positive and negative charge in the two systems. As shown in Fig. 1, the calculated energy required to extend a typical ketonic static dipole from 1.22 Å to the 2.02 Å proposed for cyclopropenones is 14 Kcal. The corresponding energy requirement in tropone is 19 Kcal. According to Bertelli [6], this amount of energy is simply not available in tropone. It should be noted especially that the energy required to increase the length of a dipole rises as the square of the fractional charge separation, so that simultaneously extending and increasing the fractional charge separation across a dipole is thermodynamically unlikely.

$$E = \frac{(f \cdot e)^2}{D} \left(\frac{1}{r_1} - \frac{1}{r_2} \right)$$

$$e = 4.8 \cdot 10^{-10} \text{ esu} \qquad r_1 = 1.22 \text{ Å}$$

$$D = 2 \qquad f = 0.5$$

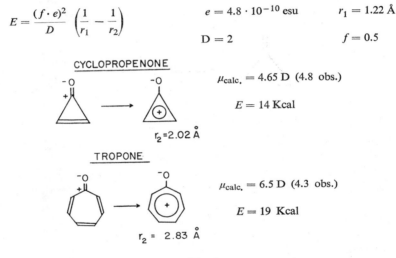

CYCLOPROPENONE

$$\mu_{calc.} = 4.65 \text{ D } (4.8 \text{ obs.})$$

$$E = 14 \text{ Kcal}$$

$$r_2 = 2.02 \text{ Å}$$

TROPONE

$$\mu_{calc.} = 6.5 \text{ D } (4.3 \text{ obs.})$$

$$E = 19 \text{ Kcal}$$

$$r_2 = 2.83 \text{ Å}$$

Fig. 1

Comparison of the energies required to extend the carbonyl dipoles in cyclopropenone and tropone to the cyclopropenium oxide and tropylium oxide forms, assuming an internal dielectric constant of 2 and a fractional charge separation of 0.5

Calculations aside, it is clear from the NMR data given in Fig. 2 that in vinyl-cyclopropenones, the cyclopropenone nucleus does exert a very strong deshielding effect on the geminal vinyl proton, -1.26 ± 0.09 p.p.m. [7]. This effect is essentially independent of the nature of the other group attached to the ring. This strong deshielding effect could, in principle, be ascribed to strong cyclopropenium ion character Ib in the ring. However, this special effect does not appear to be involved.

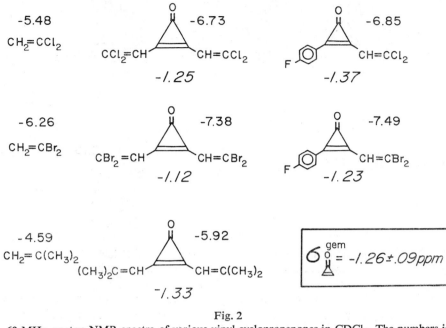

Fig. 2

60 MHz proton NMR spectra of various vinyl cyclopropenones in $CDCl_3$. The numbers in italics denote the chemical shift differences between the cyclopropenones and reference olefins [7]. Primary chemical shifts are given in p.p.m. downfield from internal tetramethylsilane standard. The cyclopropenones shown are all readily prepared by reaction of the trichlorocyclopropenium ion [8–9] with suitable olefins and aromatic substrates, as described elsewhere [9–10]

The data in Fig. 3 show clearly that the carbonyl group of the cyclopropenone nucleus is not required to obtain the observed effects. The 3,3-dichloro analogue of any given cyclopropenone shows a vinyl proton resonance position essentially identical to that of the cyclopropenone, with, if anything, a slightly stronger deshielding effect in the 3,3-dichloride. The possibility that this result is simply fortuitous seems unlikely. First and foremost, it appears reasonable that the *sigma* inductive effect of a $=CCl_2$ group should be approximately equal to that of a C=O group, but certainly not larger. Second, the observed deshielding effects in the 3,3-dichlorides do not arise from a 'chlorocyclopropenium chloride' ionic contribution, analogous to form Ib for cyclopropenones. The 3,3-dichlorides, though readily hydrolysable [10], are covalent, being quite soluble in cyclohexane and $CDCl_3$, the solvent in which our NMR data were obtained. Third, short-range special anisotropic effects of the C–Cl bonds in the 3,3-dichlorides do not appear to be involved. The data in Fig. 2 show that the nature of a group attached to the 2-position of a cyclopropenone has little effect on the *gem*-vinyl proton resonance of a group attached at the 1-position. By extension, the same should hold true for substituents at the 3-position. Molecular models show that divinyl-cyclopropenones (and -cyclopropenes) have very open structures, with no necessarily close, non-bonded intramolecular interactions [11]. This open structure arises from the $\sim 150°$ C_1–C_2–C_α exterior angle in cyclopropenoid compounds [12–13].

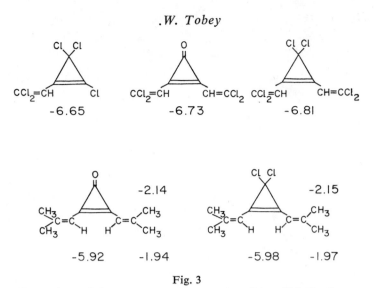

Fig. 3

Comparison of the proton resonance spectra of two divinylcyclopropenones with those of the corresponding 3, 3-dichloro analogues

Since the simplest charge-separated resonance contributor to the ground state of a vinyl cyclopropenone, Ic, shows charge build-up only at the β-vinyl carbon, it could be argued that failure to observe stronger deshielding of the *gem*-vinyl proton in vinyl-cyclopropenones than in the 3, 3-dichloro analogues is without meaning.

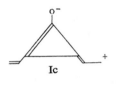

Ic

The data in the lower part of Fig. 3 refute this argument. Although the data are for methyl substituents on the vinyl backbone, it is clear that at both the α- and β-vinyl positions the 3, 3-dichlorocyclopropene nucleus exerts the same essential deshielding effects as the parent cyclopropenone.

Finally, the question can be asked: If unusually strong positive charge were present in the cyclopropenium nucleus, would unusually strong deshielding of the vinyl protons occur? In other words, are the vinyl proton NMR spectra of vinyl-cyclopropenones a suitable probe of charge build-up in the cyclopropenone nucleus? As can be seen in Fig. 4, both protonation of the carbonyl group and conversion of the ketone to the corresponding chlorocyclopropenium ion do cause appreciable deshielding.

In summary, then, the fact that no cyclopropenium oxide form Ib is possible in the 3,3-di-chloride analogue of a cyclopropenone ring and that strong deshielding of the vinyl protons in the side chains is still observed, suggests strongly that an exceptional amount of positive charge need not (and should not) be invoked to explain the observed deshielding of the side-chain protons in cyclopropenones. In fact, the -1.26 p.p.m. *geminal* vinyl proton deshield-ing effect of the cyclopropenone ring is very similar to that caused by several other planar, anisotropic, uncharged groups. Specifically, the σ^z_{gem} values [7] for Ph, C=O and C=C groups are -1.43, -1.10 and -1.26 p.p.m., respectively [14].

354

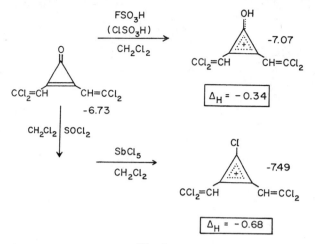

Fig. 4

The effects of protonation and conversion to the chlorocyclopropenium ion on the vinyl proton resonance positions in *bis*-(2,2-dichlorovinyl) cyclopropenone

The entire argument presented above can be verified using F^{19} resonance data, as shown below. Bertelli and others at this Symposium have argued that proton NMR chemical shifts are the subtle outcome of so many large counteracting effects that even the qualitative interpretation of results such as those presented above must be held open to some question [15]. The interpretation of F^{19} chemical shifts is less ambiguous. Taft and co-workers have found F^{19} resonance positions in *p*-substituted fluorobenzenes to be an extremely sensitive probe of σ- and π-electron withdrawing properties of substituent groups [16]. The F^{19} resonance position of fluorobenzene, the standard reference compound, lies at $+113.4$ p.p.m. above the internal standard, $CFCl_3$. *p*-Nitrofluorobenzene resonates at $+103.9$ p.p.m. so that this strongly σ- and π-electron withdrawing group causes a -9.5 p.p.m. downfield shift in the *p*-F^{19} resonance position. Similarly, *p*-fluorobenzaldehyde resonates -9.0 p.p.m. below fluorobenzene.

As shown in Fig. 5, *bis-p*-fluorophenylcyclopropenone shows F^{19} resonance centered at $+104.7$ p.p.m. or -8.7 p.p.m. below fluorobenzene. Although this suggests that the *p*-cyclopropenone ring is a strong σ- and π-electron withdrawing group, its effect can not be considered as due to any peculiar build-up in the ring of positive charge derived from structure Ib. This is shown by the fact that the covalent 3,3-dichloro analogue resonates at $+106.9$ p.p.m. above $CFCl_3$, still -6.5 p.p.m. below fluorobenzene. This more sensitive experiment can, however, be interpreted to show that delocalization of the positive end of the dipole around the cyclopropenone ring, as in Ib, does augment somewhat the fundamentally strong deshielding properties of the cyclopropene ring.

Introduction of strong formal positive charge into this ring will cause even more dramatic F^{19} deshieldings, as shown by the other data in Fig. 2. For example, protonation of *bis-p*-fluorophenylcyclopropenone with FSO_3H or $ClSO_3H$ in CH_2Cl_2 drives the F^{19} resonance position of -10.9 p.p.m. further downfield (-19.6 p.p.m. below fluorobenzene).

Ultimately, the picture one comes to is that the cyclopropenone ring derives most of its NMR deshielding properties from the anisotropy of the C=C double bond, assisted by the

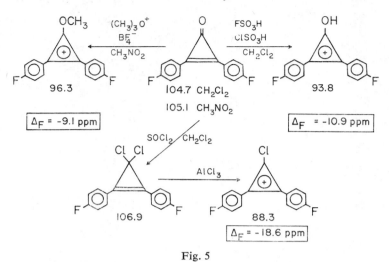

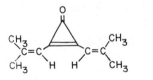

Fig. 5

56.4 MHz F^{19} spectra of *bis-p*-fluorophenylcyclopropenone and its derivatives. All shifts shown are in p.p.m. upfield from internal $CFCl_3$ standard. The Δ_F values in boxes are the chemical shifts caused by the various conversions shown

increased σ-electron withdrawing power of the vinyl carbon atoms, which lie somewhere between the sp_2 and sp hybridized states. Electronically, then, a divinyl-cyclopropenone should be similar to a simple triene. This is the case!

Firstly, as shown in Fig. 6, divinylcyclopropenones show an intense triene $\pi \to \pi^*$ transition near 285 nm, which shifts to longer wavelength on replacement of methyl substituents by chlorine. Secondly, like other triene absorptions, the band shows very little solvent dependence, λ_{max} shifting only 2.5 nm towards the blue on passing from cyclohexane to ethanol. Thirdly, and most fascinating of all, the UV spectra of *bis*-(2,2-dichlorovinyl) cyclopropenone and the 3,3-dichloride derived from it are essentially superimposable, both in λ_{max} and $log_{10}\varepsilon$ (Fig. 7). A related observation has been recorded by Osawa, Kitamura and Yoshida [17], who found that the UV spectrum of diphenyl-cyclopropenone is essentially superimposable on that of *trans*-stilbene.

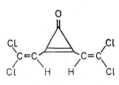

	C_2H_5OH	CYCLOHEXANE
λ_{MAX}	285 (SH 305)	287.5 (SH 307)
ε	4.47	4.45

	C_2H_5OH	CH_2Cl_2
λ_{MAX}	307.5	310
ε	4.39	4.58

Fig. 6

Divinylcyclopropenones; UV spectra

356

If cyclopropenones derived their fundamental electronic structures from exceptional fractional charge separation, such as seen in Ib, it seems unlikely that λ_{max} and $\log_{10} \varepsilon$ values for the *gem*-dichlorides, or olefins which do not bear a cyclopropenone carbonyl at all, would be so similar, or that the solvent-induced changes would be so minimal. Assuming this is the case, the similarity of the UV spectra in Fig. 7 is readily explained, as shown in

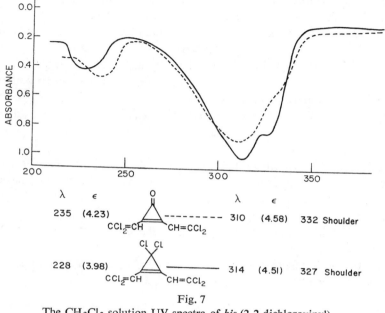

λ	ϵ		λ	ϵ	
235	(4.23)		310	(4.58)	332 Shoulder
228	(3.98)		314	(4.51)	327 Shoulder

Fig. 7

The CH_2Cl_2 solution UV spectra of *bis*-(2,2-dichlorovinyl) cyclopropenone and its 3,3-dichloro analogue

Fig. 8. The strongly-allowed $\pi \rightarrow \pi*$ transition takes place between states of B_1 symmetry, common to both the cyclopropenone and its 3,3-dichloro analogue. This transition skips over an A_2 state, which in the case of cyclopropenones shows an electron density node in the carbonyl group.

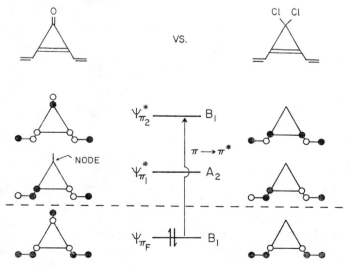

Fig. 8
Molecular orbital energy levels for divinyl-cyclopropenes and their 3,3-dichloro analogues; the dots represent the signs of the wave functions at the various carbon centres

357

Turning now to the question of the origin of cyclopropenone basicity, it seems desirable to ascribe this property more to the polarizability of the molecule than to its intrinsic polarity, which, as indicated earlier, is not necessarily unusual. It seems straightforward that the ability of a cyclopropenone molecule to delocalize a formal positive charge placed on the ring (by protonation) over three carbon atoms via Ib (i.e., the molecule is highly polarizable) should make the molecule more basic than acetone. The latter, to a first approximation, can only delocalize positive charge onto one carbon atom.

In this regard, it should be recalled that tropone, which has a dipole moment of only 4.3 D, and which, as shown by Bertelli [6], should not be considered 'tropylium oxide' in its ground state, has a pK_0 of -0.4 [1]. This additional basicity can logically be ascribed to the fact that the molecule is easily polarized to the tropylium ion under the influence of an attacking proton. Such induced polarization delocalizes charge over seven carbon atoms. Ground-state polarity will only develop in a molecule until the competing electron-withdrawing properties of the two ends of the dipole 'match'. Under attack by a proton, no such competition exists.

Krebs and Schrader [2] have investigated the $n \to \pi^*$ transitions of several cyclopropenones and have interpreted their results in terms of the high polarities and basicities of the molecules. Fig. 9 shows data on several additional molecules, the β-haloalkyl-cyclopropenones being from our current work. The low intensity cyclopropenone $n \to \pi^*$ transitions occur at characteristically short wavelengths: 267–283 nm in cyclohexane v. 310 nm for α,β-unsaturated ketones, and 340 nm for 1,1-dimethylcyclopropanone. Furthermore, the 25 nm blue shift on going from a non-polar to a polar, hydrogen-bonding solvent is twice as great as the shift for typical ketone $n \to \pi^*$ transitions.

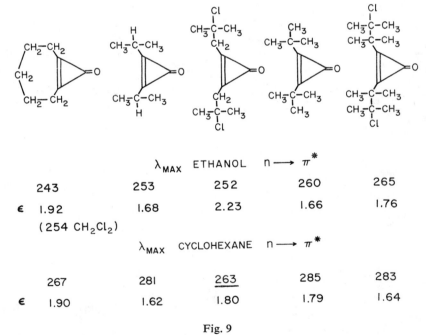

Fig. 9
Locations and $\log_{10}\varepsilon$ intensities of the $n \to \pi^*$ transition bands in dialkylcyclopropenones and β-chloro-alkylcyclopropenones [2, 18]

Krebs and Schrader have ascribed the abnormally short wavelength (high energy) of the cyclopropenone $n \to \pi^*$ transition to a lowering of the energy of the cyclopropenone ground electronic states through a strong contribution from polar form Ib, which (by an unspecified mechanism) is expected to result in a raising of the π^* excited state energy levels relative to those in simple ketones.

A comparison of the pertinent molecular orbital energy levels in the two types of ketones, as shown in Fig. 10, suggests that a more direct explanation is possible. In isolated ketones the (forbidden) $n \to \pi^*$ transition occurs between adjacent purely ketonic $n(B_2)$ and $\pi^*(B_1)$ states. In cyclopropenones the corresponding states contain both ketonic and ethylenic components, the ground states being strongly stabilized (and excited states being strongly destabilized) by mixing. Furthermore, the two states involved in the $n \to \pi^*$ transition are separated by an intervening first-excited π^* state of A_2 symmetry. If any mixing of this $\pi^*(A_2)$ state with the second $\pi^*(B_1)$ state occurs, the $n \to \pi^*$ transition will occur at even higher energy. The point is that no special contribution by cyclopropenium oxide form Ib need be invoked to explain the observed results.

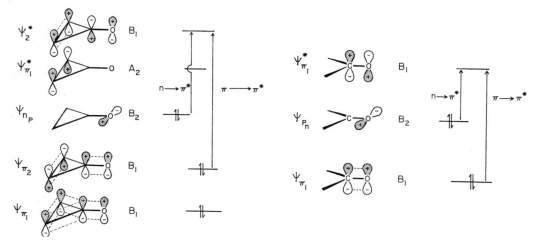

Fig. 10
Comparison of the molecular orbital diagrams for cyclopropenones and simple ketones

Lastly, the data in Fig. 9 suggest that the admittedly dramatic hypsochromic cycloprope-none $n \to \pi^*$ solvent shift, ascribed by Krebs and Schrader primarily to the abnormally strong hydrogen-bonding properties of the carbonyl group due to its unusually high negative charge [2], is obviously sensitive to other factors. As the alkyl side chains on the cyclopropenone become increasingly σ-electron-donating, the $n \to \pi^* \lambda_{max}$ moves to longer wavelength in both ethanol and cyclohexane. With one exception, however, the solvent shift between the two solvents is essentially unaltered. This is inconsistent with the solvent shift being due solely to hydrogen bonding at the carbonyl oxygen of a highly polarized cyclopropenone. Alkyl groups of any reasonable size will not interfere sterically with hydrogen bonding at the carbonyl group of a cyclopropenone—they are simply too far away. Therefore, we would expect a larger solvent shift with more electron-releasing groups. As inductive stabilization of the supposedly highly positive ring charge increases, higher negative charge on oxygen is permitted and increased basicity is expected, contrary to observation. A similar dilemma is reported by Ciabattoni in accounting for the relative stabilities of *t*-butylcyclopropenium ions [18; private correspondence with Prof. J. Ciabattoni].

The answer lies in the fact that solvation of the cyclopropenone ring is also critically important. Clear evidence for this effect is shown by the *bis*-β-chloroalkyl-cyclopropenone bearing $-CH_2-C(CH_3)_2Cl$ groups. This compound behaves, when it is dissolved in cyclohexane, as if it were dissolved in CH_2Cl_2. The other β-chloro compound falls smoothly in order. As shown below, the negatively polarized halogens in the $-CH_2-C(CH_3)_2Cl$ group can easily take up positions over the positively polarized ring, which they would tend to do in non-polar solvents, thus 'dissolving the compound in itself'. In polar media, this special conformation-dependent interaction would be expected to be swamped out by the high dielectric constant, as is observed.

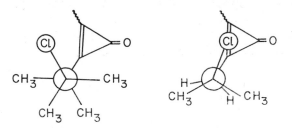

In the compound bearing $-C(CH_3)_2-C(CH_3)_2Cl$ groups, steric eclipsing of the methyls would be required for such self-solvation, and the phenomenon is understandably not observed.

ACKNOWLEDGEMENTS

This work was carried out with the experimental help of Mrs Katherine S. Whittemore, Mrs Marilyn Z. Lourandos and Dr Stanley D. McGregor, while at the Dow Chemical Co., Eastern Research Laboratory. I am particularly indebted to Drs R. G. Czerepinski, D. T. Dix and F. Johnson of the same laboratory for their helpful comments and suggestions on data interpretation, as well as to Prof. P. D. Bartlett of Harvard University and Prof. J. Ciabattoni of Brown University.

REFERENCES

1 A. Krebs (1965) *Angew. Chem. (Int. Edn)*, 4 : 10.

2 A. Krebs & B. Schrader (1967) *Ann.*, 709 : 46.

3 J. M. Pochan, J. E. Baldwin & W. H. Flygare (1969) *J. Am. Chem. Soc.*, 91 : 1896.

4 C. P. Smyth (1955) *Dielectric Behavior and Structure*, McGraw-Hill, New York, pp. 289, 314.

5 M. Sundaralingam & L. H. Jensen (1963) *J. Am. Chem. Soc.*, 85 : 3302.

6 D. J. Bertelli & T. G. Andrews Jr (1969) *ibid.*, 91 : 5280.

7 S. W. Tobey (1969) *J. Org. Chem.*, 34 : 1281.

8 S. W. Tobey & R. West (1964) *J. Am. Chem. Soc.*, 86 : 1459.

9 R. West, A. Sadô & S. W. Tobey (1966) *ibid.*, 88 : 2488.

10 S. W. Tobey & R. West (1964) *ibid.*, 86 : 4215.

11 S. W. Tobey, K. S. Whittemore & M. Z. Lourandos (1968) *Abstracts of Papers, 156th National Meeting of the American Chemical Society, Atlantic City, N. J.*, No. 168.

12 M. K. Kemp & W. H. Flygare (1967) *J. Am. Chem. Soc.*, 89 : 3925.

13 *Ibid.* (1969) 91 : 3163.

14 C. Pascual, J. Meier & W. Simon (1966) *Helv. Chim. Acta*, 49 : 164.

15 D. J. Bertelli, these Proceedings, pp. 326 ff.

16 R. W. Taft, E. Price, I. R. Fox, I. C. Lewis, K. K. Anderson & G. T. Davis (1963) *J. Am. Chem. Soc.*, 85 : 709, 3146.

17 E. Osawa, K. Kitamura & Z. Yoshida (1967) *ibid.*, 89 : 3814.

18 J. Ciabattoni & E. C. Nathan III (1969) *ibid.*, 91 : 4766.

Discussion

E. D. Bergmann:

Would it not be possible by dipole moment measurements to ascertain that indeed the chlorine atoms in β-haloalkyl-cyclopropenones are near to the carbonyl group, due to dipole–dipole interaction?

Second, would you give us your opinion on the assignment of the infra-red absorption bands of the cyclopropenones, about which, as you know, there has been some divergence of opinion?

W. S. Tobey:

In answer to the first question, I think it is reasonable to assume a preferred conformation, if that is the correct word, wherein the chlorine atoms are located over the rings. We have not done any dipole moment measurements, which might indeed give some answer. What we propose to do, and we have not yet succeeded, is to make cyclopropenones bearing β-hydroxyalkyl groups. If our explanation of the solvent-shift data is correct, these cyclopropenones dissolved in cyclohexane will behave as if they were dissolved in an alcohol.

There is a synthetic problem: When we attempt hydrolysis of the corresponding chloride, dehydrochlorination to the vinyl side chain occurs.

In answer to the second question, Krebs and co-workers have done a very complete coordinate analysis, as you know, on the infra-red stretching frequencies for cyclopropenone. I am totally in agreement with their findings, namely, that the two bands that are located near 1,640 and 1,850 cm^{-1} are a common feature of the cyclopropenone entity. Vibrations of one of the parts of such a molecule cannot occur without vibrations of the other; they are as mechanically coupled as it is possible to be. I do believe, as they point out, that the 1,640 cm^{-1} band is more carbonyl, the other more double-bond.

E. Heilbronner:

Please do not be alarmed. We have examined about twenty-five compounds containing the cyclopropyl group by photo-electron spectroscopy, and they are rather peculiar. I just want to make two comments with respect to the locations of the π absorptions in your β-chloroalkyl cyclopropenones. First, there is an interaction between the carbonyl lone pairs of the order of 0.75 eV. That is, there are two n orbitals which are separated by that amount, and this is certainly due to through-bond interaction. These linear combinations of n orbitals can interact with σ skeleton orbitals in the three-membered rings that have the same symmetry. Both sets of orbitals will be destabilized by this interaction, although there remains a gap of some size due to the energy difference between the carbon–carbon and lone-pair orbitals. Now, I do not know whether in your case you will have an interaction of this kind, but if you do, alkyl substitution on the cyclopropenone nucleus would have interesting effects. We know that an alkyl group on such a σ system will shift its energy upwards by 0.3 to 0.4 eV; this would move the ring and lone-pair orbitals closer in energy, and thus yield strong electronic shifts of the type you see.

Second, if we take an indicator, let us say a bromine atom, and we add 1, 2, 3, 4 or 5 carbon atoms, we always have a methyl group at the end. The bromine knows this, and its lone pairs continue to shift by a small amount to higher energies. If you introduce a double bond in the alkyl chain,

361

the uppermost bromine lone pair is shifted to lower energy by about 0.3 eV, even though the distance is 1, 2, 3 or 4 methylene groups. So, in contrast to what is very often believed, certainly the σ orbitals individually are very sensitive to such effects, even though, when taken as a whole, the effect dies out very rapidly.

The point is, in cyclopropyl derivatives we have seen that the three σ bonds in the three-membered ring are very, very peculiar: they interact with all sorts of different things. It is most probable that the σ frame of your cyclopropenone ring is interacting both with the carbonyl lone pairs and with the bromine lone pairs through the σ system, in addition to the π effects you have pointed out.

W. S. Tobey:

My only comment is that you promised that there was no cause for alarm. If I understand your comments, you have changed the explanation of the cause of the effects quite appreciably!

Note added during correction of proofs:

The effects Mr Heilbronner mentioned occur through the σ system and would be expected to be relatively insensitive to solvent effects. The abnormal position of the $-CH_2-C(CH_3)_2$ cyclo-propenone $n \rightarrow \pi^*$ transition in cyclohexane does suggest specific intramolecular (through-space) solvation, in addition to the σ effects Mr Heilbronner has drawn attention to.

362

Conjugated Cyclic Chlorocarbons
Structural and Bonding Aspects

by R. WEST

Department of Chemistry, University of Wisconsin, Madison

I. INTRODUCTION

THE CHEMISTRY of the chlorocarbons — fully or highly chlorinated organic compounds — is now undergoing a renaissance, in which many of the most interesting findings centre about the cyclic conjugated perchloropolyenes. Monocyclic members of this family are shown in Fig. 1. Of these species only hexachlorobenzene was known before 1964. Within the last six years, the stable species $C_3Cl_3^+$, $C_7Cl_7^+$ and C_8Cl_8 have been isolated (the cations as salts), and the transient existence of C_4Cl_4, $C_5Cl_5^+$ and $C_5Cl_5^-$ has been convincingly demonstrated [1].

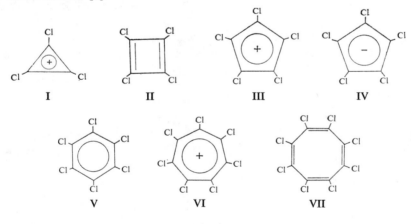

Fig. 1
Monocyclic conjugated chlorocarbons

The family of species in Fig. 1, like the cyclic polyenes from which they are derived, can be classified as aromatic if they contain $4n+2$, and anti-aromatic if they contain $4n$ π electrons. Thus, I and IV–VI can be considered aromatic, and so are expected to be stabilized, but II–III and VII are anti-aromatic and should be destabilized, at least when planar.

This work was supported by a grant from the US Public Health Service – National Institutes of Health.

However, the chlorine substituents may perturb the carbocyclic π system either by withdrawing electronic charge inductively through the C–Cl *sigma* bonds, or by direct participation of the non-bonding pairs on the chlorine in π interaction with carbon. Fragmentary evidence suggests that both effects may be significant in certain delocalized chlorocarbons [1].

At the University of Wisconsin, we have been concerned with the chemistry of all of the species shown in Fig. 1, except for C_4Cl_4, which has been studied by Scherer and Meyers [2]. The chemistry of $C_3Cl_3^+$ and $C_7Cl_7^+$ will be treated in this paper, with emphasis on properties relating to the bonding in these species. The properties of a related bicyclic compound of theoretical interest, perchlorofulvalene, will also be described.

II. Trichlorocyclopropenium Ion

The starting material for preparation of $C_3Cl_3^+$ is tetrachlorocyclopropene, C_3Cl_4, whose synthesis has been fully described elsewhere [3]. Tetrachlorocyclopropene is now commercially available in the United States, where it is being tested for use as a vapour-phase fumigant. When treated with powerful Lewis acids, which can serve as chloride acceptors, liquid C_3Cl_4 is converted into crystalline salts of the $C_3Cl_3^+$ cation [4]:

$$C_3Cl_4 + MCl_3 \rightarrow C_3Cl_3^+ \cdot MCl_4^- \qquad M = Al, Fe, Ga; \text{ also } SbCl_5.$$

Studies of the infra-red and Raman spectra of $C_3Cl_3^+$ confirm the expected planar triangular structure, with D_{3h} symmetry [4–5].

The vibrational data, summarized in Table 1, were used to investigate the bonding in $C_3Cl_3^+$ by means of normal coordinate analysis. There are six constants in the Urey-Bradley force field used for the calculation, and only five fundamental frequencies were observed, so that one constant must be assumed; but, fortunately, the chlorine atoms are

Table 1

Vibrational Frequencies for $C_3Cl_3^+$

cm^{-1}	IR	Raman	Assignment
200	s		E
459	—	s, pol	A_1
735	s	vw	E
1,312	vs	s, dep	E
1,348	s	—	—
1,791	—	m, pol	A

so far apart that the non-bonded interaction constant can be taken as zero [6]. The results of the Urey-Bradley calculation are shown in Table 2, together with those from a partial normal coordinate analysis for the corresponding brominated ion, $C_3Br_3^+$. In Table 3 the C–C and C–Cl stretching force constants are compared with those for other known aromatic species. Note that both K_{C-C} and K_{C-Cl} are decidedly higher than for chlorobenzene; indeed, K_{CC} for $C_3Cl_3^+$ is markedly higher than for any other known aromatic

364

Table 2

Force Constants for $C_3X_3{}^+$

(millidynes/Å; [5])

	$C_3Cl_3{}^+$	$C_3Br_3{}^+$
K_{CC}	6.31	6.46
K_{CX}	2.99	2.15
H_{CCC}	−0.248	(−0.248)
H_{CCCl}	0.385	(0.275)
F_{CCl}	0.808	(0.674)
C_{ClCl}	(0)	(0)

Values in parentheses are assumed

species. Consistent with this is the X-ray finding by Sundralingham that the C–C distance in triphenylcyclopropenium ion is shorter than that in benzene [7].

Table 3

C–C and C–Cl Stretching Force Constants
for Aromatic Species

	K_{CC}	K_{CX}	ρ^*
C_6H_6	5.59	4.67	0
	5.15	4.79	0.35
$C_5H_5{}^-$	5.39	4.79	0.21
C_6Cl_6	4.81	2.30	0.37
$C_3Cl_3{}^+$	6.32	2.99	0

* Resonant constant

Why should the C–C bond in cyclopropenium ions be so strong? Simple Hückel calculations predict a π-bond order of 0.667 for cyclopropenium, identical to that for benzene. The σ bonding in cyclopropenium ions is probably external to the ring and 'bent' even more strongly than in cyclopropane or cyclopropene (Fig. 2); if so, the σ bonds should be weaker than those in benzene. But bending of the σ bonds may bring the carbon atoms closer

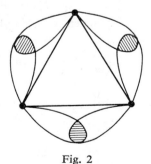

Fig. 2

Schematic drawing of orbitals forming C–C σ bonds in cyclopropenium ion,
indicating overlap external to the three-membered ring

together, allowing for much increased overlap of *p* orbitals on adjacent carbon atoms. According to this model, the unprecedented bond strength in $C_3Cl_3^+$ arises from the increased π bonding in cyclopropenium ions, which more than makes up for the decreased σ-bond strength.

In I, contributions from the chlorine atoms to the π bonding may also aid in raising both C–Cl and C–C bond strengths and force constants. Nuclear quadrupole resonance measurements of Lucken and Mazeline [8; cf. 9] give a value of 0.35 for the asymmetry parameter η for the ^{35}Cl nuclei in I. This asymmetry parameter can be related directly to the partial double-bond character, which is found to be 0.16. Both the asymmetry parameter and the bond order are higher than for any other known carbon-chlorine compound. The data suggest that about half of the positive charge on the ring (3×0.16) is delocalized through the π system onto the chlorine atoms [8; cf. 9].

Space precludes a full discussion of the chemistry of the trichlorocyclopropenium ion. However, one class of derivatives of exceptional theoretical interest will be mentioned. When treated with 2,6-disubstituted phenols, $C_3Cl_3^+$ is converted first to the *tris*(hydroxyaryl)-cyclopropenium ion (VIII), which easily loses a proton upon treatment with base, forming a diarylquinocyclopropene (IX) [10]. The latter undergoes reversible oxidation to a novel 3-radialene derivative, a triquinocyclopropane (X) [11]:

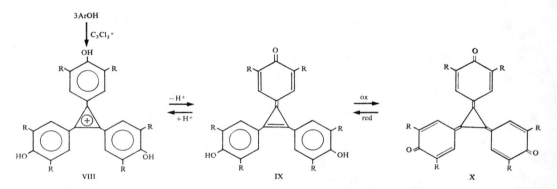

The triquinocyclopropanes are intensely violet-coloured substances, remarkable in that their electronic absorption spectrum extends throughout the visible and into the near infra-red range; the strongest band occurs at about $\lambda = 770$ nm, with log $\varepsilon = 4.7$. The electronic spectrum of X (R = *tert*-butyl) is shown in Fig. 3. The anion radical [12] and dianion corresponding to X have also been prepared; both the neutral species and the dianion have singlet ground states in agreement with simple MO calculations (Fig. 4) [11].

366

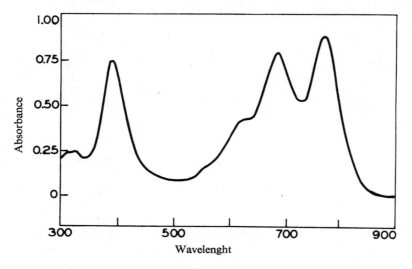

Fig. 3

Electronic spectrum of triquinocyclopropane (X) (R=*tert*-butyl), 1.5×10^9 M in benzene

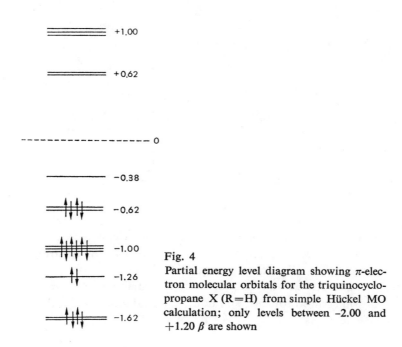

Fig. 4

Partial energy level diagram showing π-electron molecular orbitals for the triquinocyclopropane X (R=H) from simple Hückel MO calculation; only levels between −2.00 and +1.20 β are shown

III. Heptachlorotropenium Ion

The octachlorobicyclo[3.2.0]heptadiene XI can be prepared from hexachlorocyclopentadiene and trichloroethylene [13]. When XI is treated with two equivalents of aluminum chloride at 155° (the temperature is quite critical) it undergoes loss of chloride and ring opening, to give the heptachlorotropenium ion (XII) as the heptachlorodialuminate [14; R. West & K. Kusuda, unpublished studies].

367

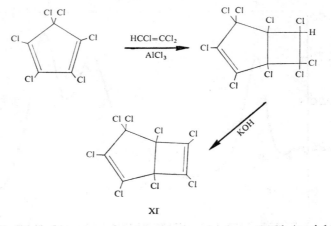

XI

Water reacts with this salt to transfer chloride ion back to $C_7Cl_7^+$, yielding octachloro-cycloheptatriene (XIII). Little is yet known about structure and bonding in $C_7Cl_7^+$, but it appears to be at least as stable as $C_3Cl_3^+$, even though it is probably non-planar, because of Cl–Cl repulsion.

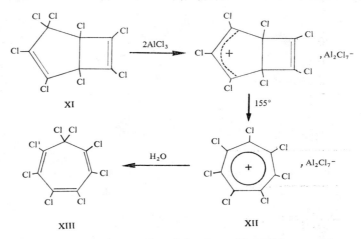

XI   XII

XIII

Solvolysis of XIII with concentrated sulphuric acid yields perchlorotropone (XIV) [14; R. West & K. Kusuda, unpublished studies] and, under more stringent conditions, perchlorotropolone (XV) [K. Scherer, unpublished investigations]. Like other cyclo-heptatrienes, XIII and XIV show some tendency to aromatize. For example, XIII under-goes thermolysis to perchlorotoluene, and both XIII and XIV react with alcohols to yield pentachlorobenzoate esters.

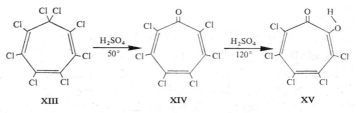

XIII   XIV   XV

The reaction chemistry of $C_7Cl_7^+$ has not been studied extensively, but the cation is known to react with phenols to give stable quinocycloheptatrienes [14; R. West & K. Kusuda, unpublished studies]:

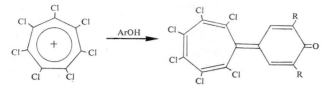

It seems probable that $C_7Cl_7{}^+$ will prove a valuable intermediate in the synthesis of other novel delocalized structures, just as $C_3Cl_3{}^+$ is proving to be.

IV. OCTACHLOROFULVALENE

Now let us consider a bicyclic chlorocarbon of unusual theoretical interest: octachloro-fulvalene (XVI). This compound was first synthesized by Mark from hexachlorocyclo-pentadiene [15], and we have improved this synthesis somewhat [16]:

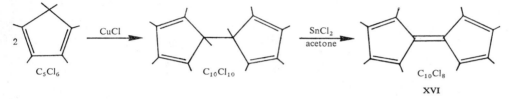

Whereas fulvalene itself is an orange-coloured material, so unstable that it cannot be iso-lated, XVI can easily be crystallized as stable, blue-violet prisms. The striking difference in colour arises from a large shift in wavelength for the lowest energy electronic excitation: from 416 nm for fulvalene to 604 nm for XVI. This strong bathochromic shift is partially due to the chlorine substituents, but probably mainly to the non-planarity of XVI. X-ray measuremements have shown that (because of chlorine repulsions) the central double bond in XVI is twisted so that the planes of the two rings form a dihedral angle of 42° [17].

Fulvalenes generally are electron-deficient molecules possessing an unfilled bonding orbital, and in octachlorofulvalene this orbital lies even lower in energy than in fulvalene itself (Fig. 5). Consistent with this is the finding that XVI forms a surprisingly stable anion radical upon electrolytic or alkali metal reduction [R. West & R. M. Smith, unpublished results].

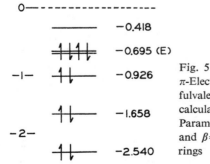

Fig. 5

π-Electron molecular orbitals for octachloro-fulvalene (XVI) from simple Hückel MO calculation; only bonding levels are shown. Parameters used were $k_{CCl}=0.4$ and $h_{Cl}=2.00$, and $\beta=0.709$ for the bond between the two rings

Octachlorofulvalene is also a strong π acid, and this fact allows a spectacular demonstration of colour changes upon charge-transfer complex formation. Solutions of XVI in non-interacting solvents, such as dichloromethane XVI, are blue. Addition of benzene or toluene changes the colour to green, mesitylene gives a yellow colour, phenanthrene gives a yellow-brown, and pyridines — red!

Spectral changes upon charge-transfer complex formation with XVI are subtle. The electronic spectrum of XVI contains bands at 390 nm (log ε = 4.60) and 604 nm (log ε = 2.40). Moderately strong π bases give with XVI a charge-transfer band between 390 and 475 nm, which 'fills in' the area between the two absorption bands of pure XVI [R. West & R. M. Smith, unpublished results]. Fig. 6 illustrates this, showing the spectrum of XVI in

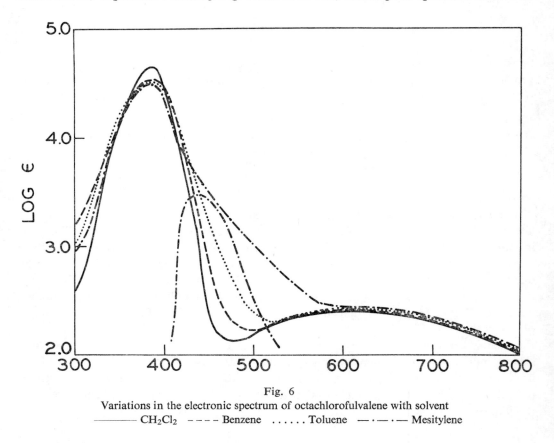

Fig. 6
Variations in the electronic spectrum of octachlorofulvalene with solvent
——— CH_2Cl_2 – – – – Benzene . . . . . . Toluene — · — · — Mesitylene

the presence of increasing amounts of mesitylene. Even though the CT maxima cannot be established precisely, the spectral observations allow us to estimate the charge-transfer energy, and therefore the strength of XVI, as a π acid (Table 4). The data indicate that XVI is comparable in acid strength to chloranil.

Table 4

Charge-Transfer Energies for Various Acceptors (nm)

π Acid	E_{CT} Mesitylene	E_{CT} Phenanthrene	Ref.
Tetracyanoethylene	—	540	[18]
Chloranil	427	473	[19]
Octachlorofulvalene	400–475	400–475	[West & Smith, unpublished results]
p-Benzoquinone	344	—	[19]
Maleic anhydride	—	354	[20]

Although XVI is a relatively good π acceptor, it is not so strong as might be predicted from MO calculations. The energy of the charge-transfer transition is commonly [21] given as

$$E_{CT} = IP - EA + C,$$

where IP is the ionization potential of the donor, EA the electron affinity of the acceptor, and C a constant denoting all other forms of energy (solvation etc.) changing upon complex formation. The electron affinity of XVI (LUMO at $-0.4\,\beta$; see Fig. 5) should, from MO theory, be even greater than for tetracyanoethylene (LUMO $-0.34\,\beta$). The reason why XVI is a poorer π acid than TCNE probably lies in the C term and reflects the non-planarity of XVI, which prevents close approach of the π system of the donor. Consistent with this explanation is the fact that the equilibrium constants for charge-transfer complex formation to XVI are two or three orders of magnitude smaller than typically found for such complexes [R. West & R. M. Smith, unpublished results]. Thus, octachlorofulvalene is a unique example of a powerful but strongly sterically hindered π acid.

The stability of XVI, compared to fulvalene, is probably due mostly to steric hindrance to attack at the central double bond by the flanking chlorine atoms. Similar steric stabilization by chlorine is observed for the highly inert compound, octachlorocycloöctatetraene, which is unchanged even when boiled with fuming nitric acid [22].

REFERENCES

1 R. West (1970) *Acc. Chem. Res.*, 3 : 130.

2 K. V. Scherer Jr & T. J. Meyers (1968) *J. Am. Chem. Soc.*, 90 : 6253.

3 S. W. Tobey & R. West (1966) *ibid.*, 88 : 2481.

4 *Ibid.* (1964) 86 : 1459.

5 R. West, A. Sadô & S. W. Tobey (1966) *ibid.*, 88 : 2488.

6 T. Shimanouchi (1949) *J. Chem. Phys.*, 17 : 848.

7 M. Sundralingham & L. H. Jensen (1966) *J. Am. Chem. Soc.*, 88 : 198.

8 E. A. C. Lucken & C. Mazeline (1968) *J. Chem. Soc. (A)*, p. 153.

9 R. M. Smith & R. West (1969) *Tetrahedron Lett.*, p. 2141.

10 R. West & D. C. Zecher (1967) *J. Am. Chem. Soc.*, 89 : 152.

11 *Ibid.* (1970) 92 : 155.

12 *Ibid.*, p. 161.

13 A. Roedig & L. Hörnig (1956) *Ann. Chem.*, 598 : 208.

14 R. West & K. Kusuda (1968) *J. Am. Chem. Soc.*, 90 : 7354.

15 V. Mark (1961) *Tetrahedron Lett.*, p. 333.

16 R. M. Smith & R. West (1970) *J. Org. Chem.*, 35 : 2681.

17 P. J. Wheatley (1961) *J. Chem. Soc.*, p. 4936.

18 M. J. S. Dewar & H. Rogers (1962) *J. Am. Chem. Soc.*, 89 : 395.

19 A. Kuboyama (1960) *Nippon Kagaku Zasshi*, 81 : 558.

20 M. Chowdhary (1962) *J. Phys. Chem.*, 66 : 353.

21 H. McConnell, J. S. Ham & J. R. Platt (1953) *J. Chem. Phys.*, 21 : 66

22 A. Roedig, R. Helm, R. M. Smith & R. West (1969) *Tetrahedron Lett.*, p. 2137.

Discussion

J. F. Labarre:

In the case of completely chlorinated hydrocarbons, you cannot, of course, use the NMR technique. But have you tried to measure the ring current magnitude by means of NQR as M. Kubo and E. A. C. Lucken suggested some years ago?

R. West:

We have studied the nuclear quadrupole resonance of many of our chlorinated compounds, including the trichlorocyclopropenium ion. Our work on $C_3Cl_3^+$, however, is much inferior to that carried out by Lucken and Mazeline at Cyanamid in Geneva, who have done a very careful study of the trichlorocyclopropenium ion. Their results are of interest in connection with the bonding in this compound. Of course, it is not easy to separate the ring current effect, which should influence the nuclear quadrupole moments, from everything else that is present in such molecules.

H. Kuroda:

We have always been looking for a new electron-acceptor compound; therefore, we must be grateful to you for introducing us to such an interesting acceptor as octachlorofulvalene. Have you tried to obtain a crystalline complex?

R. West:

Thank you, Mr Kuroda, you may take with you the sample of octachlorofulvalene, if you like! We have not obtained crystalline charge-transfer complexes, but we have not tried very hard. We have, however, studied very powerful π donors with octachlorofulvalene and shown that ground state electron charge transfer takes place, just as it does with tetracyanoethylene or tetracyanoquinodimethane.

E. D. Bergmann:

Your explanation of the blue colour of octachlorofulvalene as due to the twist of the two halves of the molecule against each other, has a parallel, viz. the explanation of the deep red colour of dibiphenyleneethene,

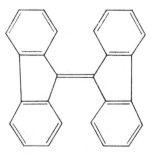

which is quite out of line with that of its vinylogs. However, I am not sure that this is the only reason. Also diphenyltetrabromofulvene

372

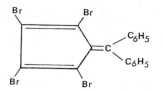

is deep blue, although here the steric interference can be overcome by rotation of the phenyl rings out of the plane of the double bond — one of them, at least, does certainly not lie in this plane.

We have studied the dipole moment of octachlorofulvalene, which is 0.95 D in benzene and 0.40 in cyclohexane [R. Lowenthal & I. Agranat, unpublished results]. It is likely that a twisted molecule of this kind has a small dipole moment, but there is obviously also an effect of solvent interaction of the type you have described. Such an effect has also been observed in dipole moment measurements of tetrachlorofulvenes in various solvents [H. Weiler-Feilchenfeld, unpublished results].

Finally, I would like to mention that perchlorofulvene

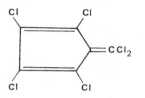

also has a finite moment (1.00) in cyclohexane.

Altogether, it would be interesting to carry out some calculations on these compounds, and perhaps even more interesting to study their X-ray diffraction, so as to get a clearer picture of the interaction between the chlorine atoms.

R. West:

I think that the differences in dipole moments can be accounted for naturally in terms of charge-transfer complex formation. What we find here corresponds very well to what Mr Agranat has found in his studies.

As for colour, I'm quite interested in your finding that tetrabromodiphenylfulvene is blue. This suggests to me that there must be considerable twisting of the central carbon–carbon double bond in this compound, even though some of the steric strain can be accommodated by deviations of the benzene rings from the molecular plane.

E. Heilbronner:

I would like to ask you first: The electronic spectrum of your trianisylcyclopropenium cation has probably one large band in the long-wave region. Is that correct and is this a clean single band?

R. West:

Yes, the band is at 362 nm, but there is a closely lying band at 343 nm.

E. Heilbronner:

I ask for a specific reason. We have carried out a theoretical investigation of the following problem: If one dissolves trianisylcarbonium ions in a non-polar solvent and if one studies their spectrum as a function of concentration, one finds a splitting of the first band for higher concentrations. This is due to the fact that in the ion-pair the counterion does not sit on the three-fold axis of the molecule, but probably shuffles all over the molecule.

373

One of the difficulties is that the triarylcarbonium ion is not necessarily planar. I assume that your trianisylcyclopropenium ion is flatter than a trianisylcarbonium ion, and might therefore be a more favourable test case for our theory.

R. West

I am trying to remember whether we have carried out these experiments. I think that nearly all of our work was done in dichloromethane, in which one might expect considerable solvation and dissociation. I am not sure if we have carried out, say, studies in cyclohexane.

B. Pullman:

If I may make one comment, too, I think that Mr West should not have any complexes about utilizing the Hückel method even for these complicated problems. Of course, we do know that Hückel's is a very simple method and applicable with much less rigour to the aromatic non-benzenoid compounds than to the aromatic compounds. Nevertheless, as far as qualitative aspects of phenomena are being concerned, I must stress the extreme value of the Hückel method in finding out the most astonishing features, of, say, the fulvenes. If I go back as far as the late forties or the early fifties, when we used to work with Mr E. D. Bergmann on these problems, I remember that it was the utilization of Hückel's method which enabled us to put into evidence some of the most outstanding features of the fulvenes.

I myself was brought up on a very remarkable book, *The Resonance Theory* of Wheland, and when one reads this book, everything seems to be very logical in chemistry. You increase the size of a conjugated system, and you get more resonance energy, you get deepening of the colour, you get nice ring currents and diamagnetic exaltation, and so on. I found this very logical, but I found it somewhat monotonous. Now, when we came across the fulvenes, everything changed: nothing was so logical any more in chemistry. You can find in the fulvenes dipole moments which you do not have in the aromatic hydrocarbons; you find that the dipole moments decrease when you increase the dimension of the molecules. You find that when you increase a conjugated system, instead of diamagnetic exaltation, you have diamagnetic depression, and so on.

Now, all these phenomena, all these features, are perfectly accounted for by the utilization of Hückel's method. It was Hückel's method which enabled us to show that we should have hypsochromic shifts upon enlarging some conjugated systems, that we should have depression of diamagnetism, and so on. So, the value of the method is undoubtedly great, and I think that an intelligent chemist today can perfectly well use the Hückel method, provided he knows what he is doing, provided he knows its limitations. He still can discover new phenomena by this method, but he must be careful not to use it quantitatively, and not to extrapolate too much.

Transition Features of
Aromatic, Non-Aromatic and Anti-Aromatic Compounds

by I. FISCHER-HJALMARS

Institute of Theoretical Physics, University of Stockholm

I. Introduction

DURING THE LAST DECADE the concept of aromaticity has raised increasing interest. Several suggestions have been made concerning suitable criteria for aromatic character. These discussions will not be repeated here, but reference is made to some of the more recent publications on the subject [1–3]. Although we are well aware of the merits and weaknesses of the various criteria, an energy criterion of aromaticity and non-aromaticity will be used in the present context. The definition of this criterion will be given in Sect. II. The molecules investigated and their geometry are discussed in Sect. III. Results from Pariser-Parr-Pople (PPP) calculations are reported in Sect. IV, and their bearing upon the concept of aromaticity is discussed.

II. Energy Criterion of Aromaticity

As discussed previously [4], the π-electron energy, $E(\pi)$, cannot easily be used for comparison of different molecules. It has been stressed by Ruedenberg [5] and others [4, 6] that the repulsion between the positively charged core atoms, $E(\text{repuls})$, must be added to $E(\pi)$ to give $E(\text{total})$. However, it has been pointed out [4–5] that the quantity $E(\text{bonding})$, obtained as the difference between $E(\text{total})$ and $E(\text{atoms})$, is even more appropriate for comparison of molecules with different numbers of π electrons, $N(\pi)$, or even for the same number when the environmental influence on W_μ is included [7]. As in [4], we define

$$E(\text{atoms}) = \Sigma[n_\mu W_\mu + (n_\mu - 1)\gamma_{\mu\mu}], \qquad (1)$$

where n_μ is the number of π electrons contributed by the atom μ. In the present study, $E(\text{bonding})$/electron is found to be the quantity most suitable for classification of the molecules:

$$E(\text{bonding})/\text{electron} = [E(\pi) + E(\text{repuls}) - E(\text{atoms})]/N(\pi). \qquad (2)$$

On account of the well-known difficulties of finding an unequivocal criterion of aromaticity, we must first find out which value the quantity in Equation (2) will take on in case of some

This investigation has been supported by grants from the Swedish Natural Science Research Council. The data machine computations have been made possible by grants from Kungl. Statskontoret.

375

typical aromatic and non-aromatic compounds. As such reference compounds we have chosen benzene, naphthalene, anthracene and phenanthrene, on one side, and the *trans* conformations of the conjugated olefinic compounds, C_nH_{n+2} — where $n = 2, 4, 6, 8, 10, 12$ — on the other. The various energy terms, as well as other properties described below, have been calculated within the modified PPP approximation developed by Roos, Skancke [7–8] and others in the Stockholm-Oslo Quantum Chemistry Groups.

III. MOLECULES INVESTIGATED AND THEIR GEOMETRY

In addition to the above-mentioned reference compounds, several molecules of inter-mediate character have been investigated, viz. C_6H_6-fulvene (F) [9], C_6H_6-dimethylenecyclo-butene (D) [9], C_8H_6-pentalene (P),* C_8H_8-heptafulvene (Hf),* $C_{10}H_8$-dimethylenebenzo-cyclobutene (Dz) [9], $C_{10}H_8$-azulene (Az) [7, 10], $C_{10}H_8$-fulvalene (Fa),* $C_{12}H_8$-ace-naphthylene (Ac)* and $C_{12}H_{10}$-heptalene (H).* The reference after each molecule indi-cates where full details are given about the computational results with the present para-metrization, as well as comparison with available experiments. Fig. 1 displays the struc-tures of the various molecules.

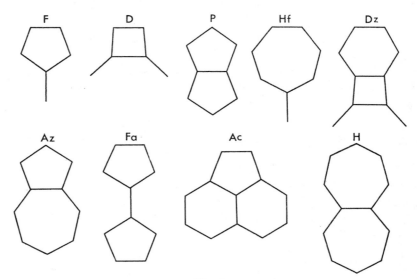

Fig. 1

F=Fulvene; D=Dimethylenecyclobutene; P=Pentalene; Hf=Heptafulvene; Dz=Dimethylenebenzo-cyclobutene; Az=Azulene; Fa=Fulvalene; Ac=Acenaphthylene; H=Heptalene

In several cases, the geometry of these molecules is not known from experiments. As described earlier [7, 9–10], the present computational scheme allows a theoretical deter-mination of bond lengths with rapid convergence and very good agreement with experi-ments. This method has been applied to all molecules treated here and has given univocal results in all but two cases, viz. pentalene and heptalene. A theoretical study of these two molecules by the PPP method has previously been carried out by Nakajima et al. [11].

---

* I. Fischer-Hjalmars, to be published.

These authors obtained self-consistency for two different models of each compound, one with almost equal bond lengths and D_{2h} symmetry, the other with alternating single and double bonds and C_{2h} symmetry. The present study has led to similar results. To distinguish between these two models, Nakajima et al. [11] studied the sum of $E(\pi)$ and $E(\sigma)$. No details about their energy calculation are given. The final result was that the C_{2h} symmetry should be more stable than D_{2h} by about 10 Kcal/mole.

The energy values from the present investigation are reproduced in the table. It is interesting to note that the values of $E(\pi)$ and E(total) indicate that D_{2h} should be the most stable species, but that the values of E(bonding) clearly predict the C_{2h} symmetry to be the stable one for both molecules. Thus, already within the π-electron frame, the present computational scheme leads to a clear distinction between the two models, predicting the π stabilization of the C_{2h} model to be 3 Kcal/mole and 8 Kcal/mole for pentalene and heptalene, respectively. Inclusion of $E(\sigma)$ will only further enhance this result. It is most gratifying that no comparison between the values of $E(\pi)$ and $E(\sigma)$ is needed, since these two quantities necessarily must originate from different sources, which are not easily put on the same footing.

Energies (in a.u.) of the Two Geometries of Pentalene and Heptalene

Energy Term	Pentalene		Heptalene	
	D_{2h}	C_{2h}	D_{2h}	C_{2h}
E(atoms)	−2.8212	−2.7340	−4.2464	−4.1485
E(repuls)	+5.7655	+5.7111	+11.3257	+11.2347
$E(\pi)$	−9.0984	−8.9607	−16.3344	−16.1578
E(total)	−3.3330	−3.2504	−5.0086	−4.9231
E(bonding)	−0.5118	−0.5164	−0.7623	−0.7746
E(bonding)$/N(\pi)$	−0.0640	−0.0646	−0.0635	−0.0646

IV. RESULTS

1. π Bonding

The results of the calculation of E(bonding)$/N(\pi)$ of Equation (2) are displayed in Fig. 2. The solid line connects the values obtained for the various olefinic compounds, and the broken lines connect the points from aromatic compounds. Since we have chosen two aromatic C_{14} compounds, anthracene (A) and phenanthrene (Ph), there are two broken lines between C_{10} and C_{14}. It should be observed that the quantity of Equation (2) depends on the number of carbon atoms involved, although this dependency becomes less pronounced with increasing molecular size. Therefore, it is only justified to compare molecules with the same number of π-electron centres, which in the present context means the same number of carbon atoms.

The values found for the intermediate molecules are gathered around the vertical line corresponding to the appropriate number of carbon atoms. In principle, there is no difference between the marks to the left and to the right of this vertical line; the only reason for the way of drawing is to make the results as easily surveyable as possible. Marks for molecules with the same number of carbon atoms are always arranged in the same left-right sequence in all the following figures, the degree of aromaticity decreasing from the left to the right.

377

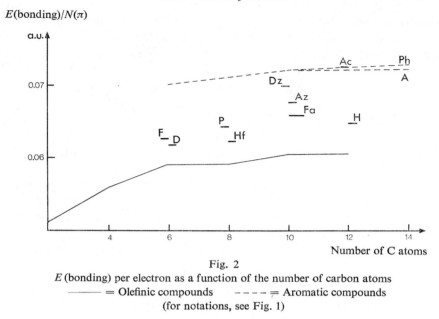

Fig. 2

E (bonding) per electron as a function of the number of carbon atoms

——— = Olefinic compounds   – – – – = Aromatic compounds

(for notations, see Fig. 1)

Fig. 2 shows that the properties of molecules with a major aromatic kernel, such as di-methylenebenzocyclobutene (Dz) and acenaphthylene (Ac), are almost completely dominated by this kernel, the small olefinic parts only leading to minor modifications. The aromaticity of azulene (Az) has been much debated. It is interesting that according to the present criterion, azulene takes an intermediate position, though closer to the aromatic than to the olefinic line.

The present energy criterion does not give any support for classifying certain molecules as anti-aromatic.

2. *Bond Lengths*

The theoretical values of bond lengths are shown in Fig. 3. Again, the solid lines refer to olefinic and the broken lines to aromatic molecules. In olefinic compounds the bond lengths can be referred to two distinct groups. The lengths of the formal single bonds are found in the narrow range between the two upper lines, and the lengths of the double bonds in the range between the two lower lines. It is seen that from C_8 and onwards these ranges are almost constant.

In aromatic molecules, all the bond lengths fall within a single range, indicated in Fig. 3 by the broken lines. However, it should be noted that this range spreads appreciably with increasing molecular size, and, at C_{14}, almost reaches both the upper and the lower olefinic range. The frequently assumed constancy of bond length in aromatic compounds is, therefore, not borne out by the present theoretical results, which are in good accord with experiments [10].

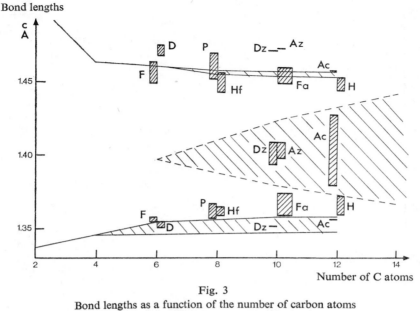

Fig. 3

Bond lengths as a function of the number of carbon atoms
(for notations, see Figs. 1–2)

The bond lengths found for the various intermediate molecules are indicated by isolated ranges, marked with the short-hand notation for the molecule in question, F, D, P,... It is seen that, with the exception of Dz and Ac, only azulene has a typical aromatic character. For this molecule only one bond, viz. 3–9, falls outside the aromatic range.

As a general result, it can be stated that the classification of molecules according to bond lengths is in conformity with the present energy criterion.

3. *Singlet Electronic Transitions*

The results of the calculations of electronic transitions contain a wealth of information which cannot be surveyed easily. Here, we shall only try to sort out those facts which are most pertinent in the present context.

Again, we shall start by looking at the reference compounds. Fig. 4 summarizes some of the results for these molecules. The solid line connects the energy levels of the lowest excited singlets of the olefinic compounds. The transition to this level from the ground state is, in all molecules, strongly allowed. The broken lines connect the levels of aromatic molecules corresponding to the transitions α, p, β and β' in Clar's notation. Here, only β and β' are strongly allowed transitions.

A point of interest, shown in Fig. 4, is that the lowest transition occurs at almost the same frequency in olefinic and in aromatic molecules of the same size. In fact, this similarity does not hold only for the lowest transition, but also for higher ones. Obviously, energies of singlet transitions cannot be used for classification of molecules according to degree of aromaticity.

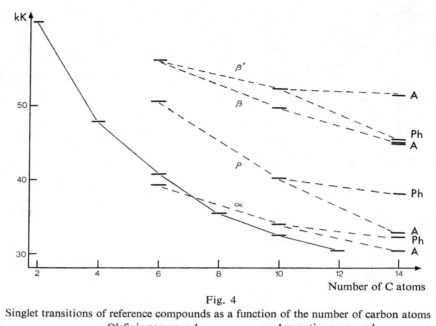

Fig. 4
Singlet transitions of reference compounds as a function of the number of carbon atoms
———— = Olefinic compounds   – – – – = Aromatic compounds

This point is further elucidated in Fig. 5. Here, we have reproduced the excited singlets of the C_{10} compounds included in the present study. Again, the degree of aromaticity, ac-

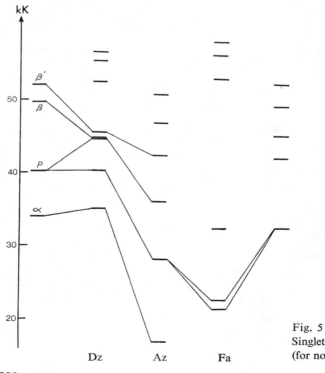

Fig. 5
Singlet transitions of C_{10} compounds
(for notations, see Fig. 1)

cording to the energy criterion, decreases from the left to the right. The connections between the levels of the various molecules have been made mainly with the aid of the information from transition density diagrams. Such diagrams will be discussed in full detail elsewhere [I. Fischer-Hjalmars, to be published].

Fig. 5 shows that the best correspondence between the excited singlets of the various molecules is found between the two extreme cases, i.e., between the typical olefinic and the typical aromatic molecule. The intermediate cases, azulene (Az) and fulvalene (Fa), exhibit spectra without any similarity to any of the reference molecules.

Similar results have been obtained for the series of C_6 compounds: benzene, fulvene, dimethylenecyclobutene and hexatriene.

4. *Triplet Electronic Transitions*

The present parameter scheme is adopted to give agreement between calculated singlet transitions and observations. Since the parametrization implicitly takes account of part of the correlation, and since correlation effects are different for singlet and triplet states, the present scheme is expected to give too low values for triplets. However, it has been shown [M. Sundbom, private communication] that this lowering varies systematically with the size of the molecule, and that the relative order of the triplets is correctly given by the theory. It is therefore of interest and significance to study the energy of the triplet levels as a function of the size of the molecules within the present series of compounds.

The results obtained for the lowest triplets are shown in Fig. 6. As before, the solid line connects the values of the olefinic compounds and the broken line the aromatics. For these transitions a clear difference is found between the two groups of reference compounds. Moreover, the intermediate compounds, Dz, Ac and Az, are found in the region between the two lines. Also, transition energies for most of the other compounds are in good accord with the bonding energies of Fig. 2. Therefore, it is of particular interest and significance

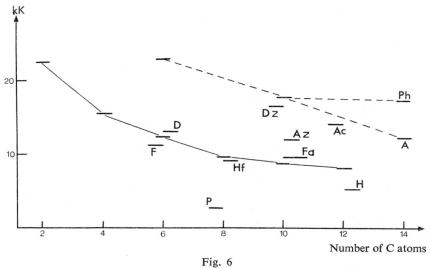

Fig. 6
Triplet transitions as a function of the number of carbon atoms
———— = Olefinic compounds   - - - - = Aromatic compounds
(for notations, see Figs. 1–2)

that two triplet levels are found well below the above-mentioned region. These values are the triplets of pentalene (P) and heptalene (H) of C_{2h} symmetry. It may be mentioned that in the case of D_{2h} symmetry, these triplets are found to be still lower.

V. Concluding Remarks

According to the present study, the reference compounds, aromatic and non-aromatic, seem to form reasonably well-defined classes. Compounds with intermediate properties, e.g. azulene, could, of course, be classified as pseudo-aromatic. A classification of certain molecules as anti-aromatic is less clear. However, the distribution of the lowest triplet levels seems to give some support also to this latter concept.

Acknowledgement

I am indebted to the members of the Quantum Chemistry Groups in Stockholm and Oslo for putting unpublished details of their work at my disposal.

References

1 D. P. Craig (1964) *Educ. Chem.*, 1 : 136.

2 D. Lloyd (1966) *Carbocyclic Non-Benzenoid Aromatic Compounds*, Elsevier, Amsterdam–London–New York.

3 A. J. Jones (1968) *Rev. Pure Appl. Chem.*, 18 : 253.

4 I. Fischer-Hjalmars (1966) *Theor. Chim. Acta*, 4 : 332.

5 K. Ruedenberg (1961) *J. Chem. Phys.*, 34 : 1861.

6 G. Del Re & R. G. Parr (1963) *Rev. Mod. Phys.*, 35 : 604.

7 B. Roos & P. N. Skancke (1967) *Acta Chem. Scand.*, 21 : 233.

8 B. Roos, *ibid.*, p. 2318.

9 A. Skancke & P. N. Skancke (1968) *ibid.*, 22 : 175.

10 I. Fischer-Hjalmars & M. Sundbom (1968) *Acta Chem. Scand.*, 22 : 607.

11 T. Nakajima, Y. Yaguchi, R. Kaeriyama & Y. Nemoto (1964) *Bull. Chem. Soc. Japan*, 37 : 272.

A New Definition of the Degree of Aromaticity

by A. JULG

Laboratoire de Chimie Théorique, Faculté des Sciences, Marseille

FEW CONCEPTS are as basic and as often used as that of aromaticity. However, it is one of the least clear and most difficult of concepts to define. Although several definitions have been proposed and discussed [1–10], the precise meaning of the word 'aromaticity' remains vague.

Recently, we proposed a definition based on the uniformization of the length of the peripheral bonds [5], which led us to estimate the degree of aromaticity (A_1) by the relation

$$A_1 = 1 - (225/n) \sum_{(rs)} (1 - d_{rs}/\bar{d})^2,$$

where n is the number of peripheral bonds (rs); d_{rs} their lengths, and $\bar{d}$ their mean length. The constant 225 has been chosen in order to obtain a value of zero for the aromaticity of the Kekulé form of benzene. One gets then for A_1:

Benzene	1	Fulvene	0.62
Naphthalene	0.90	Azulene	1.00
Anthracene	0.89		

The values for the aromatic compounds and for fulvene are more or less satisfactory. However, it is surprising that the value for azulene is the same as that for benzene, that is to say, much higher than for naphthalene. Similarly, according to this definition, borazole, in which all the B–N bonds are identical, should be as aromatic as benzene. Actually, our definition represents only one aspect of the general phenomenon of aromaticity and should therefore be broadened.

During the last few years attention has been drawn to the relation between aromaticity, on the one hand, and some magnetic phenomena, on the other (e.g. ring currents, coupling constants in NMR [2] and the Faraday effect [6]). This relation is due to the fact that aromaticity is closely linked to the ability of the electrons to circulate along the rings. It is natural to think that this mobility is closely connected with the electronic charge. Indeed, the higher the charge of an atom in a molecule, the greater is the tendency of the atom to attract electrons and to retain them, i.e. to slow down their displacement and, therefore, to decrease the aromaticity of the system. Incidentally, we will note that the bonding σ electrons contribute to this circulation only slightly. Thus, the π electrons are the only ones concerned and the π charges intervene primarily in the phenomena. This effect is particularly clear with molecules like furan, pyrrole and thiophene, where the π charges of the hetero-atom, obtained by the same method of computation, are, respectively: 1.92 [11], 1.78 [12] and 1.64 [13]. Similarly, boroxane is much less aromatic, in the magnetic sense, than borazole itself [6], because the charge on oxygen is higher than that on nitrogen, which is less electronegative.

To express numerically the effect due to the difference in the behaviour of the ring atoms, we will assume, as a first approximation, that the resistance opposed to the circulation of electrons above the atom i can be given as a function of the π charge on this atom (q_i). With these conditions, going from atom i to atom j, the resistance to the electronic circulation can be expressed as the charge gradient

$$\frac{q_i - q_j}{d_{ij}} = \frac{\Delta q_{ij}}{d_{ij}},$$

where d_{ij} is the distance between atoms i and j.

Since aromaticity must be independent of the indexing order of the atoms, the resistance to the circulation will be an even function of this gradient. Therefore, the aromaticity correction due to the difference in charge between the atoms i and j can be written, if we limit the gradient expansion to the first term,

$$1 - B(\Delta q_{ij}/d_{ij})^2.$$

That is, for all the bonds (ij) of the ring

$$A_2 = \prod_{(ij)} \left[1 - B(\Delta q_{ij}/d_{ij})^2\right].$$

For a molecule with several rings, the product obtained for each of them has to be used. We note that for small values of the gradient, A_2 can be expressed simply:

$$1 - B \sum_{(ij)} (\Delta q_{ij}/d_{ij})^2,$$

or, also, if the interatomic distances are small enough:

$$1 - B' \sum_{(ij)} (\Delta q_{ij})^2.$$

This expression can be compared with the expression given by Labarre et al. [6], who introduce the mean value of the absolute value of the gradient:

$$\frac{1}{n} \sum_{(ij)} |\Delta q_{ij}|.$$

Although these authors obtain satisfactory results from this expression, using the square of the gradient is certainly better than using the absolute value, the latter having no physical meaning.

To conclude, in order to take into acount simultaneously the uniformization of the lengths of the peripheral bonds and the resistance to the circulation of the electrons along the rings, the aromaticity degree has to be expressed in the following form: $A = A_1 \cdot A_2$.

This definition gives the following results:

	A_2	A
Benzene	1.00	1.00
Naphthalene	$1 - 0.000 \, B$	0.90
Anthracene	$1 - 0.000 \, B$	0.89
Azulene	$1 - 0.148 \, B$	$1 - 0.15 \, B$
Fulvene	$1 - 0.000 \, B$	0.62

d is measured in Å, and Δq in electrons

Thus, we see that azulene is less aromatic than naphthalene if, with the chosen units, B is larger than 2/3. For $B = 1$, for instance, A_2 (azulene) = 0.85. For the heterocycles, by using for the A_1 computation only the C–C bonds [5], we get with B = 1:

Thiophene	$0.93 \, (1 - 0.28 \, B) = 0.67$
Pyrrole	$0.91 \, (1 - 0.58 \, B) = 0.38$
Furan	$0.89 \, (1 - 0.93 \, B) = 0.06$
Pyridine [14]	$1.00 \, (1 - 0.025 \, B) = 0.97$

For borazyle [15], we get approximately 0.01.

Hence, the introduction of the factor A_2 not only decreases the aromaticity of azulene, but also greatly spreads the values for the heterocycles. The value B = 1, which leads to an aromaticity practically equal to zero for furan, seems very reasonable, and we will adopt it.

Therefore, the general expression for the aromaticity degree that we propose is:

$$A = \left[1 - 225/n \sum_{(rs)} (1 - d_{rs}/d)^2\right] \prod_{(ij)} \left[1 - (\Delta q_{ij}/d_{ij})^2\right].$$

REFERENCES

1 A. Streitwieser (1961) *Molecular Orbital Theory for Organic Chemists*, J. Wiley, New York, p. 288.

2 J. A. Elvidge & L. M. Jackman (1961) *J. Chem. Soc.*, p. 859.

3 J. A. Elvidge (1965) *Chem. Comm.*, 8: 160.

4 M. J. S. Dewar (1966) *Tetrahedron Suppl.*, 8: 75.

5 A. Julg & P. François (1967) *Theor. Chim. Acta*, 7: 249.

6 J. F. Labarre & F. Crasnier (1967) *J. Chim. Phys.*, 64: 1664.

7 J. F. Labarre, M. Graffeuil, J. P. Faucher, M. Pasdeloup & J. P. Laurent (1968) *Theor. Chim. Acta*, 11: 423.

8 G. M. Badger (1969) *Aromatic Character and Aromaticity*, Cambridge Univ. Press.

9 J. Krussewki & T. M. Krygowski (1970) *Tetrahedron Lett.*, 4: 319.

10 These Proceedings, pp. 9 ff.

11 L. Pujol & A. Julg (1964) *Theor. Chim. Acta*, 2: 125.

12 A. Julg & P. Carles (1967) *Theor. Chim. Acta*, 7: 103.

13 A. Julg, M. Bonnet & Y. Ozias (1970) *ibid.*, 7: 49.

14 L. Pujol & A. Julg (1965) *Tetrahedron*, 21: 717.

15 J. L. Tassetto (1968) Diplôme Études Supérieures, Marseille.

Concluding Remarks

by E. D. BERGMANN

Department of Organic Chemistry, the Hebrew University of Jerusalem

WHEN I TRY TO SUMMARIZE this third Jerusalem Symposium, I cannot help wondering whether we should have tried to define aromaticity instead of accepting the instinctive understanding of the notion that the organic chemists appear to possess. However, we cannot overlook the fact that the mind of the scientist seeks classification, perhaps in the hope that one may extract from the classification more than the experimental input into it.

We might have formulated the title of this Symposium better as 'Aromatics, Pseudo-Aromatics and Anti-Aromatics' and not as 'Aromaticity, Pseudo-Aromaticity, Anti-Aromaticity', because to the chemist the accurate description of compounds, known and yet unknown, is more important than the theories employed for such description. It seems to me that we can define at least two clear cases, one aromatic and one non-aromatic: benzene and ethylene, and our question is: can one classify all compounds containing one or more double bonds, in the formal sense, as belonging to one of these two categories? Obviously, this is not the case. Mr Agranat in his introductory lecture has emphasized that we have gradually changed the definition of aromatic character from a chemical one ('substances inclined more to substitution reactions and retaining their type therein' — R. Robinson), emphasizing the energy content of the molecule in the excited state, to a physical one ('substances having a low ground-state enthalpy'). There are compounds in which there is a correlation between both definitions; the difficulties arise when they do not coincide. Mr Heilbronner has suggested to drop the concept of aromaticity altogether, and, amongst several others, Mr Lloyd has supported this suggestion, introducing three formally new concepts instead: benzenoid, referring to structure; regenerative or meneidic, referring to reactivity; and Hückelian, referring to ground-state properties. I feel that there is a danger that we enter into problems of semantics rather than of substance; in the last analysis, classification is not an end in itself. Obviously, we have not found an unequivocal description of aromatic character, and there are all possible intermediary transitions from ethylene to benzene. We will, I believe, retain the structural formulae which we use, simply because of their simplicity; but in each case, the double bond will mean something different. The alternative would be to describe each molecule by a number of equations or physical properties, which is less convenient and which, again, would only allow us to define arbitrarily those properties that would characterize an 'aromatic' compound. In any event, I am not sure whether we are going to witness the disappearance of the word 'aromatic'.

386

Concluding Remarks

In surveying what we have learnt, I would like to mention two theoretical papers of general interest. In the first one, Mr Binsch expressed the conviction that the definition of aromaticity presents an insoluble problem, but that the concept has historically fulfilled an important function. He elaborated on the definition of a conjugated π-electron system as aromatic if it shows neither first-order nor second-order double-bond fixation, relating this to ring currents. He arrived at the conclusion that both pentalene and heptalene suffer second-order double-bond fixation, leading to a dynamic bond alternation around the rings. They are then anti-aromatic.

Mr Mulder has shown us a new mathematical derivation of the Woodward-Hoffmann rules for two Kekulé structures; he proved to us that the concept of the symmetry of the orbitals is not needed in these rules. These calculations introduced both anti-aromatic and aromatic transition states in which the two Kekulé structures have equal weight. The connection with Craig's rules becomes quite evident.

In the introductory lecture, we have been given a brief survey of the physical criteria applied so far to the definition of aromatic character: the anisotropy of the diamagnetic susceptibility and the exaltation of the latter property; the resonance energy (e.g., in Dewar's treatment); the ring current; the bond alternation, as related to NMR coupling constants and to X-ray diffraction data in the solid state; the energy barrier for restricted rotation; the dipole moment; the deviation of peripheral bonds of the molecules from the bond length in benzene; and, on the theoretical side, the already-mentioned first- and second-order bond fixation and Craig's rules, to mention only a few. In any event, it is more or less agreed that if aromatic character exists at all, it is a property of the π-electron system only.

Two additional physical methods have explicitly been discussed at this meeting. Mr Pacault's study of optical anisotropy — parallel to magnetic anisotropy, but being a molecular property — showed that there is a vast difference between aromatic and aliphatic compounds, and it remains to be seen what one can learn from this method for the description of the less clearly defined intermediate stages between these two extremes. Mr Labarre has given us a very comprehensive picture of the application of the Faraday effect to our problem: The excess of magnetic rotation due to conjugation *versus* the sum of the free valence indices is a straight line, the slope for unsubstitued arenes being double that for olefins. In the former, the electrons are more easily set in motion by an external magnetic field. There is thus a relation to the Pauling-Pople ring current. Substitution changes that; e.g., fluorobenzene does not appear to be an aromatic compound. Thus, Mr Labarre suggested to abandon aromaticity and to replace it by strobilism, that is, the ability of π electrons to move around.

Not every participant in this Symposium has stated his views *expressis verbis*. Mr Heilbronner, in his review of an impressive amount of experimental material on photoelectronic spectroscopy and on electronic spectra, found himself unable to define the aromaticity of compounds considered as aromatic by chemists, such as various annulenes. The main points he made were that there is π-bond interaction also in non-conjugated unsaturated molecules, and that, secondly, delocalization is a very general phenomenon.

A clear, but rather opposite stand has been taken by Mr Del Re in his attempt to define aromaticity in five-membered heterocycles:

He demanded that one must find out which properties are to be attributed to the presence of a *cis*-diene system in these compounds, which to conjugation and which to delocalization of the lone pair of electrons of the hetero-atom, including transfer of charge into the ring; cyclic delocalization infers aromatic character. Benzene can be considered as a special case of a *cis*-diene in which the X is the vinylene group — it has been made clear that we have to refer to a standard compound if we want to speak at all of the aromatic character of a given substance.

Mr Lloyd, too, as mentioned already, rejected the notion of aromatic character even in the classical sense, because chemical properties cannot be taken as real criteria, even if they are associated with a completely planar system. He reminded us that the metal complexes of 1,3-diketones such as acetylacetone are easily substituted at the C_2 atom, and he added a fascinating description of the dihydroazepinium salts, in which the ring contains a fully saturated part, but also an unsaturated portion, which is even more easily given to the typical benzene reactions than benzene itself and its derivatives.

There are a few other types of molecules and reactions which should be mentioned here — first, perhaps the beautiful iodination reaction of azulenes, cycloheptatrienes, ferrocenes and azapentalenes, described by Mr Treibs. We have heard a report by Mr Cais on the metallocenes which are so easily substituted in the aromatic manner — ferrocene acetylates one million times faster than benzene — but which should not be considered as benzenoid and as derivatives of the $4n + 2\pi$-electron system of the cyclopentadiene anion; they form a class in themselves.

Another group of compounds of great interest, to our problem and in general, are the helicenes. Mr Martin's lecture on the synthesis of hexahelicene and its homologues, up to the 13-helicene, has been a chemist's delight; an organic chemist must be impressed by the observation that the photocyclization of polycyclic diarylethylenes should give helicenes in preference to the equally possible extended planar polycyclics. Mr Wagnière's theoretical studies on hexa- and hepta-helicene were a most welcome supplement. He showed that configuration interaction must be introduced to bridge the gap between the discordant results of previous calculations. This gives the correct sequence of transitions, but they appear too far in the blue region of the spectrum; it was suggested that the consideration of π interaction between the superimposed benzene rings of these molecules might improve the results.

Let me mention here Mr Adam's calculations on polycyclic hydrocarbons, which also introduced the configuration interaction in order to arrive at the correct values for the electronic spectra and ionization potentials of these compounds, although the results for such heterocyclic compounds as pyrimidines and purines have not been satisfactory.

A further contribution to this series of interesting compound types were Mr Boekelheide's masterful syntheses of molecules with substituents in the cavity of the π-electron cloud of the pyrene system. One observation that seems to me of particular interest is the photochemical formation of a *cis*-difluoro compound from its precursor:

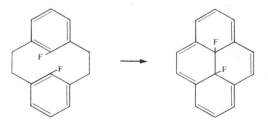

The very stable thieno [3,4-*d*] thiepin has been described to us by Mr Schlessinger as an extremely delocalized system (with 12 π electrons). This is, as physical and chemical data show, due to the contribution of charge-separated forms to the ground state

The X-ray diffraction data in the solid state are complex; but the structure is very similar to that of azulene, and the central linkage is definitely a single bond.

Finally, an interesting group of substances we learned about were the oxazolones of Mr Boyd,

zwitterionic compounds which have a fascinating chemistry. They undergo, *inter alia*, facile electrophilic substitution at C_4 with acylium ions and with diazonium salts; they are equally capable of 1,3-dipolar addition reactions.

I think we have been fortunate that two lectures, those of Mr Berthier and of Mr Meyer, have emphasized the particular position of benzene by a study of the isomers of benzene, especially the recently discovered ones. Mr Berthier dealt with the theory of the planar isomers, fulvene, dimethylene-cyclobutene and trimethylene-cyclopropane. Using *ab initio* calculations, he found that for the cyclobutene and cyclopropane derivatives good agreement between theory and experiment (e.g., for the dipole moment) could be achieved when doubly-excited configurations were taken into account. However, this is not so for fulvene. Practically the same moment results from SCF calculations for π electrons and for σ and π electrons, but the value (1.1 D) is double that experimentally found by Brown (0.5 D). It was, therefore, an agreeable surprise that Mrs Skancke has been able to calculate, referring to π electrons alone, not only the correct spectra and the correct bond lengths for fulvene and dimethylene-cyclobutene, but also the correct dipole moment for fulvene. However, the moment of the other benzene isomer came out twice as high as the experimental value. On the other hand, Mrs Pullman reported that an elaborate calculation by the CNDO method, both for σ and π electrons and invoking doubly-excited configurations, gives for the moment of fulvene a value of 0.59 D, which is in good agreement with the experimental one.

For the three non-planar benzene isomers, we obtained from Mr Meyer interesting results and predictions by both all-valence-electron and all-electron calculations. Prismane appears as a bridged system of three single ethane molecules; for Dewar benzene, a small dipole moment has been predicted, and an even larger moment for benzvalene, which thus would be a polar hydrocarbon, although it cannot be considered as a zwitterion.

These results were nicely supplemented by Mrs Serre's paper on the dehydrobenzenes, especially the ordinary benzyne. The energy difference between it and benzene is of the order of magnitude of 160 Kcal; it has a dipole moment, and it can only be hoped that the chemists will find means to verify at least the latter prediction.

We have heard rather little about anti-aromaticity, the phenomenon that π conjugation, or resonance, destabilizes a system of $4n$ π electrons. Mr Rabinovitz made it at least probable that the indenylium ion

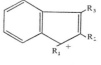

389

is such a case. He showed that not only the substituents R_1 and R_3, as in open allylic systems, but also R_2 — contrary to such allylic compounds — are identical in their NMR behaviour. A more clear-cut case is that of cyclobutadiene, derivatives of which Mr Gompper has described in such beautiful experiments. The molecule has been stabilized by the 'push-pull' mechanism, introducing in a suitable way positive and negative substituents into the molecule.

If one considers the thermal stability of these compounds and the calculated resonance energy of 0.76β, one might conclude, if one uses the 'old' nomenclature, that this substituted molecule is almost aromatic. Its reactions, especially the cycloadditions, are of great beauty, both from the theoretical and the preparative point of view.

Perhaps the greatest amount of new information we collected during this Symposium was concerned with the fulvenes (if I want to avoid the term pseudo-aromatic) — those hydrocarbons that are polar and have measurable and measured dipole moments. They are, of course, interesting, because the moment indicates a tendency to develop a cyclopentadienide and a tropylium structure, respectively, in the best-known cases, the fulvenes and the heptafulvenes, which are both $4n+2\ \pi$ systems.

Mr Nakajima used for his comprehensive study of fulvenes and related compounds the principle of first- and second-order bond fixation, and arrived at very reasonable values for double-bond fixation and bond length. In pentalene, heptalene and the — unknown — 1,2,5,6-dibenzopentalenes, we have bond fixation, but not in the known isomer of the latter nor in Hafner's hydrocarbons, in which the bonds are all of nearly equal length:

Equally, the anions of heptalene, pentalene and the anions and cations of fulvalene and heptafulvalene were treated with success. The method also gave results in good agreement with the experimental values of Dauben for the exaltation of magnetic succeptibility, except for the case of aceheptylene (cyclopenta [e, f] heptalene).

Mr Kuroda has studied essentially the same large group of fulvenes and fulvenoid molecules by the variable-bond length or variable-β, γ procedure; he examined in particular the influence of the assumptions on the relation between the two-centre integrals and the bond length. On the strength of his many calculations, Mr Kuroda proposed that aromatic character should be expected if the degree of bond-order fluctuation is smaller than 0.3.

Let me shortly survey the experimental work in this series that was reported to us. Mr Hafner told us of the electrophilic and nucleophilic substitution in the five- and seven-membered ring, respectively, of azulene and of the nonafulvenes, which expectedly resemble the pentafulvenes; he added a new type of pentafulvene, which is derived from the unknown isoindene:

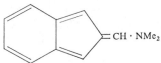

It is interesting that these compounds show a bathochromic shift, compared with the classical benzofulvenes.

Mr Prinzbach has given us a present in the form of a group of triapenta- and pentahepta-fulvalenes which are not stabilized electronically by aryl groups or annelation, but sterically by the introduction of *tert*-butyl groups. He has determined the physical properties, especially the dipole moments, of these compounds and studied their interesting addition reactions, which let them appear — in his opinion — as pure polyenes. From Mr Kitahara we have had a report on the substitution and addition reactions of heptafulvenes, stabilized in the dipolar form by cyano groups at the exocyclic carbon atom C_8. Among these reactions, the most interesting appears to me the substitution at this C_8 atom.

Mr Sutherland has given us new insights into the fine structure of the 6-amino-substituted fulvenes. In particular, he determined by NMR techniques the rotational barriers around the C=C double bond (which, of course, in the zwitterionic form, is not double) and around the C–N bond. With the same technique, he arrived at some interesting results for the dibenzoheptafulvene system and similar heterocyclic seven-membered rings, using, very elegantly, the α-hydroxyisopropyl group at the bridge as a probe.

A welcome addition to this series was Mr Battiste's study of the azatria-pentafulvalenium salts, for which a beautiful synthesis has been developed. The degree of charge delocalization between the rings and the related degree of double-bond character of the central bond were determined by the NMR spectrum of suitable derivatives. The rotational barrier around the central bond is, indeed, low.

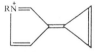

Mr Bertelli has made the most impressive use of the NMR spectrum, especially of the coupling constants, for the determination of the conformation of 8,8-diphenyl-heptafulvene and heptalene, and has supported his results by CNDO/2 calculations. His conclusion is that the heptafulvene system is planar, while heptalene is not.

Another method, namely electron spin resonance, was applied by Mr Vincow to this series of compounds. He determined the entropy of cleavage of ditropenyl and found the C_7H_7 radical to be highly resonance-stabilized. He further reported the interesting result that the cation and anion radicals of heptafulvene both show considerable π-electron delocalization.

I think that we should include here Mr Tobey's most original study of the cyclopropenone system, which has almost always been considered as an oxatriafulvene, but, from what we have heard, wrongly so. He has used the physical properties of various divinylcyclopropenones, which have become rather easily available by reaction of suitably activated olefins with trichlorocyclopropenylium chloroaluminate. The equally available *p*-fluorophenyl-cyclopropenones show in the NMR spectrum a similar shift for the fluorine atom as fluorophenyl-nitrobenzene; this shift is completely different for a *p*-fluorophenylcyclopropenylium ion.

A fitting conclusion for our meeting was Mr West's paper on conjugated cyclic chlorocarbons, colourful in the material and spiritual sense. His paper paid particular attention to the trichlorocyclopropenylium ion $C_3Cl_3{}^+$, the heptachlorotropylium ion $C_7Cl_7{}^+$ and the perchlorofulvalene. Whilst calculations predict for $C_3Cl_3{}^+$ a π-bond order of 0.667

391

(as in benzene), the C–C distance is shorter than in the latter, probably due to bent σ bonding. $C_7Cl_7^+$ is at least as stable as $C_3Cl_3^+$, but may not be planar. Octachlorofulvalene is blue, and we have witnessed its chamaeleon-like colour changes. The colour is ascribed to the twist of the two halves of the molecule against each other, as has been done before in the case of dibiphenyleneethene.

Thus, I think, we all have learnt a great deal, and we have received inspiration and challenges; we have had a fairly good equilibrium between theory and experiment. It is true, we have not solved the problem of what is aromaticity (and, therefore, also not of what is pseudo- and anti-aromaticity), but we all agree that aromaticity can be defined only artificially, by convention, if we do not want to go to the extreme of abandoning the notion altogether. However, I hope that you will agree with me that classification and theory are not ends in themselves. If they generate new experimental work, new compounds, new processes, new methods — they are good; if they are sterile — they are bad. Speaking for myself, I have been greatly stimulated to further work, and if this is true not only for me, perhaps we will meet again at a Jerusalem Symposium on, shall we say, π-bond interaction or π-electron delocalization, or on strobilism.

I also hope that you have found the air of Jerusalem scientifically stimulating; it has always been a city of thinkers and prophets, who also believed that their predictions were experimentally verifiable.

Index

Index

Index

398